Floristic Studies in the Light of Biodiversity Knowledge and Conservation

Floristic Studies in the Light of Biodiversity Knowledge and Conservation

Editor

Robert Philipp Wagensommer

MDPI

Basel • Beijing • Wuhan • Barcelona • Belgrade • Novi Sad • Cluj • Manchester

Editor
Robert Philipp
Wagensommer
Free University of
Bozen-Bolzano
Brixen-Bressanone, Italy

Editorial Office
MDPI
St. Alban-Anlage 66
4052 Basel, Switzerland

This is a reprint of articles from the Special Issue published online in the open access journal *Plants* (ISSN 2223-7747) (available at: www.mdpi.com/journal/plants/special_issues/floristic_studies).

For citation purposes, cite each article independently as indicated on the article page online and as indicated below:

Lastname, A.A.; Lastname, B.B. Article Title. *Journal Name* **Year**, *Volume Number*, Page Range.

ISBN 978-3-0365-9563-4 (Hbk)
ISBN 978-3-0365-9562-7 (PDF)
doi.org/10.3390/books978-3-0365-9562-7

Cover image courtesy of Robert Philipp Wagensommer

Contents

About the Editor

Robert Philipp Wagensommer

Robert P. Wagensommer is an Associate Professor at the Free University of Bozen-Bolzano in Italy. R.P. Wagensommer is a botanist. His research is focused on Flora, Systematics, Taxonomy, Red Lists, and the Conservation of threatened plant species, especially in the Mediterranean area. He also works on STEM Education.

Preface

Although they play a key role in ecosystem functioning and human life, plants are generally overlooked by policy makers and humans in general. In addition, in recent years we have been witnessing an increase in biologists unable to recognize even common plants, and the number of plant experts (i.e., botanists) able to identify plants is decreasing. In light of these considerations, the Special Issue "Floristic Studies in the Light of Biodiversity Knowledge and Conservation" was launched in the journal *Plants*, and eight research articles, two reviews, and an editorial were published in it. I am grateful to all the authors that contributed to this Special Issue, the reviewers for their valuable work, and the academic editors of the journal *Plants* who have helped me greatly during the editorial process. Thanks to this teamwork, it has been possible to publish research from different countries, underlining that the knowledge of the flora of a territory is essential to slow down the effects of anthropic pressure on biodiversity and likewise essential for most of the research activities of modern botany. For example, management actions, planning strategies, and biodiversity conservation are efficient and effective only with an in-depth knowledge of how many and which taxa constitute the floristic richness of a territory.

Robert Philipp Wagensommer
Editor

plants

MDPI

Editorial

Floristic Studies in the Light of Biodiversity Knowledge and Conservation

Robert Philipp Wagensommer

Faculty of Education, Free University of Bozen-Bolzano, 39042 Brixen-Bressanone, Italy; robertphilipp.wagensommer@unibz.it

Floristic studies are often considered "simply" traditional research. However, the knowledge of the flora of a territory is essential to slow down the effects of anthropic pressure on biodiversity and essential for most of the research activities of modern botany, such as molecular research on the taxonomy of critical genera, which requires in-depth floristic knowledge for the correct identification of the analyzed taxa. Management actions, planning strategies, and biodiversity conservation are efficient and effective only with an in-depth knowledge of how many and which taxa constitute the floristic richness of a territory. Therefore, floristic research has not only been important in the past, but still has an important role in botanical studies today, even in countries with a long tradition of floristic studies and where the flora is considered already well known. Indeed, even in these countries, new taxa are still found [1,2] or described as new to science [3–5].

Although they play a key role in ecosystem functioning and human life, plants are generally overlooked by policy makers and humans in general [6]. An important task of flora experts is the correct identification of plants. This is even more important today, considering the increase in biologists unable to recognize even common plants, and the dwindling number of plant experts (i.e., botanists) able to identify plants [7].

In light of these considerations, eight research articles [8–15] and two reviews [16,17] were published in the journal *Plants*, in the Special Issue "Floristic Studies in the Light of Biodiversity Knowledge and Conservation".

The eight research articles cover different taxonomic groups from Europe and Asia. Perrino et al. [8] evaluated the effects of ecology and climate on the composition of the essential oils of two species of wild officinal plants of the Lamiaceae family from Mediterranean Italy, i.e., *Satureja cuneifolia* Ten. and *Thymus spinulosus* Ten., the latter endemic to southern Italy and Sicily. The authors also presented a survey of the vegetation in which the two species grow. The results suggest a potential use of both species as food, pharmacy, cosmetics and perfumery. Glišić et al. [9] investigated the floristic composition and diversity of seven urban habitat types in 24 Serbian cities with different climatic characteristics. Significant differences were observed in habitats based on diagnostic species and representation of life forms, demonstrating that habitat type influences species composition much more than climate. Molnár et al. [10] presented a survey of the Orchidaceae diversity in Azerbaijani cemeteries, highlighting that Azerbaijani cemeteries can be important refuges for rare and threatened orchids. Furthermore, *Epipactis turcica* Kreutz was reported for the first time for the flora of Azerbaijan. Dar et al. [11] has published an analysis of the vegetation composition and distribution of *Aeluropus lagopoides* (L.) Trin. ex Thwaites communities in five different regions of Saudi Arabia, with a floristic survey, emphasizing environmental factors influencing species distribution in highly saline ecosystems such as Sabkhas. The authors concluded that, given the economic potential of *Aeluropus lagopoides* as a forage plant and sand stabilizer, the conservation of its habitats is of great importance, adding that this grass could be integrated as a promising forage candidate in saline-affected areas, even in the dry season. Apostolova et al. [12] presented a survey on the vascular and cryptogamic floristic diversity of ancient mounds in Bulgaria, demonstrating that as

Citation: Wagensommer, R.P. Floristic Studies in the Light of Biodiversity Knowledge and Conservation. *Plants* **2023**, *12*, 2973. https://doi.org/10.3390/plants12162973

Received: 4 August 2023
Revised: 14 August 2023
Accepted: 14 August 2023
Published: 17 August 2023

many as 98% of the more than one thousand vascular plant taxa recorded were native species. In addition, the lichen *Arthopyrenia salicis* A.Massal. was reported for the first time from Bulgaria. Iamonico [13] explored the plant diversity of a protected area in the city of Rome, reporting the inventory of the vascular flora, with information on its biological, ecological and biogeographical composition, and notes of the physiognomy of vegetation and landscape. Furthermore, *Denisophytum bessac* (Choiv.) E.Gagnon & G.P.Lewis has been reported as an alien species new to Europe. Bartolucci et al. [14] confirmed the presence of *Ranunculus gracilis* E.D.Clarke in Italy, highlighting the morphological characteristics that distinguish it from similar species, and assessing its conservation status in Italy. They concluded that floristic research and the study of herbarium collections are of crucial importance in the conservation of vascular plant biodiversity, and are necessary to collect data to plan the correct conservation strategies. Bonsanto et al. [15] has published a survey on the rare and little known species *Scabiosa garganica* Porta & Rigo ex Wettst., with a morphological analysis on the populations of the Gargano, including the *Locus classicus* of the species, and an ecological characterization. The new morphological framework has made it possible to highlight the taxonomic autonomy of the species and has helped to clarify its relationship with the related species *Scabiosa holosericea* Bertol. and *S. taygetea* Boiss. & Heldr. Furthermore, a lectotype was designated for the name *Scabiosa garganica*, and the conservation status of the species was assessed.

The two reviews deal with plants from different geographical regions, namely America and Africa on one side, and the Mediterranean Basin on the other. Gómez-Maqueo and Gamboa-deBuen [16] presented the biological diversity of the genus *Ceiba* (Malvaceae), that includes 18 recognized species, distributed in America and Africa, and introduced in various countries, especially in Asia, due to their ornamental interest and potential uses for their fiber. The authors pointed out that *Ceiba* species are also considered potentially useful for restoring ecosystems affected by human activity. On the other hand, Accogli et al. [17] presented the halophytic species traditionally gathered in south-eastern Italy, with particular regard to their ecology and distribution, traditional uses, medicinal properties, commercialization and first cultivation attempts.

In conclusion, it is possible to state with certainty that even today floristic studies are of great importance, especially in geographic areas in which the flora is not well known [18], in protected areas [19], and in biodiversity hotspots [20], but also in territories where the flora is well known [21], where these studies still contribute to the knowledge and conservation of biodiversity.

Acknowledgments: I want to thank all the authors that contributed to this Special Issue, the reviewers for their valuable work, and the academic editors of the journal, who have helped me greatly during the editorial process.

Conflicts of Interest: The author declares no conflict of interest.

References

1. Wagensommer, R.P.; Fröhlich, T.; Fröhlich, M. First record of the southeast European species *Cerinthe retorta* Sibth. & Sm. (Boraginaceae) in Italy and considerations on its distribution and conservation status. *Acta Bot. Gall. Bot. Lett.* **2014**, *161*, 111–115. [CrossRef]
2. Wagensommer, R.P.; Bartolucci, F.; Fiorentino, M.; Licht, W.; Peccenini, S.; Perrino, E.V.; Venanzoni, R. First record for the flora of Italy and lectotypification of the name *Linum elegans* (Linaceae). *Phytotaxa* **2017**, *296*, 161–170. [CrossRef]
3. Perrino, E.V.; Silletti, G.N.; Erben, M.; Wagensommer, R.P. *Viola cassinensis* subsp. *lucana* (Violaceae), a new subspecies from the Lucanian Apennine, southern Italy. *Phyton* **2018**, *58*, 109–115. [CrossRef]
4. Wagensommer, R.P.; Venanzoni, R. *Geranium lucarinii* sp. nov. and re-evaluation of *G. kikianum* (Geraniaceae). *Phytotaxa* **2021**, *489*, 252–262. [CrossRef]
5. Turco, A.; Medagli, P.; Wagensommer, R.P.; D'Emerico, S.; Gennaio, R.; Albano, A. A morphometric study on *Ophrys* sect. *Pseudophrys* in Apulia (Italy) and discovery of *Ophrys japigiae* sp. nov. (Orchidaceae). *Plant Biosyst.* **2022**, *156*, 560–571. [CrossRef]
6. Balding, M.; Williams, K.J.H. Plant blindness and the implications for plant conservation. *Conserv. Biol.* **2016**, *30*, 1192–1199. [CrossRef]
7. Crisci, J.V.; Katinas, L.; Apodaca, M.J.; Hoch, P.C. The End of Botany. *Trends Plant Sci.* **2020**, *25*, 1173–1176. [CrossRef]

8. Perrino, E.V.; Valerio, F.; Jallali, S.; Trani, A.; Mezzapesa, G.N. Ecological and Biological Properties of *Satureja cuneifolia* Ten. and *Thymus spinulosus* Ten.: Two Wild Officinal Species of Conservation Concern in Apulia (Italy). A Preliminary Survey. *Plants* **2021**, *10*, 1952. [CrossRef]
9. Glišić, M.; Jakovljević, K.; Lakušić, D.; Šinžar-Sekulić, J.; Vukojičić, S.; Tabašević, M.; Jovanović, S. Influence of Habitat Types on Diversity and Species Composition of Urban Flora—A Case Study in Serbia. *Plants* **2021**, *10*, 2572. [CrossRef]
10. Molnár V., A.; Löki, V.; Verbeeck, M.; Süveges, K. Orchids of Azerbaijani Cemeteries. *Plants* **2021**, *10*, 2779. [CrossRef]
11. Dar, B.A.; Assaeed, A.M.; Al-Rowaily, S.L.; Al-Doss, A.A.; Abd-ElGawad, A.M. Vegetation Composition of the Halophytic Grass *Aeluropus lagopoides* Communities within Coastal and Inland Sabkhas of Saudi Arabia. *Plants* **2022**, *11*, 666. [CrossRef] [PubMed]
12. Apostolova, I.; Sopotlieva, D.; Valcheva, M.; Ganeva, A.; Shivarov, V.; Velev, N.; Vassilev, K.; Terziyska, T.; Nekhrizov, G. First Survey of the Vascular and Cryptogam Flora on Bulgaria's Ancient Mounds. *Plants* **2022**, *11*, 705. [CrossRef] [PubMed]
13. Iamonico, D. Biodiversity in Urban Areas: The Extraordinary Case of Appia Antica Regional Park (Rome, Italy). *Plants* **2022**, *11*, 2122. [CrossRef] [PubMed]
14. Bartolucci, F.; De Santis, E.; Conti, F. Nomenclatural Synopsis, Revised Distribution and Conservation Status of *Ranunculus gracilis* (Ranunculaceae) in Italy. *Plants* **2022**, *11*, 3094. [CrossRef]
15. Bonsanto, D.; Biscotti, N.; Wagensommer, R.P. New Morphological, Distribution, and Ecological Data on *Scabiosa garganica* (Caprifoliaceae), a Poorly Known Species of the Italian Flora, with Evaluation of Its Conservation Status and Typification of the Name. *Plants* **2023**, *12*, 1915. [CrossRef]
16. Gómez-Maqueo, X.; Gamboa-deBuen, A. The Biology of the Genus *Ceiba*, a Potential Source for Sustainable Production of Natural Fiber. *Plants* **2022**, *11*, 521. [CrossRef]
17. Accogli, R.; Tomaselli, V.; Direnzo, P.; Perrino, E.V.; Albanese, G.; Urbano, M.; Laghetti, G. Edible Halophytes and Halo-Tolerant Species in Apulia Region (Southeastern Italy): Biogeography, Traditional Food Use and Potential Sustainable Crops. *Plants* **2023**, *12*, 549. [CrossRef]
18. Figueiredo, E.; Smith, G.F.; César, J. The Flora of Angola: First Record of Diversity and Endemism. *Taxon* **2009**, *58*, 233–236. [CrossRef]
19. Jiménez, J.E.; Juárez, P.; Díaz, A. Checklist of the vascular flora of Reserva Biológica San Luis, Costa Rica. *Check List* **2016**, *12*, 1859. [CrossRef]
20. Noroozi, J.; Naqinezhad, A.; Talebi, A.; Doostmohammadi, M.; Plutzar, C.; Rumpf, S.B.; Asgarpour, Z.; Schneeweiss, G.M. Hotspots of vascular plant endemism in a global biodiversity hotspot in Southwest Asia suffer from significant conservation gaps. *Biol. Conserv.* **2019**, *237*, 299–307. [CrossRef]
21. Turco, A.; Albano, A.; Medagli, P.; D'Emerico, S.; Wagensommer, R.P. Orchidaceae in Puglia (Italy): Consistency, Distribution, and Conservation. *Plants* **2023**, *12*, 2223. [CrossRef] [PubMed]

plants

Article

Ecological and Biological Properties of *Satureja cuneifolia* Ten. and *Thymus spinulosus* Ten.: Two Wild Officinal Species of Conservation Concern in Apulia (Italy). A Preliminary Survey

Enrico V. Perrino [1,*], Francesca Valerio [2], Shaima Jallali [1], Antonio Trani [1] and Giuseppe N. Mezzapesa [1]

1 CIHEAM, Mediterranean Agronomic Institute of Bari, Via Ceglie 9, 70010 Valenzano, Italy; jellelishaima@yahoo.com (S.J.); trani@iamb.it (A.T.); mezzapesa@iamb.it (G.N.M.)
2 National Research Council of Italy, Institute of Sciences of Food Production, Via Amendola 122/O, 70126 Bari, Italy; francesca.valerio@ispa.cnr.it
* Correspondence: perrino@iamb.it or enricoperrino@yahoo.it; Tel.: +39-080-4606407

Abstract: This study evaluated the effects of ecology (plant community, topography and pedology), as well as of climate, on the composition of essential oils (EOs) from two officinal wild plant species (Lamiales) from Apulia, namely *Satureja cuneifolia* Ten. and *Thymus spinulosus* Ten. Few scientific data on their chemical composition are available, due to the fact that the first has a limited distribution range and the second is endemic of southern Italy. Results for both species, never officially used in traditional medicine and/or as spices, showed that the ecological context (from a phytosociological and ecological point of view) may influence their EO composition, and hence, yield chemotypes different from those reported in the literature. *S. cuneifolia* and *Th. spinulosus* can be considered good sources of phytochemicals as natural agents in organic agriculture due to the presence of *thymol* and *α-pinene*. Overall, the obtained trend for EOs suggests a potential use of both species as food, pharmacy, cosmetics and perfumery. Hence, their cultivation and use represent a positive step to reduce the use of synthetic chemicals and to meet the increasing demand for natural and healthier products.

Keywords: correlation; ecology; essential oils; Lamiaceae; vegetation

Citation: Perrino, E.V.; Valerio, F.; Jallali, S.; Trani, A.; Mezzapesa, G.N. Ecological and Biological Properties of *Satureja cuneifolia* Ten. and *Thymus spinulosus* Ten.: Two Wild Officinal Species of Conservation Concern in Apulia (Italy). A Preliminary Survey. *Plants* **2021**, *10*, 1952. https://doi.org/10.3390/plants10091952

Academic Editor: Calvin O. Qualset

Received: 25 August 2021
Accepted: 14 September 2021
Published: 18 September 2021

1. Introduction

Medicinal and aromatic plants (MAPs) have a long history of being used for multiple purposes, including for food and therapeutics. Before the advances of modern medicine, ancient civilizations discovered a wealth of useful therapeutic agents from within the plant and fungi kingdoms [1]. Knowledge of these medicinal preparations was passed down through generations and was occasionally recorded in the herbal literature [2]. Hence, the use of these wild officinal plants became a significant aspect of populations' cultural heritage and transformed into different traditions that kept on being passed down [3,4]. Much research in the field of biology, chemistry and medicine is directed at the identification and characterization of plant secondary metabolites with a pharmacological activity, which may be candidates for the synthesis of new drugs [5].

MAPs are defined as plants possessing aromatic and/or medicinal characteristics that can be extracted in the form of essential oils (EOs) through their secondary metabolites of bioactive properties [6]. These extracts are synthesized for several reasons, including plant protection against pathogens, insects, pollinators attraction and for allelopathic activity [7–9].

EOs are natural complexes of volatile and semi-volatile organic compounds, responsible for specific aroma, flavor and fragrance of certain plant species [10]. They can be stored by all plant organs (flowers, leaves, stems, twigs, seeds, fruits, roots, wood, or bark) and are synthesized in different histological structures such as secretory cells, cavities, canals or glandular trichomes [8,11].

The chemical structure of plant EOs falls into two distinct classes: terpenes and phenylpropanoids. Terpenes are formed by the combination of isoprene units such as monoterpenes (two units), sesquiterpene (three units), diterpenes (four units), and phenylpropanoids, which are aromatic and non-terpenoid compounds responsible in part for the fragrance of the plant [11,12]. These volatile oils represent a wide chemical diversity and are responsible for many plant applications, such as antimicrobial, antiparasitic, insecticidal, herbicidal, antioxidant and other medicinal properties.

The Lamiaceae family (formerly known as Labiatae) is the largest family of the order Lamiales, containing more than 245 genera and 7886 species [13], which grow in different natural ecosystems worldwide, with a particularly high concentration in the Mediterranean region, including Italy, providing a great level of biodiversity [14]. Most of the species belonging to this family are aromatic and possess EOs, making them valuable in cosmetics, perfumery, agriculture and medicine [15,16].

EOs can present a wide variability in their yield and chemical composition due to a series of intrinsic and extrinsic factors that affect the plants in their environment. In fact, it is well documented that the same species, under different environmental and geographical conditions, can produce EOs with different chemical profiles and biological properties [17–19]. In addition, some of these taxa are rare or endemic and have not yet been investigated for their ecology, chemical composition and biological properties.

The aim of the present research was to investigate two wild species of conservation concern (SCC; species for which we have concerns about its ability to remain on a landscape for a long time), *Satureja cuneifolia* Ten. (wild savory) and *Thymus spinulosus* Ten. (little thorn thyme), in order to: (a) perform ecological studies to understand the relationships existing between plant associations (sociology) and their surrounding habitat; (b) verify and eventually explain the ecological context influences on the composition of EOs; (c) determine chemical composition; (d) evaluate the potential for commercial purpose.

2. Study Area

Samplings were carried out in 2020 in Apulia. For *Satureja cuneifolia*, two sites, San Basilio (Mottola, province of Taranto) and Difesa di Malta (Fasano, province of Bari), were detected, while for *Thymus spinulosus*, three sites were identified, one of which, San Basilio (Mottola), was the same as *S. cuneifolia*, and the other two sites being Scannapecora (Altamura, province of Bari) and S. Egidio in S. Giovanni Rotondo (province of Foggia) (Figure 1).

The investigated sites have a Mediterranean macro-bioclimate, except S. Egidio, which is cooler, resulting Temperate. The bioclimate is oceanic, particularly the semicontinental transition at S. Egidio and Scannapecora, while the ombrotype is always the subhumid and dry type at Difesa di Malta (Table 1) (Phytoclimatic map—Italian Ministry for the Environment, Land and Sea, http://www.pcn.minambiente.it/viewer, accessed on 26 May 2021) [20] (Table 1).

Table 1. Phytoclimate at the investigated sites of *S. cuneifolia* (*Sc*) and *Th. spinulosus* (*Ts*).

Site (IC)	Macroclimate	Bioclimate	Ombrotype	Italian Distribution
S. Egidio (Ts3)	Temperate	Oceanic-semicontinental transition	Subhumid	Hilly areas of Apennines with Adriatic exposure
Scannapecora (Ts2)	Mediterranean			Coastal areas of the middle Adriatic, of the inland plains of the pre-Apennines and of Sicily
San Basilio (Sc2, Ts1)		Oceanic		Areas of the middle and lower Adriatic, the Ionian and the major islands; moderate presence also in the middle and upper Tyrrhenian regions
Difesa di Malta (Sc1)		Mediterranean oceanic	Dry	

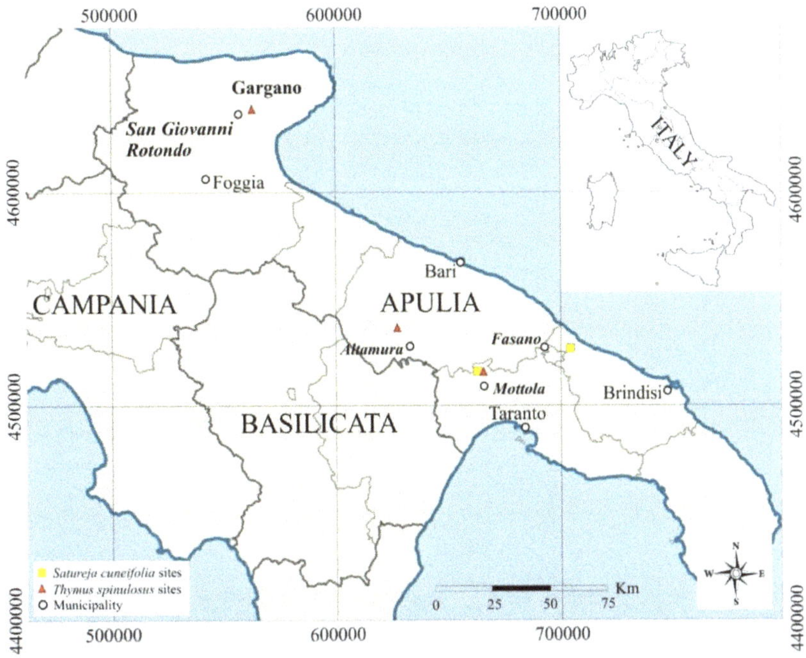

Figure 1. Site locations of *Satureja cuneifolia* and *Thymus spinulosus*.

Within the San Basilio site in the province of Taranto, the most widespread pedotype is Lithic Ruptic–Inceptic Haploxeralf fine, predominantly clayey, very thin, and very rocky with substrate within 50 cm [21]. At Scannapecora in the province of Bari, the geological type is that of Skeletal limestones of neritic and carbonate platform facies (Upper Cretaceous), on Tyrrhenian carbonate reliefs with material defined by calcareous sedimentary rocks and climate from Oceanic to Suboceanic Mediterranean, partially mountainous. The Difesa di Malta and S. Egidio sites share the same ecopedology as the Scannapecora site but different geopedology and geolithology, respectively, with Terrigenous-skeletal limestones such as "Panchina" (Pleistocene) and Colluvial, terraced alluvial, fluviolacustrine and fluvioglacial deposits (Pleistocene) (Geological, Geolithological and Ecopedologic maps—Italian Ministry for the Environment, Land and Sea, http://www.pcn.minambiente.it/viewer, accessed on 26 May 2021) [20] (Table 2).

Table 2. Geopedology, geolithology and ecopedology in the investigated sites of *S. cuneifolia* (*Sc*) and *Th. spinulosus* (*Ts*).

Site (IC)	Geo-Pedology and Geolithology	Ecopedology
S. Egidio (Ts3)	Colluvial, terraced alluvial, fluviolacustrine and fluvioglacial deposits (Pleistocene)	Tyrrhenian carbonate reliefs with material defined by calcareous sedimentary rocks and climate from Oceanic to Suboceanic Mediterranean partially mountainous
Scannapecora (Ts2)	Skeletal limestones of neritic and carbonate platform facies (Upper Cretaceous)	
San Basilio (Sc2, Ts1)		
Difesa di Malta (Sc1)	Terrigenous-skeletal limestones such as "Panchina" (Pleistocene)	Hilly reliefs with undifferentiated tertiary sedimentary rocks and sub-continental Mediterranean to continental Mediterranean climate

3. Materials and Methods

3.1. Vegetation Analysis

The field inspections of *Satureja cuneifolia* and *Thymus spinulosus* were carried out in 2020. A total of five vegetation surveys were conducted for both species, following the phytosociological method of the Zurich–Montpellier Sigmatist School [22], updated by Géhu and Rivas-Martínez [23], using the coefficient of abundance–dominance of identified species, well known to field botanists, and concerning the covering percentage of the relief area (value 5: species with more 75% coverage; value 4: species with coverage from 50% to 75%; value 3: species with coverage from 25% to 50%, value 2: species with coverage from 5% to 25%: value 1: species with coverage from 1% to 5%; value +: species with coverage less than 1%). Identification code, taxon, location, date, and geographical position (expressed in WGS84—World Geodetic System 1984) are reported in Table 3, while phytosociological and topographical data (identification code, altitude (m a. s.), aspect, slope (°), relevé area (m^2), stoniness (%), rockiness %), cover total (%), soil depth detected (cm), macroclimate, geolithological and eco-pedological types), endemicity, number of species identified, and number of individuals collected for laboratory analyses are reported in Tables 4 and 5. Other identified and recorded species not used for the phytosociological classification are not reported here. A total of 5 specimens for each species and for each relève were collected and deposited at the official herbarium of Bari University (Italy) (*Herbarium Horti Botanici Barensis*—BI).

Table 3. Location, date retrieved, and geographical position in the investigated sites of *S. cuneifolia* (*Sc*) and *Th. spinulosus* (*Ts*).

IC	Taxon	Location	Date Retrieved	Geographic Position (WGS84)
Sc1	*Satureja cuneifolia* Ten.	Difesa di Malta (Fasano–Brindisi)	03.06.20	40.7986 N 17.4946 E
Sc2		San Basilio (Mottola–Taranto)	26.05.20	40.6877 N 16.9694 E
Ts1	*Thymus spinulosus* Ten.	San Basilio (Mottola–Taranto)	05.06.20	40.6877 N 16.9693 E
Ts2		Scannapecora (Altamura–Bari)	03.06.20	40.9051 N 16.4422 E
Ts3		S. Egidio (S. Giovanni Rotondo–Foggia)	12.06.20	41.7132 N 15.7658 E

Table 4. *Phagnalo saxatilii-Saturejetum cuneifoliae* Biondi & Guerra 2008.

Identification Code		Sc1	Sc2
Altitude (m a. s. l.)		49	274
Aspect		NE	N
Slope (°)		3	4
Relevé area (m^2)		20	25
Stoniness (%)	endemic	20	10
Rockiness (%)		60	50
Cover total (%)		80	70
Number of species		23	41
Number of individuals collected for laboratory analysis		15	50
Geolithology		§	§§
Ecopedology		†	††
Macroclimate		≠	≠
Bioclimate		◊	◊◊
Ombrotype		#	##

Table 4. *Cont.*

Identification Code		Sc1	Sc2
Other species		10	10
Characteristics association ***Phagnalo saxatilii-Saturejetum cuneifoliae***			
Satureja cuneifolia Ten.		3	3
Phagnalon saxatile (L.) Cass.		+	-
Characteristics Alliance ***Cisto cretici-Ericion manipuliflorae*****, Order** ***Cisto-Micromerietalia julianae*****, Class** ***Ononido-Rosmarinetea***			
Helianthemum jonium Lacaita & Grosser		1	1
Micromeria graeca (L.) Benth. ex Rchb.		1	+
Asphodeline lutea (L.) Rchb.		-	1
Characteristics Alliance ***Hippocrepido glaucae-Stipion austroitalicae*****, Order** ***Scorzoneretalia villosae***			
Convolvulus cantabrica L.		-	2
Euphorbia nicaeensis All. subsp. *japygica* (Ten.) Arcang.		-	2
Teucrium capitatum L. subsp. *capitatum*		-	2
Petrorhagia saxifraga (L.) Link subsp. *gasparrinii* (Guss.) Greuter & Burdet		-	1
Eryngium amethystinum L.		-	+
Linum tommasinii (Rchb.) Nyman		-	+
Phleum hirsutum Honck. subsp. *ambiguum* (Ten.) Cif. & Giacom.		-	+
Scorzonera villosa Scop. subsp. *columnae* (Guss.) Nyman	E	-	+
Thesium humile Vahl		-	+
Thymus spinulosus Ten.	E	-	+
Characteristics Class ***Festuco-Brometea***			
Centaurea deusta Ten.		-	2
Anacamptis morio (L.) R.M. Bateman, Pridgeon & M.W.Chase		-	+
Galium corrudifolium Vill.		-	+
Poterium sanguisorba L.		-	+
Transgressive Class ***Helianthemetea guttati***			
Stachys romana (L.) E.H.L. Krause		1	1
Trifolium scabrum L.		1	1
Lagurus ovatus L. subsp. *ovatus*		+	1
Hypochaeris achyrophorus L.		+	+
Alkanna tinctoria Tausch subsp. *tinctoria*		-	2
Catapodium rigidum (L.) C.E. Hubb.		+	-
Aegilops ovata auct.		-	+
Festuca danthonii Asch. & Graebn. subsp. *danthonii*		-	+
Medicago truncatula Gaertn.		-	+
Plantago coronopus L.		-	+
Stipellula capensis (Thunb.) Röser & H.R. Hamasha		-	+
Transgressive Class ***Lygeo sparti-Stipetea tenacissimae***			
Daucus carota L. subsp. *carota*		1	+
Asphodelus ramosus L. subsp. *ramosus*		+	-
Convolvulus elegantissimus Mill.		+	-
Pallenis spinosa (L.) Cass. subsp. *spinosa*		+	-
Ferula communis L. subsp. *communis*		-	+
Reichardia picroides (L.) Roth		-	+
Legend E = Endemic species **Geopedology and Geolithology** §—Terrigenus-skeletal limestones like "Panchina" (Pleistocene) §§—Skeletal limestones of neritic and carbonate platform facies (Upper Cretaceous) §§§—Colluvial, terraced alluvial, fluviolacustrine and fluvioglacial deposits (Pleistocene) **Ecopedology** †—Hilly reliefs with undifferentiated tertiary sedimentary rocks and sub-continental Mediterranean to continental Mediterranean climate			

Table 4. *Cont.*

Identification Code	Sc1	Sc2
††—Tyrrhenian carbonate reliefs with material defined by calcareous sedimentary rocks and climate from Oceanic to Suboceanic Mediterranean, partially mountainous		
Macroclimate		
≠—Mediterranean		
≠≠—Temperate		
Bioclimate		
◊—Mediterranean Oceanic		
◊◊—Oceanic		
◊◊◊—Oceanic-semicontinental transition		
Ombrotype		
#—Dry		
##—Subhumid		

Identification of the taxa was carried out according to Flora d'Italia [24] and Flora Europea [25], with nomenclature standardized by "An updated checklist of the vascular flora native to Italy" [26] and "An updated checklist of the vascular flora alien to Italy" [27], while the syntaxonomic framework was conceived by several contributions [28–30], reported in the phytosociological tables and summarized in the syntaxonomic scheme.

The aerial parts of the studied plant species were harvested in the same sites and date of vegetation surveys. The number of individuals collected (from 15 to 100) for the laboratory analysis is reported in the phytosociological tables (Tables 4 and 5).

Table 5. Physico-chemical characteristics of *S. cuneifolia* (*Sc*) and *Th. Spinulosus* (*Ts*) for each investigated soil site.

Soil Parameters	Unit	Sc1	Sc2	Ts1	Ts2	Ts3
Sand (2–0.05 mm)	$g \cdot kg^{-1}$	307	576	783	200	161
Silt (0.05–0.002 mm)	$g \cdot kg^{-1}$	649	383	193	759	787
Clay (<0.002 mm)	$g \cdot kg^{-1}$	44	41	23	41	51
Texture Class (USDA)		Silt loam	Sandy loam	Loam sand	Silt loam	Silt loam
Total Carbonate	$g \cdot kg^{-1}$	306	698	965	19	<0.1
pH H_2O	-	8.1	8.2	7.9	7.9	6.9
pH $CaCl_2$	-	7.5	7.4	7.5	7.4	6.4
Electrical Conductivity (1:2)	$dS \cdot m^{-1}$	0.2	0.2	0.3	0.2	0.2
Total Nitrogen	$g \cdot kg^{-1}$	6.0	3.5	3.7	6.8	6.7
Available P_2O_5	$mg \cdot kg^{-1}$	103	43	88	27	24
Organic Carbon	$g \cdot kg^{-1}$	63	34	42	62	69
Organic Matter	$g \cdot kg^{-1}$	109	59	72	107	119
C/N	-	10.5	10	11.2	9.2	10.3
Ca exchangeable	$mg \cdot kg^{-1}$	9772	4842	4146	10,450	7711
K exchangeable	$mg \cdot kg^{-1}$	253	93	87	515	855
Mg exchangeable	$mg \cdot kg^{-1}$	240	131	105	231	368
Na exchangeable	$mg \cdot kg^{-1}$	68	40	15	85	81
Available Fe	$mg \cdot kg^{-1}$	13.1	7.8	8.9	20.3	31.9
Available Mn	$mg \cdot kg^{-1}$	7.9	15.1	7.6	8.1	19.9
Available Zn	$mg \cdot kg^{-1}$	2.5	20.5	8.9	1.6	3.0
Available Cu	$mg \cdot kg^{-1}$	0.7	1.0	0.9	1.9	1.9
Available Ni	$mg \cdot kg^{-1}$	0.27	0.27	0.14	0.16	0.21
Available Co	$mg \cdot kg^{-1}$	0.03	0.06	0.03	0.03	0.11

3.2. *Essential Oils Analysis*

The essential oils (EOs) of the two wild plant species were extracted by hydrodistillation [31] using a Clevenger type apparatus for 4–5 h. Air-dried plant material (100 g) was covered with 500 mL of distilled water in a 1L volume distillation flask (extraction ratio

1:5 w/vol). EOs were collected in amber glass vials, weighed, and stored at 4 °C. EO yields (% *w/w*) were determined as grams of EOs per 100 g of dry weight plant material.

Identification of the compounds present in the EO extracts was carried out by gas chromatography coupled with mass spectrometry (GC–MS) using a Clarus 680 GC interfaced with a Clarus SQ8C single quadrupole mass spectrometer (PerkinElmer, Waltham, MA, USA) equipped with an Elite-5 MS (PerkinElmer, Waltham, MA, USA) fused silica capillary column (30 m × 0.25 mm and 0.25 μm film thickness). Mass spectra of the target compounds were obtained by an electron impact ionization system with standardized ionization energy of 70 e V. Helium 5.5 was used as carrier gas at a constant flow rate of 1ml/min. Mass transfer line and injector temperatures were set at 280 °C and the oven temperature was programmed from 50 °C to 160 °C at 5 °C/min, then raised to 250 °C at 10° C/min, and held at the final temperature for 5 min. Diluted samples (1:10, *v/v*, in hexane) were injected in split mode with a split ration of 1:100. Data were collected in full scan mode in the range 40–300 amu. A solvent delay of 4 min was applied. Qualitative results include compound identification and area percentage of related peaks in the total ions chromatogram (Table 5).

Compounds identification was performed by both Kovats retention indexes (RI) [32,33] and mass spectra (MS) searches in NIST and Wiley databases (Table 6).

Table 6. EO composition as area percentage of *S. cuneifolia* (*Sc*) and *Th. spinulosus* (*Ts*) with the identification method used. RT: retention time; RI: Kovats retention index; RI: experimental retention index; RI ref.: bibliographic retention index according to [32].

Compound Name	RT	RI	RI ref.	Sc1	Sc2	Ts1	Ts2	Ts3
α-Thujene	6.54	925	927	-	-	1.84	1.59	1.84
α-Pinene	6.74	932	936	36.8	38.82	0.95	0.73	0.83
Camphene	7.16	949	950	1.66	0.49	0.52	0.28	0.28
2,4(10)-Thujadiene	7.25	952	943	1.28	0.36	-	-	-
Sabinene	7.75	971	973	0.74	1.71	0.13	0.09	0.08
β-Pinene	7.89	977	977	0.36	0.38	0.87	0.72	0.94
α-Myrcene	8.16	987	989	0.84	1.43	1.33	1.1	0.83
2,3-Dehydro-1,8-cineole	8.19	989	989	1.78	0.99	-	-	-
3-Octanol	8.34	994	993	0.09	0.07	-	-	-
α-Phellandrene	8.63	1005	1004	0.81	0.71	0.18	0.14	0.18
α-Terpinene	8.94	1016	1017	0.14	0.11	2.24	1.84	2.9
p-Cymene	9.15	1023	1024	1.55	0.63	17.87	17.51	17.5
Limonene	9.29	1028	1029	5.08	6.39	0.51	0.41	1.43
β-Ocimene. (Z)	9.41	1033	1037	2.8	3.62	2.48	1.14	0.9
β-Ocimene. (E)	9.74	1044	1047	1.17	1.52	15.4	11.68	10.97
γ-Terpinene	10.10	1057	1060	0.25	0.19	-	-	-
Sabinene hydrate, *cis*-	10.43	1069	1066	0.12	0.1	0.45	0.49	0.52
Terpinolene	10.88	1085	1087	0.61	0.3	0.14	0.14	0.14
Linalool	11.25	1098	1099	6.35	6.36	1.83	2.41	2.9
Hotrienol	11.34	1101	1106	0.12	0.13	0.11	-	-
Thujone, cis	11.45	1104	1105	0.1	0.05	0.06	0.49	0.05
Chrysanthenone	11.89	1120	1124	0.31	0.1	-	-	-
α-Campholenal	12.00	1124	1124	1.05	0.33	0.05	0.05	0.19
Pinocarveol, *trans*-	12.47	1141	1140	1.78	0.78	-	-	-
trans-3-caren-2-ol	12.58	1146		4.1	1.9	-	-	-
Pinocarvone	13.05	1160	1160	0.33	0.12	-	-	-
Borneol	13.31	1170	1166	6.9	1.4	1	0.5	0.5
α-Terpineol	13.97	1193	1190	11.03	17.11	0.29	0.22	0.19
Carveol, *trans*-	14.60	1217	1217	0.92	0.35	-	-	-
Carvacrol, methyl ether	15.16	1238	1243	-	-	2.05	3.19	0.54
Bornyl acetate	16.42	1284	1283	0.21	0.07	-	-	-
Thymol	16.58	1289	1290	-	-	42.87	48.77	45.88
Carvacrol	16.77	1297	1300	-	-	0.94	0.62	2
Caryophyllene, (E)	20.01	1422	1420	0.07	0.08	2.07	1.22	2.22

Table 6. *Cont.*

Compound Name	RT	RI	RI ref.	Sc1	Sc2	Ts1	Ts2	Ts3
Aromadendrene	20.48	1441	1440	0.21	0.35	-	-	-
γ-Curcumene	21.39	1478	1480	0.16	0.18	-	-	-
α-Bisabolene, (Z)	21.88	1497	1503	1.52	2.6	-	-	-
γ-Bisabolene	22.14	1509	1508	0.2	0.39	0.96	1.05	0.73
Caryophyllene oxide	23.75	1581	1580	3.25	2.44	-	0.14	-
Guaia-1(10),11-diene	24.81	1637	-	0.31	0.25	0.38	0.37	0.51
Other components				5	7.19	2.48	3.11	4.95

The chemical composition (%) of the main compounds of *Satureja cuneifolia* and *Thymus spinulosus* is reported in Table 7. Statistical analysis was not applicable due to the low number of samples per species and site, limited amount of material per sample, and few and variable numbers of individuals per specimen (15–70), also due in part to the fact that the study was carried out on two species of conservation concern, which limits the harvest of material.

Table 7. *Hippocrepido glaucae-Stipion austroitalicae* Forte & Terzi in Forte, Perrino and Terzi 2005.

Identification Code	endemic	Ts1	Ts2	Ts3
Altitude (m a. s.l.)		274	561	626
Aspect		N-NE	S-SW	SE
Slope (°)		3	10	12
Relevé area (m^2)		50	50	50
Stoniness (%)		5	20	10
Rockiness (%)		60	40	45
Cover total (%)		70	70	85
Number of species		38	44	64
Number of individuals collected for laboratory analysis		70	80	60
Geolithology		§§	§§	§§§
Ecopedology		††	††	††
Macroclimate		≠	≠	≠≠
Bioclimate		◊◊	◊◊◊	◊◊◊
Ombrotype		##	##	##
Other species		12	9	22
Characteristics Association *Sideritido syriacae-Stipetum austroitalicae*				
Sideritis italica Mill.		-	-	2
Characteristics Association *Acino suaveolentis-Stipetum austroitalicae*				
Thymus spinulosus Ten.	E	2	2	1
Euphorbia nicaeensis All. subsp. *japygica* (Ten.) Arcang.	E	1	+	-
Clinopodium suaveolens (Sm.) Kuntze		-	1	-
Characteristics Alliance *Hippocrepido glaucae-Stipion austroitalicae*				
Stipa austroitalica Martinovský subsp. *austroitalica*	E	+	3	2
Petrorhagia saxifraga (L.) Link subsp. *gasparrinii* (Guss.) Greuter & Burdet		1	1	+
Linum tommasinii (Rchb.) Nyman		+	-	-
Hippocrepis glauca Ten.		-	-	+
Characteristics Order *Scorzoneretalia villosae*				
Teucrium capitatum L. subsp. *capitatum*		1	1	1
Thesium humile Vahl		+	1	-
Phleum hirsutum Honck. subsp. *ambiguum* (Ten.) Cif. & Giacom.		1	-	3
Convolvulus cantabrica L.		1	-	+
Scorzonera villosa Scop. subsp. *columnae* (Guss.) Nyman	E	-	+	2
Koeleria splendens C. Presl	E	-	1	+
Eryngium amethystinum L.		-	+	1
Satureja montana L. subsp. *montana*		-	4	-
Anthyllis vulneraria L. subsp. *rubriflora* (DC.) Arcang.		-	1	-
Teucrium capitatum L. subsp. *capitatum*		1	1	1

Table 7. *Cont.*

Identification Code		Ts1	Ts2	Ts3
Characteristic Class *Festuco-Brometea*				
Eryngium campestre L.		+	2	-
Centaurea deusta Ten.		+	-	+
Festuca circummediterranea Patzke		-	2	+
Poterium sanguisorba L.		-	+	1
Filipendula vulgaris Moench		-	-	+
Euphorbia spinosa L.		-	-	1
Helianthemum oelandicum (L.) Dum. Cours. subsp. *incanum* (Willk.) G. López		-	-	1
Melica ciliata L. subsp. *ciliata*		-	-	+
Potentilla calabra Ten.	E	-	-	+
Trifolium arvense L.		-	-	+
Trifolium campestre Schreb.		-	-	+
Transgressive Class *Helianthemetea guttati*				
Lagurus ovatus L.		1	1	2
Aegilops ovata auct.		+	1	+
Hypochaeris achyrophorus L.		+	1	+
Stachys romana (L.) E.H.L. Krause		+	1	+
Trifolium scabrum L.		1	1	-
Alkanna tinctoria Tausch subsp. *tinctoria*		1	+	-
Ononis reclinata L.		+	+	-
Stipellula capensis (Thunb.) Röser & H.R. Hamasha		+	+	-
Avena barbata Pott ex Link		+	-	+
Trifolium stellatum L.		+	-	+
Linum strictum L.		-	+	1
Brachypodium distachyon (L.) P. Beauv.		-	+	+
Helianthemum salicifolium (L.) Mill.		1	-	-
Onobrychis aequidentata (Sm.) d'Urv.		-	+	-
Briza maxima L.		-	-	1
Hippocrepis ciliata Willd.		-	-	1
Transgressive Class *Lygeo sparti-Stipetea tenacissimae*				
Asphodelus ramosus L. subsp. *ramosus*		+	2	2
Reichardia picroides (L.) Roth		1	-	+
Daucus carota L.		+	-	+
Carlina corymbosa L.		-	+	+
Convolvulus elegantissimus Mill.		-	+	+
Petrosedum ochroleucum (Chaix) Niederle		-	+	+
Ferula communis L. subsp. *communis*		+	-	-
Plantago bellardii All. subsp. *bellardii*		-	1	-
Dactylis glomerata L. subsp. *hispanica* (Roth) Nyman		-	-	+
Pallenis spinosa (L.) Cass. subsp. *spinosa*		-	-	+
Phlomis herba-venti L. subsp. *herba-venti*		-	-	+
Transgressive Class *Ononido-Rosmarinetea*				
Micromeria graeca (L.) Benth. ex Rchb.		-	1	1
Satureja cuneifolia Ten.		-	3	-
Helianthemum jonium Lacaita & Grosser		-	2	-
Rhamnus saxatilis Jacq.		-	+	-

Legend
E = Endemic species
Geopedology and Geolithology
§—Terrigenus-skeletal limestones like "Panchina" (Pleistocene)
§§—Skeletal limestones of neritic and carbonate platform facies (Upper Cretaceous)
§§§—Colluvial, terraced alluvial, fluviolacustrine and fluvioglacial deposits (Pleistocene)
Ecopedology
†—Hilly reliefs with undifferentiated tertiary sedimentary rocks and sub-continental Mediterranean to continental Mediterranean climate

Table 7. *Cont.*

Identification Code	Ts1	Ts2	Ts3
††—Tyrrhenian carbonate reliefs with material defined by calcareous sedimentary rocks and climate from Oceanic to Suboceanic Mediterranean, partially mountainous			
Macroclimate			
≠—Mediterranean			
≠≠—Temperate			
Bioclimate			
◊—Mediterranean Oceanic			
◊◊—Oceanic			
◊◊◊—Oceanic-semicontinental transition			
Ombrotype			
#—Dry			
##—Subhumid			

3.3. Soil Analysis

Soil samples were collected from each vegetation sample of topsoil in the following way: in 5–9, subsamples (depending on the habitat extension; each subsample ~ 100 g) were taken from the first 0–20 cm of depth.

Soil samples were sieved to <2 mm and stored for physical and chemical analysis. Particle size distribution was determined by the pipette method, whereas textural class was categorized according to the USDA classification [34]. Soil pH was measured both in water and saline solution using 1:2.5 ratio (*w/v*). Electrical conductivity (EC) was measured on a soil to distilled water ratio (1:2, *w/v*). Soil organic carbon (OC) was determined according to the Walkley and Black method as described by Nelson and Sommers [35]. Total nitrogen (N) was measured following the Kjeldahl method as described in Bremner [36]. Available P was measured in sodium bicarbonate alkaline soil extracts and determined colorimetrically [37]. The total carbonate content in the soils was determined using a Dietrich–Fruhling calcimeter. Exchangeable bases were extracted by $BaCl_2$ (1 M) and triethanolamine solution, while available microelements were extracted by DTPA and triethanolamine solution. After extractions, both were analyzed by ICP-OES.

4. Results and Discussions

4.1. Satureja Cuneifolia Ten.

S. cuneifolia (synonym: *Satureja montana* subsp. *cuneifolia* (Ten.) O. Bolós & Vigo) is shown in Figure 2.

Herbarium samples are shown in Figure 3: The sample from San Basilio, in the municipality of Mottola (Taranto), garrigue, 26 May 2020, *legit* and *determinavit* determinative *E.V. Perrino* 42457 (*Herbarium Horti Botanici Barensis*—BI), is shown in Figure 3a. The sample from Difesa di Malta, in the municipality of Fasano (Brindisi), garrigue, 3 June 2020, *legit* and *determinavit E.V. Perrino* 42458 (*Herbarium Horti Botanici Barensis*—BI), is shown in Figure 3b.

(a) (b)

Figure 2. *S. cuneifolia* (**a**) in flowering; (**b**) in its habitat, *Phagnalo saxatilii-Saturejetum cuneifoliae*. San Basilio, 26 March 2020.

(a) (b)

Figure 3. *S. cuneifolia* herbarium samples: (**a**) San Basilio (Mottola) (BI 42457); (**b**) Difesa di Malta (Fasano) (BI 42458).

The genus *Satureja* L. comes from the Latin "*satureia*" and was first named by Roman writer Pliny [38]. It means "*herb of satyrs*" and, for this reason, its cultivation was banned in monasteries [39]. *S.* belongs to the Lamiaceae family, sub-family *Nepetoideae*, and the tribe *Mentheae* [40] including about 200 wild species of herbs and shrubs, often aromatic, widely distributed in the Mediterranean area, Asia and boreal America [41,42]. More than 30 of them grow in eastern parts of the Mediterranean area [43,44], up to 1200 m above sea level, in arid, sunny, stony and rocky habitats [45], 5 of which are in Italy: *S. cuneifolia* Ten.,

S. montana L. subsp. *montana*, *S. montana* L. subsp. *variegata* (Host) P.W. Ball, *S. subspicata* Bartl. ex Vis. subsp. *liburnica* Šilic, and *S. thymbra* L. [26]. *S. cuneifolia* Ten. is a SE European entity reported in Italy, Albania, Greece, Croatia and Montenegro [46,47], with its western limit of distribution in Italy, where the species occurs only in Apulia, throughout the Region, in Salento, Murge and Gargano [48–52], and in the nearby Basilicata, in the Murge di Matera [53]; it was recorded by mistake at Laino Borgo in Calabria [24,26]. As highlighted by some authors [54], this taxon is one of the many plant species with eastern distribution that exhibit a disjunct Italian distribution more or less restricted to the Apulian region, such as *Bromus parvispiculatus* H. Scholz [55], *Carex phyllostachys* C. A. Mey. [56,57], *Cerinthe retorta* Sm. [58], *Linum elegans* Spruner ex Boiss. [59], *Scrophularia lucida* L. [60,61], and *Ophrys oestrifera* complex [62].

S. cuneifolia grows from sea level to 600 m of altitude, and is a diagnostic taxon of three plant communities: (a) *Phagnalo saxatilii-Saturejetum cuneifoliae* Biondi & Guerra (2008) associated with *Phagnalon saxatile* (L.) Cass.; (b) *Sedo ochroleuci-Saturejetum cuneifoliae* Di Pietro & Misano (2010) associated with the endemism *Petrosedum ochroleucum* (Chaix) Niederle subsp. *mediterraneum* (L. Gallo) Niederle; (c) *Stipo austroitalicae-Seslerietum juncifoliae* Di Pietro & Wagensommer (2014) subass. *typicum* [54]. These vegetations fall into the *Ononido-Rosmarinetea* class, with the first two (a, b) described for central of Apulia in the "Gravine" gorges [49,63], while the third (c) for the Gargano [50,54]. The *Stipo austroitalicae-Seslerietum juncifoliae* subass. *typicum* grows on small rocky outcrops emerge from a matrix composed of limestone debris, found within the steep slopes of the gorges, which cut across the southern side of the Gargano promontory [54]; it could be considered a distinct aspect of the habitat "Calcareous rocky slopes with chasmophytic vegetation" (code 8210) according to Directive 92/43 EEC [64,65].

Satureja species have been traditionally used in the treatment of many diseases such as nausea, indigestion, cramps, diarrhea, infectious diseases and muscle pains [66,67], in relation to the EOs secreted by the glands found on their leaf surface, as with other aromatic plants belonging to some genera of the Lamiaceae family.

Previous studies from the Mediterranean area showed variability in chemical composition in EOs of *S. cuneifolia*, attributed to many factors such as climate, regional, local and ecological conditions, growth stages, harvesting period, and genetics [67–71]. In three different localities of Dalmatia (Croatia), corresponding to as many stages of development, the yield of the EOs ranged from 0.1% to 0.6%, and all samples had a low percentage of thymol (0–3.5%) and carvacrol (0–16.3%), but were relatively rich in α-pinene (1.3–20.7%), limonene (1.8–17.4%), γ-terpinene (0–5.6%), and p-cymene (1.8–14.8%) [69]. In a research study carried out in the southwestern part of Montenegro (Lovćen National Park), a high percentage of linalool (20.3%) has been found, the most abundant compound, followed by much lower values of α-terpineol (3.8%) and borneol (3.6%), while limonene (1.1%) and α-pinene (0.7%) were under-represented, and carvacrol and thymol were even absent [67]. In high altitude conditions and precisely from Sogut mountain in Turkey, *S. cuneifolia* collected during the flowering stage in August recorded elevated values of carvacrol (45%), p-cymene (21.6%), and then, lower but appreciable values thymol (9%), γ-terpinene (4.3%), and borneol (2.5%) [72]. A further study [70] conducted in 12 wild sites far east of the Italian peninsula at lower altitudes than Turkey showed significant percentage changes in the main compounds observed: α-pinene (0.3–26.4%), linalool (0.2–45.8%) and borneol (1.2–41.1%), with linalool and α-pinene not always present. An accurate reading of the Italian data also highlights that linalool and α-pinene reached high values (>35% and >14%, respectively) only in four samples, while borneol was the most present chemotype.

S. cuneifolia from Croatia and Montenegro is very different from that collected in the Mediterranean parts of Turkey and eastern Italy (Apulia). EOs of Croatian and Montenegro origin have a low content (even absent) of both thymol and carvacrol, as observed in Italian sites, while, in general, EOs of plants from Turkey have a mean percentage of thymol accompanied by a high percentage of carvacrol or vice versa. In Italy, the values are even more extreme and variable because not only have thymol and carvacrol not been recorded, but they have considerable fluctuations within the same compound in comparable environments. The authors [70] explain that this intraspecific polymorphism indicates that homogeneous environments may host different genotypes.

The data collected about *S. cuneifolia* show that the vegetation belongs to *Phagnalo saxatilii-Saturejetum cuneifoliae* at both investigated sites (Table 4), where there is the same Mediterranean macroclimate, but different bioclimate and ombrotype, respectively: Dry Mediterranean Oceanic under the influence of the sea at Difesa di Malta, and Subhumid Oceanic in the hinterland at San Basilio. In addition to the climatic differences, there are also geolithological and ecopedological differences, with a more recent Pleistocene geology near the sea, where we find Terrigenous-skeletal limestones such as "Panchina", and older in the hinterland (Upper Cretaceous) with skeletal limestones of neritic and carbonate platform facies. Differences were confirmed by Hilly reliefs with undifferentiated tertiary sedimentary rocks at Difesa di Malta and Tyrrhenian carbonate reliefs with material defined by calcareous sedimentary rocks at San Basilio. This means that the two sites only have the same macroclimate, but different bioclimate, geology, and pedology characteristics, and partially different plant communities. In fact, within the same plant association, we observe less human-made interference in San Basilio than Difesa di Malta, and a greater biodiversity due to the many species referable to the *Hippocrepido glaucae-Stipion austroitalicae* alliance and *Scorzoneretalia villosae* order, enhanced with two endemic taxa: *Scorzonera villosa* subsp. *columnae* and *Thymus spinulosus* Ten. The vegetation detected in San Basilio deserves more analysis from a syntaxonomic point of view, through ad hoc surveys in the surrounding areas, to better define this distinct aspect with *S. cuneifolia*.

The soil parameters show some differences between the two sites (Sc1, Sc2). In the more disturbed aspect, less natural, in a cultivated context, as Difesa di Malta (Sc1), a fine-silt loam soil has been recorded, characterized by poor total carbonate, very high phosphorus availability and high total nitrogen and organic carbon, while at San Basilio (Sc2), in a more sandy soil, lower and moderate values were recorded, except for the total carbonate which, here, has a much higher concentration (Table 5), confirming the geopedological and geolithological features (Table 4).

The exchangeable cation and available microelement (macro and micronutrients, respectively) soil contents showed that all samples are rich in exchangeable cation but less rich in available microelements, with differences in types and percentage at both sites. The available microelements have comparable concentrations at both sites, except for Zn and Mn, which are more abundant in natural contexts, unlike the Fe that prevails in agricultural land (Table 5).

The environmental, geological, ecological, vegetational and pedological differences have a slight effect on the phytochemical properties of *S. cuneifolia*, remembering that the San Basilio site is located at a higher altitude (274 m), has a major plant biodiversity, and older soils, compared to the maritime site of Difesa di Malta (49 m), which is situated in a more agricultural context and in a more recent soil. In particular, the correlation between EO compositions was meaningful and positive with the monoterpenoid alkene and monoterpenoid alkene alcohol, which are the most abundant classes in *S. cuneifolia* EOs. Additionally, monoterpenoid alkenes have a significant positive correlation with soil pH, and significant negative correlation with total nitrogen and soil organic carbon (Table 5).

In all, 36 compounds were identified at the Difesa di Malta (Sc1) and San Basilio (Sc2) sites, but with some quantitative differences (Table 6). Most of the compounds are present with similar percentages at both sites. In particular, α-Pinene, the most abundant phytochemical, is present with the highest percentage at both sites (36.8% and 38.8%, respectively), the same slight difference occurs for limonene (5.1% and 6.4%), but significant differences arise for α-Terpineol (11% and 17.1%), trans-3-caren-2-ol (4.1% and 1.9%), and Borneol (6.9% and 4.1%). The rest of the compounds for both sites did not exceed 5%.

Finally, even other minor compounds, not mentioned, were present with slightly different percentages.

4.2. *Thymus Spinulosus Ten.*

Th. spinulosus (synonym: *Th. paronychioides* Celak.) is shown in Figure 4.

(a) **(b)**

Figure 4. *Th. spinulosus* (**a**) in flowering; (**b**) in its habitat, *Hippocrepido glaucae-Stipion austroitalicae*. Scannapecora, 03.06.2020.

Herbarium samples are shown in Figure 5: The samples from San Basilio, in the municipality of Mottola (Taranto) (garrigue, 5 June 2020, *legit* and *determinavit E.V. Perrino* 42454 (*Herbarium Horti Botanici Barensis*—BI) and San Egidio, in the municipality of San G. Rotondo (Foggia), garrigue, 12 June 2020, *legit* and *determinavit E.V. Perrino* 42455 (*Herbarium Horti Botanici Barensis*—BI) are shown in Figure 5a. The samples from Scannapecora, in the municipality of Altamura (Bari), garrigue, 3 June 2020, *legit* and *determinavit E.V. Perrino* 42456 (*Herbarium Horti Botanici Barensis*—BI) are shown in Figure 5b.

The genus *Thymus*, described by Carl Linnaeus in *Species Plantarum* [73], is one of the most important genera within the family Lamiaceae, due to its high number of species, commercial uses and medicinal features [74]. It includes 220 accepted taxa distributed in Europe, Northwest Africa, Ethiopia, Asia and southern Greenland [75–77]. The center of diversity of the genus is the Mediterranean region, where the typical endemism grows especially in the west sector (Iberian peninsula, Northwest Africa) [78]. In Italy, 20 species of *Th.* are reported, with 5 endemims of the southern regions (*Th. paronychioides* Čelak., *Th. picentinus* (Lacaita) Bartolucci, *Th. praecox* Opiz subsp. *parvulus* (Lojac.) Bartolucci, Peruzzi & N.G. Passal., *Th. richardii* Pers. subsp. *nitidus* (Guss.) Jalas, and *Th. spinulosus* Ten.) [26], where they occur mainly in rocky habitats, from hills to mountain tops. *Th. spinulosus* is reported in Campania, Apulia, Basilicata, Calabria and Sicily, while its presence is uncertain in Molise, no longer recorded in Latium, and reported by mistake in Abruzzo [26]. It grows in arid stony slopes, breached, clearings of deciduous oaks, from sea level up to about 1100 m of altitude and flowering in May–June. In Apulia, it is widespread in the perennial grasslands, especially in the pseudo-steppes of *Stipa austroitalica* Martinovský [29,54,79], which is considered is habitat (Eastern sub-Mediterranean dry grasslands (*Scorzoneretalia villosae*)" (code 62A0)) under Directive 92/43 EEC [64,65], while it is very rare in typically rocky vegetation [54].

Many species of *Thymus* are popular in the traditional medicine of many countries and nations as a source of valuable crude drug, where they are used for acute and chronic bronchitis, and the leaf extracts are prescribed as expectorant and analgesic [80]. In medicine and perfumery, *Th. serpyllum* and *Th. vulgaris* are widely used [81,82], and in general, many species of this genus show antimicrobial, anti-inflammatory, antioxidant, cytotoxic, spasmolytic, and antinematodal activities. *Thymus linearis* Benth. is used in Kashmir regions for gastrointestinal problems and fever issues [83]. *Thymus capitatus* (L.) Hoffmanns & Link (=*Thymbra capitata* (L.) Cav.) were used in Southern Italy to make fumigations in the treatment of colds together with dried figs and chestnuts [84]. Traditionally, the plants of the genus *Th.* are used as a source of EOs and are promising in pest management and control of harmful insects, especially thanks to thymol and carvacrol content, as well as citral, geraniol and nerolidol [85–87]. *Th.* still need to be explored as supplementary sources of botanical insecticides, to face the growing concerns arising from the use of chemical pesticides [88–90], with special reference to insecticide resistance [91]. *Th. spinulosus* is a less explored taxon for both phytochemistry and biological activities [92], and there are few works in the literature detailing information on its use in the kitchen such as the preparation called "sanguinaccio" in the Campania region, based on pig blood with dried fruit, sultanas and cooked wine [93]. The Sicilian data, obtained from four populations located above 1000 m of altitude, but in different soil conditions (siliceous and calcareous), gave oils with different composition, 62 components in total, which is inconsistent with the previous chemical composition of EOs described in the literature for this endemic taxon, suggesting a new chemotype, characterized by myrcene-limonene among monoterpenes and γ-muurolene, caryophyllene and germacra-1(10),4-dien-6-ol among sesquiterpenes as the main constituents [87].

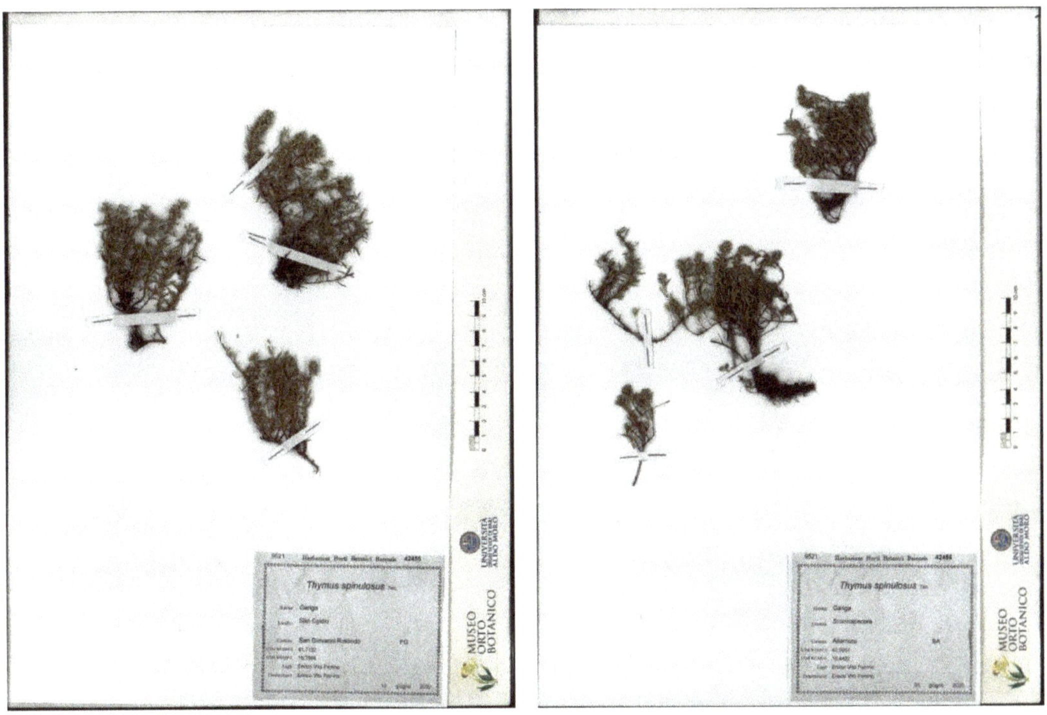

(a) (b)

Figure 5. *Th. spinulosus* herbarium samples: (**a**) San Egidio (S. G. Rotondo) (BI 42455); (**b**) Scannapecora (Altamura) (BI 42456).

The macroclimate in the two eastern sites (Ts1, Ts2) is Mediterranean, while in the Gargano, it is colder and temperate (Ts3). The bioclimate is Oceanic-semicontinental transition and subhumid Ombrotype in the two sunniest sites placed at higher altitudes (Ts2, Ts3), and Dry Mediterranean Oceanic type in cooler expositions, as observed at San Basilio (Ts1). The Gargano site differs from the other two sites also due to its geolithological characteristics, more recent (Pleistocene) for colluvial, terraced alluvial, fluviolacustrine and fluvioglacial deposits (Ts3) compared to those of the upper Cretaceous corresponding to skeletal limestones of neritic and carbonate platform facies (Ts1, Ts2). Additionally, the geopedology is different, with Tyrrhenian carbonate reliefs and material defined by calcareous sedimentary rocks on Gargano (Ts3), and Hilly reliefs with undifferentiated tertiary sedimentary rocks in the other two sites (Ts1, Ts2). This means that the three sites only have the same macroclimate, but the Gargano site (Ts3) differs from the other two due to different bioclimate, geology, and pedology characteristics, and partially different plant communities.

The phytosociological data (Table 7) show within the same alliance *Hippocrepido glaucae-Stipion austroitalicae*, there are two different plant associations, *Acino suaveolentis-Stipetum austroitalicae* (Ts1, Ts2) and *Sideritido syriacae-Stipetum austroitalicae* (Ts3). At S. Egidio (Ts3), in a good environmental condition, in the absence of human disturbances, with 626 m altitude, 12° slope, a medium level of stoniness (10%) and a high level of rockiness (45%), and 85% coverage of 64 plant species, we observed within the *Sideritido syriacae-Stipetum austroitalicae* association identified by *Sideritis italica*, a richness of species of *Hippocrepido glaucae-Stipion austroitalicae, Scorzoneretalia villosae* and *Festuco-Brometea* communities, with a good coverage of *Stipa austroitalica* subsp. *austroitalica, Scorzonera villosa* subsp. *columnae*, both endemics, and *Phleum hirsutum* subsp. *ambiguum*, in addition to other species such as *Petrorhagia saxifraga* subsp. *gasparrinii, Hippocrepis glauca, Teucrium capitatum* subsp. *capitatum, Convolvulus cantabrica, Koeleria splendens,* and *Eryngium amethystinum*. Some transgressive species of *Helianthemetea guttati* Rivas Goday & Rivas-Mart. 1963 and *Lygeo sparti-Stipetea tenacissimae* Rivas-Mart. 1978 classes show natural catenal contact with these other plant communities. In the two easternmost sites (Ts1, Ts2), always in natural conditions, but at different topographic conditions, such as altitudes (Ts1 = 274 m, Ts2 = 561 m), exposition (N–NE, S–SW) slope degree (2° and 10°), level of stoniness (5% and 20%) and rockiness (60% and 40%), we detected the same *Acino suaveolentis-Stipetum austroitalicae* association, characterized by three species, two of which are endemic and present at both sites (*Thymus spinulosus* and *Euphorbia nicaeensis* subsp. *japygica*), while *Clinopodium suaveolens* was present only at Scannapecora (Ts2). The taxa detected are comparable to those of Gargano (Ts3), except for the characteristics of the Festuco-Brometea class, which are less represented at both sites, explained by the lower plant biodiversity (Ts1 = 38 taxa, Ts2 = 44 taxa) than Gargano (Ts3 = 64 taxa), and only for Scannapecora (Ts2), also with good coverage of Ononido-Rosmarinetea transgressive species, which this plant association makes a mosaic.

The GC–MS results showed that EO composition consists of a total of 27 identified compounds for all three sites (Table 6). The environmental differences are reflected only partially in the chemical composition of EOs. Only low differences were observed on the abundance and patterns among the three sites, since they share 25 compounds; only 1 (Hotrienol) was exclusive to Ts1 (Hotrienol) and 1 to Ts2 (Caryophyllene oxide). The most interesting phytochemicals, having the highest presence in all samples, were thymol, p-Cymene and β-Ocimene. In particular, at San Basilio (Ts1), in the same climatic condition as Altamura (Ts2), but with lower altitudes and different topographic, geological and geopedological conditions, 28 compounds were detected, with the most abundant ones similar to the Altamura (Ts2) site. Slight differences in percentages were observed: thymol (42.9% and 48.8%, respectively), followed by p-Cymene (17.9% and 17.5%) and β-Ocimene (15.4% and 11.7%). The other compounds did not exceed 3% and 4%.

At the Gargano (Ts3) site, in a different climatic, environmental and partly different plant community, 28 compounds were extracted, comparable anywhere with the results of the other two sites (Ts1 and Ts2): thymol (45.9%), p-Cymene (17.5%) and β-Ocimene (11.0%). The abovementioned data differ from the Sicilian populations [86], and highlight a different chemotype, probably due to the lower altitude compared to the Sicilian stations, which exceed 1000 m above sea level, and maybe for the different latitude, while it seems that the climate, the pedological and lithological characteristics of the soil and the vegetation do not play a relevant role. In any case, the qualitative difference seems to be particularly relevant, since Sicilian populations show a number of components almost double that of Apulia.

4.3. Syntaxonomical Scheme of Surveyed Vegetation

Festuco-Brometea Br.-Bl. et Tx. ex Soó 1947
Scorzoneretalia villosae Kovačević 1959
Hippocrepido glaucae-Stipion austroitalicae Forte et Terzi 2005 in Forte, Perrino and Terzi 2005
Hippocrepido glaucae-Stipienion austroitalicae Biondi and Galdenzi 2012
Acino suaveolentis-Stipetum austroitalicae Forte et Terzi in Forte, Perrino et Terzi 2005
Sideritido syriacae-Stipetum austroitalicae Biondi and Guerra 2008
Ononido-Rosmarinetea Br.-Bl. in A. Bolòs y Vayreda 1950
Cisto-Micromerietalia julianae Oberd. 1954
Cisto cretici-Ericion manipuliflorae Horvatić 1958
Phagnalo saxatilii-Saturejetum cuneifoliae Biondi and Guerra 2008

5. Conclusions

The composition of *Satureaja cuneifolia* EOs is only partially affected by differences in environmental conditions, specifically for the geological, ecological, vegetational and pedological aspects. The soil in the two examined sites is rich in macro- and less rich in microelements, with differences in types and percentages. In particular, the microelements have similar concentrations, except for Zn and Mn. Among the total of 36 compounds detected and identified in the EOs extracted from the analyzed samples at both sites (Difesa di Malta—Sc1; San Basilio—Sc2), the most abundant compounds, often with strong differences between sites, were α-Terpineol, trans-3-caren-2-ol, and Borneol. At Difesa di Malta (Sc1), in more disturbed conditions and in an agriculture context, with a low plant biodiversity referable to an unrepresentative aspect of *Phagnalo saxatilii-Saturejetum cuneifoliae*, to a significant decrease in trans-3-caren-2-ol and borneol, there is a significant increase in α-Terpineol, compared to the better naturalness of San Basilio. Clearly, this means that the environment may play a strong role.

The composition of the *S. cuneifolia* EOs found in the present research does not match with those of previous studies conducted in the Mediterranean area [47,67,69], which showed different chemotypes also among themselves. In Croatia, the chemical composition shows a high percentage of carvacrol (17.7%), γ-terpinene (14.8%), p-cymene (9.8%), linalool (6.6%) and limonene (6.2%) [94]. In another work conducted in Salento (Apulia), but in different areas than the present one, there was a prevalence of linalool (9.6–32.7%), borneol (12.9– 24.0%) and α-pinene (9.5–11.7%) [70]. Recently, in Montenegro, it has been observed that the main constituents were linalool (20.3%), trans-(E)- caryophyllene (6.1%), germacrene D (5.8%), nerolidol (5.2%) and spathulenol (5.0%) [47]. Again, this confirms an environmental effect.

The climatic and geological factors, soil properties, and topographical and vegetational characteristics do not seem to play a crucial role on the chemical composition in EOs of *Thymus spinulosus* in Apulia, as observed in previous studies conducted in a Sicilian population where the environmental conditions and the different soil types affect the composition of EOs. The results of research conducted in Apulia indicate a total of 35 compounds for the three investigated sites, with the highest presence in all samples of thymol, p-Cymene and β-Ocimene, always with comparable percentages, unlike the Sicilian populations, with 62 components in four sites, geographically close and located at comparable altitudes, with variable values of the main components: myrcene (1.0–15.7%), limonene (2.3–13.2%), γ-muurolene (7.3–15.9%), caryophyllene (8.3–11.1%), and germacra-1(10),4-dien-6-ol (1.3–11.3%).

Our results may highlight the following responses in the chemical composition of EOs of *Th. spinulosus*: (a) the Sicilian and Apulian populations are two different chemotypes and the Apulian ones are more stable in composition, regardless of environmental, climatic and vegetational factors; (b) latitude and altitude could be diagnostic factors, even if it would be interesting to complete the picture with other Italian populations such as those of Calabria; (c) it is confirmed that the genus *Th.* has several chemotypes. Furthermore, some studies underscore that other factors are also implicated in determining the EO composition of *Th.*, such as genetic and reproductive characteristics [94], the ecological functions of the EOs, e.g., protection from herbivores, interaction with microorganisms in the decomposition process, patterns of vegetation through allelopathic action [95], the physiological stage of the plant [87,96], and the negative impacts of herbicides [97].

Finally, adopting specific measures for selective management of the shrub vegetation is recommended, including those identified here, with a high concentration of species rich in EO, especially in the Mediterranean climate, in order to contribute to a reduction in the risk of fire [98] and safeguard the endemic and threatened flora, some of them still without studies of their chemical properties.

As general remarks, for both studied species, we may stress, once more, that the high bioactivity of *Th. spinulosus* indicates it has promising potential in organic agriculture, since it may provide thymol as a possible natural agent against phytopathogenic microorganisms; the richness of α-Pinene (insecticidal) in *S. cuneifolia* also makes the species a potential plant as natural pesticide in organic agriculture.

Author Contributions: Conceptualization, E.V.P.; methodology, E.V.P., F.V. and G.N.M.; software, E.V.P., F.V., G.N.M. and A.T.; validation, E.V.P.; formal analysis, E.V.P., F.V., G.N.M. and A.T.; investigation, E.V.P., F.V. and G.N.M.; data curation, E.V.P., F.V., G.N.M. and A.T.; writing—original draft preparation, E.V.P.; writing—review and editing, E.V.P., F.V., G.N.M., S.J. and A.T.; visualization, E.V.P., F.V., G.N.M., S.J. and A.T.; supervision, E.V.P.; pictures, E.V.P. All authors have read and agreed to the published version of the manuscript.

Funding: This research did not received funding, but it was performed within the master's program in "Mediterranean Organic Agriculture" assigned to Shaima Jallali at the CIHEAM, Mediterranean Agronomic Institute of Bari.

Institutional Review Board Statement: Not applicable.

Informed Consent Statement: Not applicable.

Data Availability Statement: The data that support the findings of this study are available from the corresponding author upon reasonable request.

Acknowledgments: The authors wish to thank Lina Al Bitar (CIHEAM) and Ramez Mohamad (CIHEAM) for the administrative support in the management of the Master. The authors are also grateful to the Botanical Garden Museum of the University of Bari, especially Rocco Mariani, the curator of the *Herbarium Horti Botanici Barensis* (BI), Gaetano Pazienza for scanning the specimens of *Satureja cuneifolia* and *Thymus spinulosus*, and Luigi Forte for having authorized their reproduction.

Conflicts of Interest: The authors declare no conflict of interest.

References

1. Cock, I.E. *Medicinal and Aromatic Plants Australia, In Ethnopharmacology Section, Biological, Physiological and Health Sciences, Encyclopedia of Life Support Systems (EOLSS)*; Developed under the Auspices of the Unesco, Eolss Publishers: Oxford, UK, 2011.
2. Petrovska, B.B. Historical review of medicinal plants' usage. *Pharm. Rev.* **2012**, *6*, 1–5. [CrossRef]
3. Ahmad, M.; Sultana, S.; Fazl-i-Hadi, S.; Ben Hadda, T.; Rashid, S.; Zafar, M.; Khan, M.A.; Khan, M.P.Z.; Yaseen, G. An ethnobotanical study of medicinal plants in high mountainous region of Chail valley (District Swat-Pakistan). *J. Ethnobiol. Ethnomed.* **2014**, *10*, 36. [CrossRef]
4. Sargin, S.A.; Büyükcengiz, M. Plants used in ethnomedicinal practices in Gulnar district of Mersin, Turkey. *J. Herb. Med.* **2019**, *15*, 100224. [CrossRef]
5. Şimşek Sezer, E.N.; Uysal, T. Phytochemical Analysis, Antioxidant and Anticancer Potential of *Sideritis niveotomentosa*: Endemic Wild Species of Turkey. *Molecules* **2021**, *26*, 2420. [CrossRef]
6. Chrysargyris, A.; Mikallou, M.; Petropoulos, S.; Tzortzakis, N. Profiling of Essential Oils Components and Polyphenols for Their Antioxidant Activity of Medicinal and Aromatic Plants Grown in Different Environmental Conditions. *Agronomy* **2020**, *10*, 727. [CrossRef]
7. Pichersky, E.; Gershenzon, J. The formation and function of plant volatiles: Perfumes for pollinator attraction and defense. *Curr. Opin. Plant Biol.* **2002**, *5*, 237–243. [CrossRef]
8. Dhifi, W.; Bellili, S.; Jazi, S.; Bahloul, N.; Mnif, W. Essential Oils' Chemical Characterization and Investigation of Some Biological Activities: A Critical Review. *Medicines* **2016**, *3*, 25. [CrossRef]
9. Dini, I. Use of Essential Oils in Food Packaging. In *Essential Oils in Food Preservation, Flavor and Safety*; Academic Press: Cambridge, MA, USA, 2016; pp. 139–147, (Chapter 14). [CrossRef]

10. Mari, M.; Bautista-Baños, A.; Sivakumar, D. Decay control in the postharvest system: Role of microbial and plant volatile organic compounds. *Postharvest Biol. Technol.* **2016**, *122*, 70–81. [CrossRef]
11. Moghaddam, M.; Mehdizadeh, L. Chemistry of essential oils and factors influencing their constituents. In *Soft Chemistry and Food Fermentation*; Grumezescu, A.M., Holban, A.M., Eds.; Academic Press: London, UK, 2019; Volume 13, pp. 379–419. [CrossRef]
12. Zuzarte, M.; Salgueiro, L. Essential Oils Chemistry. In *Bioactive Essential Oils and Cancer*; de Sousa, D., Ed.; Springer: Cham, Switzerland, 2015. [CrossRef]
13. The Plant List, 2013 Version 1.1. Published on the Internet. Available online: http://www.theplantlist.org/ (accessed on 20 May 2021).
14. Kokkini, S.; Karousou, R.; Hanlidou, E. Herbs of the Labiatae. In *Encyclopedia of Food Sciences and Nutrition*; Elsevier Science Ltd.: Amsterdam, The Netherlands, 2003; pp. 3082–3090.
15. Ozkan, G.; Simsek, B.; Kuleasan, H. Antioxidant activities of *Satureja cilicica* essential oil in butter and in vitro. *J. Food Eng.* **2007**, *79*, 1391–1396. [CrossRef]
16. Carović-Stanko, K.; Petek, M.; Grdiša, M.; Pintar, J.; Bedekovic, D.; HerakCustic, M.; Satovic, Z. Medicinal plants of the family Lamiaceae as functional foods—A review. *Czech J. Food Sci.* **2016**, *34*, 377–390. [CrossRef]
17. Maksimović, M.; Vidic, D.; Miloš, M.; Šolić, M.E.; Abadžić, S.; Siljak-Yakovlev, S. Effect of the environmental conditions on essential oil profile in two Dinaric *Salvia* species: *S. brachyodon* Vandas and *S. officinalis* L. *Biochem. Syst. Ecol.* **2018**, *35*, 473–478. [CrossRef]
18. Yeddes, W.; Aidi Wannes, W.; Hammami, M.; Smida, M.; Chebbi, A.; Marzouk, B.; Saidani Tounsi, M. Effect of environmental conditions on the chemical composition and antioxidant activity of essential oils from *Rosmarinus officinalis* L. growing wild in Tunisia. *J. Essent. Oil Bear. Plants* **2018**, *21*, 972–986. [CrossRef]
19. Perrino, E.V.; Valerio, F.; Gannouchi, A.; Trani, A.; Mezzapesa, G. Ecological and Plant Community Implication on Essential Oils Composition in Useful Wild Officinal Species: A Pilot Case Study in Apulia (Italy). *Plants* **2021**, *10*, 574. [CrossRef] [PubMed]
20. Italian Ministry for the Environment, Land and Sea (MATTM) Geoportale Nazionale. 2021. Available online: http://www.pcn.minambiente.it/viewer/ (accessed on 26 May 2021).
21. Bellini, F.; Dursi, F.; Perrino, E.V.; Wagensommer, R.P. Caratterizzazione floristica di aree di proprietà pubblica per la reintroduzione delle specie vegetali in situ. Il parco naturale regionale Terra delle Gravine. In *Grastepp tra Gravine e Steppe—Azioni per la Conservazione Della Biodiversità nel Parco Nazionale dell'Alta Murgia e nel Parco Naturale Regionale Terra Delle Gravine—Rapporto Finale. APQ FAS Biodiversità Delibera Cipe 19/2004*; Ufficio Parchi e Tutela della Biodiversità della Regione Puglia: Bari, Italy, 2014; pp. 117–127.
22. Braun—Blanquet, J.; Conard, H.S.; Fuller, G.D. *Plant Sociology*; McGraw—Hill: New York, NY, USA, 1932; p. 439.
23. Géhu, J.M.; Rivas-Martínez, S. *Notions Fondamentales de Phytosociologie*; Dierschke, H., Ed.; Berichte der Internationalen Symposien der Internationalen Vereinigung fur Vegetationskunde: Cramer, Vaduz, 1981; pp. 5–33, (Syntaxonomie).
24. Pignatti, S. *Flora d'Italia*; Edagricole: Bologna, Italy, 1982; Volume 1–3.
25. Tutin, T.G.; Heywood, V.H.; Burges, N.A.; Moore, D.M.; Valentine, D.H.; Walters, S.M.; Webb, D.A. *Flora Europea*, 1st ed.; Cambridge University Press: Cambridge, UK, 1964; Volume 1–5.
26. Bartolucci, F.; Peruzzi, L.; Galasso, G.; Albano, A.; Alessandrini, A.; Ardenghi, N.M.G.; Astuti, G.; Bacchetta, G.; Ballelli, S.; Banfi, E.; et al. An updated checklist of the vascular flora native to Italy. *Plant Biosyst.* **2018**, *152*, 179–303. [CrossRef]
27. Galasso, G.; Conti, F.; Peruzzi, L.; Ardenghi, N.M.G.; Banfi, E.; Celesti-Grapow, L.; Albano, A.; Alessandrini, A.; Bacchetta, G.; Ballelli, S.; et al. An updated checklist of the vascular flora alien to Italy. *Plant Biosyst.* **2018**, *152*, 556–592. [CrossRef]
28. Mucina, L.; Bültmann, H.; Dierssen, D.; Theurillat, J.P.; Raus, T.; Čarni, A.; Šumberová, K.; Willner, W.; Dengler, J.; García, R.G. Vegetation of Europe: Hierarchical floristic classification system of vascular plant, bryophyte, lichen, and algal communities. *Appl. Veg. Sci.* **2016**, *9*, 3–264. [CrossRef]
29. Rivas-Martínez, S. *Global Bioclimatics. Clasificación Bioclimática de la Tierra*. 2004. Available online: http://www.globalbioclimatics.org/book/bioc/bioc1.pdf (accessed on 21 November 2020).
30. Biondi, E.; Blasi, C. *Prodromo della Vegetazione d'Italia*; Check-list sintassonomica aggiornata di classi, ordini e alleanze presenti in Italia; Onlus, Ed.; Società Botanica Italiana: Firenze, Italy, 2013; Available online: http://www.prodromo-vegetazione-italia.org (accessed on 20 November 2020).
31. Meyer—Warnod, B. Natural essential oils: Extraction processes and application to some major oils. *Perfum. Flavorist* **1984**, *9*, 93–104.
32. Kováts, E. Gas chromatographische charakterisierung organischer verbindungen. Teil 1: Retentions indices aliphatischer halogenide, alkohole, aldehyde und ketone. *Helv. Chim. Acta* **1958**, *41*, 1915–1932. [CrossRef]
33. Babushok, V.I.; Linstrom, P.J.; Zenkevich, I.G. Retention indices for frequently reported compounds of plant essential oils. *J. Phys. Chem. Ref. Data* **2011**, *40*, 3–47. [CrossRef]
34. Soil Survey Staff. *Keys to Soil Taxonomy*, 11th ed.; Department of Agriculture: Washington, WA, USA, 2010.
35. Nelson, D.W.; Sommers, L.E. Total carbon, organic carbon, and organic matter. In *Methods of Soil Analysis*; Sparks, D.L., Ed.; Part 3, Chemical Methods, SSSA Book Series n. 5; Soil Science Society of America/American Society of Agronomy: Madison, WI, USA, 1996; pp. 961–1010.
36. Bremner, J.M. Nitrogen—Total. In *Methods of Soil Analysis*; Sparks, D.L., Ed.; Part 3, Chemical Methods, SSSA Book Series n. 5; Soil Science Society of America/American Society of Agronomy: Madison, WI, USA, 1996; pp. 1085–1121.

37. Olsen, S.L.; Sommers, L.E. *Phosphorus. Methods of Soil Analysis*, 2nd ed.; Chemical and Microbiological Properties, SSSA Book Series; American Society of Agronomy: Madison, WI, USA, 1982; Number 9; pp. 403–427.
38. Bektas, T.; Cilkiz, M. A pharmacological and phytochemical overview on *Satureja*. *Pharm. Biol.* **2016**, *54*, 375–412. [CrossRef]
39. Zavatti, M.; Zanoli, P.; Benelli, A.; Rivasi, M.; Baraldi, C.; Baraldi, M. Experimental study on *Satureja montana* as a treatment for premature ejaculation. *J. Ethnopharmacol.* **2011**, *133*, 629–633. [CrossRef]
40. Momtaz, S.; Abdollahi, M. A Systematic Review of the Biological Activities of *Satureja* L. Species. *Pharmacologyonline* **2008**, *2*, 34–54.
41. Rustaiyan, A.; Feizbakhsh, A.; Masoudi, S.; Ameri, N. Comparison of the Volatile Oils of *Satureja atropatana* Bung. and *Satureja mutica* Fisch. et C.A. Mey. from Iran. *J. Essent. Oil. Res.* **2004**, *16*, 594–596. [CrossRef]
42. Epling, C.C. Studies on South American Libiatae. III. Synopsis of the Genus *Satureia*. *Annals Miss. Bot. Gard.* **1927**, *14*, 47–86. [CrossRef]
43. Cantino, P.D.; Harley, R.M.; Wagstaff, S.J. Genera of *Labiatae* status and classification. In *Advances in Labiatae Science*; Harley, R.M., Reynolds, T., Eds.; Royal Botanic Gardens Press: Kew, UK, 1992; pp. 511–522.
44. Slavkovska, V.; Jancic, R.; Bojovic, S.; Milosavljevic, S.; Djokovic, D. Variability of essential oils of *Satureja montana* L. and *Satureja kitaibelii* Wierzb. ex Heuff. from the central part of the Balkan peninsula. *Phytochemistry* **2001**, *57*, 71–76. [CrossRef]
45. Greuter, W.; Burdet, H.M.; Long, G. *Med-Checklist 3: A Critical Inventory of Vascular Plants of the Circum-Mediterranean Countries, Dicotyledones (Convolvulaceae-Labiatae)*; Conservatoire et Jardin botanique de la Ville Genève: Genève, Swiss, 1986; p. 395.
46. Bezić, N.; Skočibišić, M.; Dunkić, V. Phytochemical composition and antimicrobial activity of *Satureja montana* L. and *Satureja cuneifolia* Ten. essential oils. *Acta Bot. Croat.* **2005**, *64*, 313–322.
47. Bojović, D.; Šoškićb, M.; Tadić, V. Comparative study of chemical composition of the essential oils from *Satureja cuneifolia* Ten. and *Satureja montana* L., lamiaceae collected at National Park Lovćen, Montenegro. *Stud. Univ. Babes-Bolyai Chem.* **2018**, *63*, 167–180. [CrossRef]
48. Perrino, E.V.; Ladisa, G.; Calabrese, G. Flora and plant genetic resources of ancient olive groves of Apulia (southern Italy). *Genet. Resour. Crop Evol.* **2014**, *61*, 23–53. [CrossRef]
49. Biondi, E.; Guerra, V. Vegetazione e paesaggio vegetale delle gravine dell'arco jonico. *Fitosociologia* **2008**, *45*, 57–125.
50. Wagensommer, R.P.; Di Pietro, R. Analisi fitosociologica su alcune specie rare e/o minacciate del Parco Nazionale del Gargano (Italia centro-meridionale) e considerazioni sintassonomiche sulle comunità casmofitiche della Puglia. *Fitosociologia* **2008**, *45*, 177–200.
51. Bianco, P.; Brullo, S.; Pignatti, E.; Pignatti, S. La vegetazione delle rupi calcaree della Puglia. *Braun Blanquetia* **1988**, *2*, 133–151.
52. Francini Corti, E. Il Pino d'Aleppo in Puglia. *Annali della Facoltà di Agraria dell'Università di Bari* **1953**, *8*, 309–416.
53. Gambetta, G.; Medagli, P. Guida alla flora del Parco e del territorio materano. In *Parco Regionale Della Murgia Materana*; Graficom srl: Matera, Italy, 2019; p. 431.
54. Di Pietro, R.; Wagensommer, R.P. A new *Sesleria juncifolia* association from south-eastern Italy and its position in the amphi-Adriatic biogeographical context. *Acta Bot. Croat.* **2014**, *73*, 171–207. [CrossRef]
55. Karl, R.; Scholz, H. Med-Checklist Notulae, 28. *Willdenowia* **2009**, *39*, 335–345.
56. Terlević, A.; Koopman, J.; Wieclaw, H.; Rešetnik, I.; Bogdanović, S. *Carex phyllostachys* (Cyperaceae), a new species in Croatia. *Acta Bot. Croat.* **2021**, *80*, 106–111. [CrossRef]
57. Wagensommer, R.P.; Perrino, E.V.; Silletti, G.N. *Carex phyllostachys* C.A. Mey. (Cyperaceae) new for Italy and phytogeographical considerations. *Phyton* **2014**, *54*, 215–222. [CrossRef]
58. Wagensommer, R.P.; Fröhlich, T.; Fröhlich, M. First record of the southeast European species *Cerinthe retorta* Sibth. & Sm. (Boraginaceae) in Italy and considerations on its distribution and conservation status. *Acta Bot. Gall. Bot. Lett.* **2014**, *161*, 111–115. [CrossRef]
59. Wagensommer, R.P.; Bartolucci, F.; Fiorentino, M.; Licht, W.; Peccenini, S.; Perrino, E.V.; Venanzoni, R. First record for the flora of Italy and lectotypification of the name *Linum elegans* (Linaceae). *Phytotaxa* **2017**, *296*, 161–170. [CrossRef]
60. Perrino, E.V.; Signorile, G.; Marvulli, M. A first checklist of the vascular flora of the "Polignano a Mare" coast (Apulia, southern Italy). *Nat. Croat.* **2013**, *22*, 295–318.
61. Fenu, G.; Bernardo, L.; Calvo, R.; Cortis, P.; De Agostini, A.; Gangale, C.; Gargano, D.; Gargano, M.L.; Lussu, M.; Medagli, P.; et al. Global and Regional IUCN Red List Assessments: 8. *Ital. Bot.* **2019**, *8*, 17–33. [CrossRef]
62. Wagensommer, R.P.; Medagli, P.; Turco, A.; Perrino, E.V. IUCN Red List Evaluation of the Orchidaceae endemic to Apulia Region (Italy) and considerations on the application of the IUCN protocol to rare species. *Nat. Conserv. Res.* **2020**, *5*, 90–101. [CrossRef]
63. Di Pietro, R.; Misano, G. Shrubland and garrigue vegetation in the « Gravine » gorges (Apulia region, south—eastern Italy). *Acta Bot. Gall.* **2010**, *157*, 195–229. [CrossRef]
64. Biondi, E.; Blasi, C.; Burrascano, S.; Casavecchia, S.; Copiz, R.; Del Vico, E.; Galdenzi, D.; Gigante, D.; Lasen, C.; Spampinato, G.; et al. Italian Interpretation Manual of the Habitats (92/43/EEC Directive). Ministry of Environment, Land and Sea Protection. 2010. Available online: http://vnr.unipg.it/habitat/ (accessed on 3 June 2021).
65. European Commission DG Environment. *Interpretation Manual of European Union Habitats (Version EUR27)*; European Commission DG Environment: Brussels, Belgium, 2007; Available online: https://ec.europa.eu/environment/nature/legislation/habitatsdirective/docs/2007_07_im.pdf (accessed on 3 June 2021).
66. Jafari, F.; Ghavidel, F.; Zarshenas, M.M. A Critical Overview on the Pharmacological and Clinical Aspects of Popular Satureja Species. *J. Acupunct. Meridian Study* **2016**, *9*, 118. [CrossRef] [PubMed]

67. Bezić, N.; Samanić, I.; Dunkić, V.; Besendorfer, V.; Puizina, J. Essential Oil Composition and Internal Transcribed Spacer (ITS) Sequence Variability of Four South-Croatian *Satureja* Species (Lamiaceae). *Molecules* **2009**, *14*, 925–938. [CrossRef]
68. Musarella, C.M.; Paglianiti, I.; Cano-Ortiz, A.; Spampinato, G. Indagine etnobotanica nel territorio del Poro e delle Preserre Calabresi (Vibo Valentia, S-Italia). *Atti Soc. Toscana Sci. Nat. Mem. Ser. B* **2019**, *126*, 13–28. [CrossRef]
69. Milos, M.; Radonić, A.; Bezić, N.; Dunkić, V. Localities and seasonal variations in the chemical composition of essential oils of *Satureja montana* L. and *S. cuneifolia* Ten. *Flavour Fragr. J.* **2001**, *16*, 157–160. [CrossRef]
70. Tommasi, L.; Negro, C.; De Bellis, L.; Miceli, A. Essential Oil Variability of *Satureja cuneifolia* Ten. Growing Wild in Southern Puglia (Italy). *J. Essent. Oil Res.* **2008**, *20*, 295–302. [CrossRef]
71. De Oliveira, T.L.C.; De Araújo Soares, R.; Ramos, E.M.; Das Graças Cardoso, M.; Alves, E.; Piccoli, R.H. Antimicrobial activity of *Satureja montana* L. essential oil against Clostridium perfringens type A inoculated in mortadella-type sausages formulated with different levels of sodium nitrite. *Int. J. Food Microbiol.* **2011**, *144*, 546. [CrossRef] [PubMed]
72. Oke, F.; Aslim, B.; Ozturk, S.; Altundag, S. Essential oil composition, antimicrobial and antioxidant activities of *Satureja cuneifolia* Ten. *Food Chem.* **2009**, *112*, 874–879. [CrossRef]
73. Linnaeus, C. *Species Plantarum*; Laurentii Salvii: Stockholm, Sweden, 1753.
74. Bartolucci, F.; Peruzzi, L.; Passalacqua, N. Typification of names and taxonomic notes within the genus *Thymus* L. (Lamiaceae). *Taxon* **2013**, *62*, 1308–1314. [CrossRef]
75. Morales, R. The history, botany and taxonomy of the genus Thymus. In *Thyme: The Genus Thymus*; Stahl-Biskup, E., Sáez, F., Eds.; Taylor & Francis: London, UK; New York, NY, USA, 2002; pp. 1–43.
76. Bartolucci, F.; Walter, J. Typification of names in the genus *Thymus* (Lamiaceae). *Phytotaxa* **2015**, *221*, 137–147. [CrossRef]
77. Bartolucci, F.; Mráz, P. A new combination in the genus *Thymus* and clarification on the type material of *T. thracicus* (Lamiaceae). *Phytotaxa* **2016**, *263*, 161–164. [CrossRef]
78. Morales, R. Synopsis of the genus *Thymus* L. in the Mediterranean area. *Lagascalia* **1997**, *19*, 249–262.
79. Perrino, E.V.; Tomaselli, V.; Costa, R.; Pavone, P. Conservation status of habitats (Directive 92/43 EEC) of coastal and low hill belts in a mediterranean biodiversity hot spot (Gargano-Italy). *Plant Biosyst.* **2013**, *147*, 1006–1028. [CrossRef]
80. Alekseeva, L.I.; Gruzdev, I.V. Polymorphism of Essential Oils in Thyme Species Growing in European Part of North East Russia and Ural. *Russ. J. Plant Physiol.* **2012**, *59*, 818–827. [CrossRef]
81. Hosseinzadeh, S.; Jafarikukhdan, A.; Hosseini, A.; Armand, R. The application of Medicinal Plants in Tradicional and Modern Medicine: A Review of *Thymus vulgaris*. *Int. J. Clin. Med.* **2015**, *6*, 635–642. [CrossRef]
82. Jarić, S.; Mitrović, M.; Pavlović, P. Review of Ethnobotanical, Phytochemical, and Pharmacological Study of *Thymus serpyllum* L. Evid. Based Complement. *Altern. Med.* **2015**, *2015*, 101978.
83. Bhat, M.N.; Singh, B.; Surmal, O.; Singh, B.; Shivgotra, V.; Musarella, C.M. Knowledge and practices of ethnobotany in developing countries: Validation vs traditional use of plant resources of Kashmir Himalaya, India. *Biology* **2021**, *10*, 851. [CrossRef]
84. Maruca, G.; Spampinato, G.; Turiano, D.; Laghetti, G.; Musarella, C.M. Ethnobotanical notes about medicinal and useful 1110 plants of the Reventino Massif tradition (Calabria region, Southern Italy). *Genet. Resour. Crop Evol.* **2019**, *66*, 1027–1040. [CrossRef]
85. Koul, O.; Walia, S.; Dhaliwal, G.S. Essential oils as green pesticides: Potential and constraints. *Biopestic. Int.* **2008**, *4*, 63–84.
86. Chan, W.K.; Tan, L.T.H.; Chan, K.G.; Lee, L.H.; Goh, B.H. Nerolidol: A sesquiterpene alcohol with multi-faceted pharmacological and biological activities. *Molecules* **2016**, *21*, 529. [CrossRef]
87. De Feo, V.; Bruno, M.; Tahiri, B.; Napolitano, F.; Senatore, F. Chemical composition and antibacterial activity of essential oils from *Thymus spinulosus* Ten. (Lamiaceae). *J. Agric. Food Chem.* **2003**, *51*, 3849–3853. [CrossRef]
88. Desneux, N.; Decourtye, A.; Delpuech, J.M. The sublethal effects of pesticides on beneficial arthropods. *Annu. Rev. Entomol.* **2007**, *52*, 81–106. [CrossRef] [PubMed]
89. Guedes, R.N.C.; Smagghe, G.; Stark, J.D.; Desneux, N. Pesticide induced stress in arthropod pests for optimized integrated pest management programs. *Annu. Rev. Entomol.* **2016**, *61*, 43–62. [CrossRef] [PubMed]
90. Mohammed, A.A.A.; Desneux, N.; Fan, Y.; Han, P.; Ali, A.; Song, D.; Gao, X.W. Impact of imidacloprid and natural enemies on cereal aphids: Integration or ecosystem service disruption? *Entomol. Gen.* **2018**, *37*, 47–61. [CrossRef]
91. Liu, X.L.; Tang, Q.L.; Li, Y.D.; Campos, M.R.; Liang, P.; Gao, X.W. Widespread resistance of the aphid *Myzus persicae* to pirimicarb across China, and insights on ace2 mutation frequency in this species. *Entomol. Gen.* **2017**, *36*, 285–299.
92. Pavela, R.; Bartolucci, F.; Desneux, N.; Lavoir, A.V.; Canale, A.; Maggi, F.; Benelli, G. Chemical profiles and insecticidal efficacy of the essential oils from four *Thymus* taxa growing in central-southern Italy. *Ind. Crops Prod.* **2019**, *138*, 111460. [CrossRef]
93. De Natale, A.; Menale, B.; Di Lecce, S.; Pollio, A. An ethnobotanical survey of wild edible plants of Campania (Italy). *Bull. Reg. Nat. Hist.* **2021**, *1*, 1–122.
94. Vernet, P.; Gouyon, P.H.; Valdeyron, G. Genetic control of the oil content in *Thymus vulgaris*: A case of polymorphism in a biosynthetic chain. *Genetica* **1986**, *69*, 227–231. [CrossRef]
95. Vokou, D.; Margaris, N.S. Volatile oils as allelopathic agents. In *Aromatic Plants: Basic and Applied Aspects*; Margaris, N.S., Koldam, A., Vokou, D., Eds.; Martinus Nijhoff: The Hague, The Netherlands, 1982; pp. 59–72.
96. Senatore, F. Influence of harvesting time on yield and composition of the essential oil of a *Thyme* (*Thymus pulegioides* L.) growing wild in Campania (Southern Italy). *J. Agric. Food Chem.* **1996**, *44*, 1327–1332. [CrossRef]

97. Piñar Fuentes, J.C.; Leiva, F.; Cano-Ortiz, A.; Musarella, C.M.; Quinto-Canas, R.; Pinto-Gomes, C.J.; Cano, E. Impact of Grass Cover Management with Herbicides on Biodiversity, Soil Cover and Humidity in Olive Groves in the Southern Iberian. *Agronomy* **2021**, *11*, 412. [CrossRef]
98. Raposo, M.A.M.; Gomes, C.J.P.; Nunes, L.J.R. Selective Shrub Management to Preserve Mediterranean Forests and Reduce the Risk of Fire: The Case of Mainland Portugal. *Fire* **2020**, *3*, 65. [CrossRef]

plants

MDPI

Article

Influence of Habitat Types on Diversity and Species Composition of Urban Flora—A Case Study in Serbia

Milan Glišić [1,2,*], Ksenija Jakovljević [2], Dmitar Lakušić [2], Jasmina Šinžar-Sekulić [2], Snežana Vukojičić [2], Milena Tabašević [2] and Slobodan Jovanović [2]

[1] Academy of Applied Studies Šabac, Dobropoljska 5, 15000 Šabac, Serbia
[2] Faculty of Biology, Institute of Botany and Botanical Garden, University of Belgrade, Takovska 43, 11000 Belgrade, Serbia; kjakovljevic@bio.bg.ac.rs (K.J.); dlakusic@bio.bg.ac.rs (D.L.); jsekulic@bio.bg.ac.rs (J.Š.-S.); sneza@bio.bg.ac.rs (S.V.); milenatabasevic@gmail.com (M.T.); sjov@bio.bg.ac.rs (S.J.)
* Correspondence: milanglisic88@gmail.com

Abstract: The aim of this study was to investigate the floristic composition and diversity of seven urban habitat types in 24 Serbian cities with different climatic affiliation. In each of the 24 cities, we selected 1 ha plots representing a habitat from one of the following groups: square, boulevard, residential area with compact and with open building pattern, city park, and sites with early and mid-succession vegetation stages. All vascular plant species that occur spontaneously in these plots were observed. Data on the main climatic characteristics were collected for each plot, and data on the life forms were obtained for each species recorded. Diagnostic species were identified for each habitat type analyzed, and alpha, beta and gamma diversity were calculated. A total of 674 taxa were recorded in the studied area. Significant differences were observed in habitats by diagnostic species and by life form representation. The lowest alpha and gamma diversity and the dominance of therophytes were observed in habitat types with intensive anthropogenic impact, whereas the highest number was recorded in mid-successional sites and residential areas with a compact building pattern. The analysis showed that habitat type influences species composition much more than climate.

Keywords: species composition; richness; urban areas; anthropogenic impact; climate

Citation: Glišić, M.; Jakovljević, K.; Lakušić, D.; Šinžar-Sekulić, J.; Vukojičić, S.; Tabašević, M.; Jovanović, S. Influence of Habitat Types on Diversity and Species Composition of Urban Flora—A Case Study in Serbia. *Plants* **2021**, *10*, 2572. https://doi.org/10.3390/plants10122572

Academic Editor: Robert Philipp Wagensommer

Received: 26 October 2021
Accepted: 13 November 2021
Published: 25 November 2021

1. Introduction

Human activities are an inseparable part of urban area and play a leading role in modifying its ecological characteristics, forming similar conditions in diverse, often remote areas. Hence, the similar urban habitats are found in the vast majority of cities, even in areas in different biogeographical regions, with different macroclimatic characteristics [1,2]. Large-scale introduction of species with the cosmopolitan type of distribution, sometimes associated with a decline in native species, may lead to a further increasing similarity in species composition between regions [3,4]. Additional homogenization is caused by the presence of invasive species, primarily archeophytes, whereas neophytes mainly lead to opposite effects [5,6]. However, both groups of invasive species have been shown to contribute to an increase in the richness of plant species in urban habitats [5,7]. Namely, according to Pyšek [8], archaeophytes and neophytes account for 15% and 25%, respectively, of the urban flora in Central Europe, although the negative effects of alien species on native diversity have also been observed [9]. Additionally, urban areas are very heterogeneous, so this also contributes to a larger number of species in cities [10,11]. This heterogeneity, caused by different disturbance regimes, induced differences in species composition [12]. Simultaneously with the certain similarities and the number of generalist species found both in and outside the cities, pronounced differences can be observed compared to the surroundings, wherefore the cities can be regarded as a kind of ecological island [6].

There are many characteristics of urban habitats that distinguish them from natural ones: higher levels of disturbance [13,14], herbicide use [15], air and soil pollution [16,17], nutrient enrichment [5,18], higher temperatures due to the urban heat island effect [19,20], higher input of alien species propagules [21], etc.

There are a number of reasons why urban flora research is attracting increasing attention: most of Europe's human population lives in urban areas [22]; the plants of urban habitats contribute to ecosystem services and affect the citizens' well-being [15,23]; urban areas can be centers for the spread of allochthonous plant species to neighboring territories [8,24]; and urban habitats can serve as a refugia for plant species, even those considered rare or endangered [11]. The number of studies dealing with urban flora has increased considerably in recent decades [25]. Comparative studies of urban flora in large areas and the implementation of standardized sampling protocols made a great contribution to the understanding of the distribution and ecology of plants in central European cities [5,6,12,26–29], indicating striking differences between urban habitats in terms of plant species diversity, induced by the types of urban habitats, climate, and specific spatial patterns [26,30,31]. However, the results from Central Europe are hard to generalize to the whole of Europe. For example, diversity of plant species in urban habitats in Southern Europe is greater compared to the parts of Central Europe with a different climate [30]. In addition, the urban flora of Southeastern Europe is poorly studied compared to other parts of Europe, with previous studies often focusing on individual cities or specific urban habitats within them [25]. Comparative and comprehensive studies of urban flora in SE Europe are particularly rare and are not based on standardized sampling methods, rather on a comparison of existing data [32]. For the above reasons, large-scale comparisons and generalizations of features and trends in the urban flora of Southeastern Europe are still lacking.

The aim of this study was therefore to investigate the floristic composition and diversity of urban habitats in Serbia using the standardized protocol established by Lososová et al. [26], to obtain comparable results. Cities in Serbia are a good model for such a study, as climatic differences between cities are considerable. Hence, our aim was also to investigate (1) how urban habitat types affect floristic composition and diversity and (2) which factor has a stronger effect on floristic composition, habitat types, or climatic characteristics.

2. Results

Significant differences in species composition were observed among selected habitat types. Diagnostic species for each habitat type are listed in Table 1. The highest average value of Φ (0.13) and the highest number of diagnostic species (67) were calculated for mid-successional sites (m). This habitat type was dominated by deciduous shrubs and trees (e.g., *Prunus spinosa, Cornus sanguinea*, and *Juglans regia*), but also by a large number of herbaceous perennial species (e.g., *Hypericum perforatum, Dipsacus laciniatus, Rumex patientia, Agrimonia eupatoria*) and grasses (e.g., *Calamagrostis epigejos, Poa trivialis, Holcus lanatus, Bromus hordeaceus*). Early successional sites and residential areas with compact building patterns were also characterized by a relatively high average value of Φ (0.10, both) and a relatively high number of diagnostic species (39, both). The diagnostic species with the highest Φ value (>0.50) in early successional sites were *Papaver rhoeas, Bromus hordeaceus, Polygonum lapathifolium*, and *Vicia cracca*. Residential areas with compact building patterns were mostly characterized by ornamental species, which spread from neighboring gardens (e.g., *Campsis radicans, Kerria japonica, Antirrhinum majus, Ipomoea purpurea, Rudbeckia hirta*). In residential areas with an open building pattern, 22 diagnostic species were observed, with domination of ornamental species (e.g., *Mirabilis jalapa, Commelina communis, Hibiscus syriacus*) and an average Φ value of 0.7. Only five diagnostic species were determined for city parks (p) and the species with the highest Φ value was *Quercus robur*, which often occurs as seedlings. Boulevards (b) hosted a large group of different plants, but only four species were determined as a diagnostic. By comparison, no positive correlation with city squares (s) was found for any of the plant species, and the lowest average value of Φ was obtained (0.01) for this habitat type.

Table 1. Diagnostic plant species for studied urban habitat types in Serbia.

Habitat Type	Taxon
Boulevard	*Sedum lineare* (0.44), *Sagina procumbens* (0.32), *Chelidonium majus* (0.31), *Acer pseudoplatanus* (0.30)
Early successional site	***Papaver rhoeas* (0.54), *Bromus hordeaceus* (0.53), *Polygonum lapathifolium* (0.51), *Vicia cracca* (0.50),** *Onopordum acanthium* (0.49), *Tanacetum vulgare* (0.49), *Tragopogon dubius* (0.47), *Sisymbrium loeselii* (0.47), *Rumex crispus* (0.45), *Linaria vulgaris* (0.45), *Phragmites australis* (0.44), *Petrorhagia prolifera* (0.42), *Silene vulgaris* (0.41), *Rumex patientia* (0.41), *Melilotus albus* (0.41), *Trifolium campestre* (0.40), *Melilotus officinalis* (0.40), *Matricaria perforata* (0.40), *Conium maculatum* (0.39), *Bromus tectorum* (0.38), *Silene latifolia alba* (0.38), *Dipsacus fullonum* (0.36), *Calamagrostis epigejos* (0.36), *Sinapis arvensis* (0.35), *Sanguisorba minor* (0.35), *Abution theophrasti* (0.34), *Echium vulgare* (0.34), *Euphorbia seguieriana* (0.34), *Descurainia sophia* (0.33), *Lepidium ruderale* (0.32), *Anchusa officinalis* (0.32), *Carduus acanthoides* (0.32), *Torilis arvensis* (0.31), *Verbascum phlomoides* (0.31), *Artemisia vulgaris* (0.31), *Eryngium campestre* (0.30), *Avena sativa* (0.30), *Coronilla varia* (0.30), *Polygonum persicaria* (0.30)
Mid-successional site	***Prunus spinosa* (0.70), *Hypericum perforatum* (0.65), *Cornus sanguinea* (0.59), *Dipsacus laciniatus* (0.57), *Rumex patientia* (0.52), *Calamagrostis epigejos* (0.52), *Agrimonia eupatoria* (0.52), *Dasypyrum villosum* (0.51)**, *Euphorbia cyparissias* (0.49), *Lathyrus tuberosus* (0.49), *Rumex crispus* (0.49), *Rubus ulmifolius* (0.49), *Linaria vulgaris* (0.49), *Galium verum* (0.49), *Verbascum nigrum* (0.48), *Prunus persica* (0.47), *Vicia cracca* (0.47), *Senecio erucifolius* (0.46), *Clinopodium vulgare* (0.46), *Coronilla varia* (0.46), *Melilotus albus* (0.45), *Cephalaria transsylvanica* (0.42), *Stachys palustris* (0.42), *Petrorhagia prolifera* (0.42), *Tanacetum vulgare* (0.41), *Salix alba* (0.41), *Dipsacus fullonum* (0.41), *Melilotus officinalis* (0.40), *Sambucus ebulus* (0.38), *Poa trivialis* (0.38), *Centaurea stoebe* (0.38), *Carduus acanthoides* (0.38), *Odontites vernus* (0.38), *Verbascum phlomoides* (0.38), *Populus alba* (0.38), *Juglans regia* (0.37), *Tragopogon dubius* (0.37), *Silene vulgaris* (0.37), *Rosa canina* (0.36), *Equisetum arvense* (0.36), *Conium maculatum* (0.36), *Ulmus* sp. (0.35), *Silene latifolia alba* (0.35), *Populus nigra* (0.34), *Euphorbia esula* (0.34), *Epilobium hirsutum* (0.34), *Echium vulgare* (0.34), *Glycyrrhiza echinata* (0.34), *Bromus arvensis* (0.33), *Xeranthemum cylindraceum* (0.33), *Scabiosa ochroleuca* (0.33), *Medicago falcata* (0.33), *Malus domestica* (0.33), *Centaurea arenaria* (0.33), *Avena fatua* auct. (0.32), *Prunus cerasifera* (0.32), *Triticum aestivum* (0.32), *Crepis biennis* (0.32), *Cirsium vulgare* (0.32), *Torilis arvensis* (0.31), *Artemisia vulgaris* (0.31), *Rumex conglomeratus* (0.30), *Picris hieracioides* (0.30), *Holcus lanatus* (0.30), *Cruciata laevipes* (0.30), *Bromus hordeaceus* (0.30), *Epilobium tetragonum* (0.30)
Park	***Quercus robur*** (0.51), *Philadelphus coronarius* (0.45), *Viola odorata* (0.33), *Symphoricarpos albus* (0.31), *Spiraea media* (0.31)
Residential area-compact	***Digitaria ciliaris* (0.71), *Campsis radicans* (0.63), *Syringa vulgaris* (0.51), *Kerria japonica* (0.50),** *Ligustrum vulgare* (0.47), *Armoracia rusticana* (0.47), *Vitis vinifera* (0.46), *Antirrhinum majus* (0.44), *Ficus carica* (0.43), *Aquilegia* sp. (0.43), *Mirabilis jalapa* (0.41), *Rhus typhina* (0.41), *Hibiscus syriacus* (0.40), *Lonicera japonica* (0.39), *Ipomoea purpurea* (0.39), *Albizia julibrissin* (0.38), *Vinca major* (0.38), *Rudbeckia hirta* (0.38), *Fragaria vesca* (0.36), *Oenothera biennis* (0.36), *Iris germanica* (0.35), *Coreopsis tinctoria* (0.35), *Prunus domestica* (0.34), *Bassia scoparia* (0.34), *Viola odorata* (0.33), *Anethum graveolens* (0.33), *Lactuca saligna* (0.33), *Geranium pusillum* (0.33), *Thuja orientalis* (0.33), *Cosmos bipinnatus* (0.32), *Tagetes patula* (0.32), *Crepis biennis* (0.32), *Alcea rosea* (0.32), *Ranunculus repens* (0.32), *Hedera helix* (0.32), *Oxalis corniculata* (0.31), *Mentha piperita* agg. (0.31), *Impatiens balfourii* (0.30), *Veronica arvensis* (0.30)
Residential area-open	***Mirabilis jalapa* (0.50)**, *Spiraea media* (0.39), *Duchesnea indica* (0.38), *Solanum tuberosum* (0.36), *Acer pseudoplatanus* (0.36), *Ajuga reptans* (0.35), *Sedum kamtschaticum* (0.35), *Commelina communis* (0.34), *Vinca major* (0.34), *Ligustrum vulgare* (0.33), *Rumex pulcher* (0.33), *Viola odorata* (0.33), *Syringa vulgaris* (0.33), *Hibiscus syriacus* (0.33), *Tagetes patula* (0.32), *Sedum rupestre* (0.32), *Geum urbanum* (0.32), *Chelidonium majus* (0.31), *Sisymbrium officinale* (0.30), *Prunella vulgaris* (0.30), *Petunia* x *atkinsiana* (0.30), *Anagallis arvensis* (0.30)
Square	-

Taxa are listed by decreasing values of Φ (in parenthesis). Only taxa with Φ > 0.3 are shown. The taxa with Φ > 0.5 are marked bold.

The habitat types also differed significantly in the representation of plant life forms (Figure 1). The largest participation of hemicryptophytes, as most represented compared to the other life forms, was found in early and mid-successional sites, and the lowest at squares and boulevards. By comparison, therophytes had the largest share in squares and the lowest in mid-successional sites. Parks and residential areas, both compact and open, were characterized by a somewhat higher proportion of phanerophytes and chamaephytes compared to other habitat types. Geophytes were generally the least represented, especially in mid-successional sites.

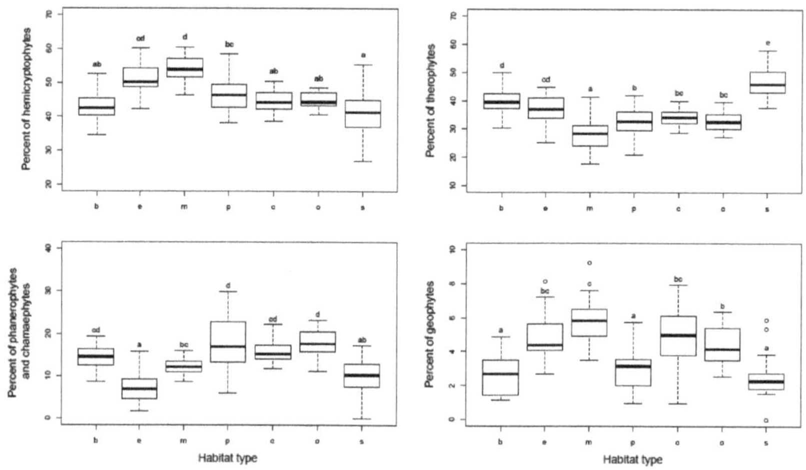

Figure 1. Proportion of hemicryptophytes, therophytes, phanerophytes/chamaephytes, and geophytes in particular urban habitat types. X-axis abbreviations: b—boulevard, e—early successional sites, m—mid-successional sites, p—city park, c—residential area (compact building pattern), o—residential area (open building pattern), s—historical city square. Homogeneous groups of urban habitat types are denoted by the same letters ($p < 0.01$).

The ordination diagram obtained by PCA indicated a predominant grouping of plots belonging to the same habitat type, with a certain overlapping between some of them (Figure 2). The city square (s) distinguished most clearly from the others, with no overlap of this habitat type with any other habitat type analyzed. The most distant from the other habitat types were the plots representing mid-successional and early successional sites. These two habitat types formed a well-separated group, with minor overlapping between each other. The differences were much less noticeable between other habitat types, particularly between city parks (p) and boulevards (b) and between the two types of residential areas (c and o).

The results of the RDA analysis indicated that species composition was affected by both habitat type and climatic variables, with the greater influence of habitat type (9.7%) in relation to climatic characteristics (2.3%), whereas shared variance was not observed (Table 3).

The total number of taxa recorded on 164 plots in the 24 investigated cities on the territory of Serbia was 674, with an average of 105 taxa observed per plot. The highest number of taxa was recorded in the residential areas with compact building pattern (c) in Niš (147), Kruševac (146), and Loznica (142), whereas the lowest number of registered taxa was observed in the city squares (s) of Smederevo and Pančevo (38) and Sremska Mitrovica (42).

There were considerable differences in alpha, beta, and gamma diversity between the habitat types studied. The lowest alpha diversity was found in the city squares (s; 53 plant taxa) and boulevards (b; 88 taxa), whereas the highest values were recorded in residential areas with compact building patterns (c; 128 plant taxa) and at the mid-successional sites (m; 127 plant taxa). Similarly, the lowest gamma diversity was found in the city squares (s; 244 plant taxa) and boulevards (b; 297 taxa), whereas the highest values were found at the mid-successional sites (m; 435 plant taxa) and residential areas with compact building pattern (e; 398; Figure 3A). The highest beta diversity was recorded within city squares (s) and the lowest in residential areas with open building patterns (c; Figure 3B).

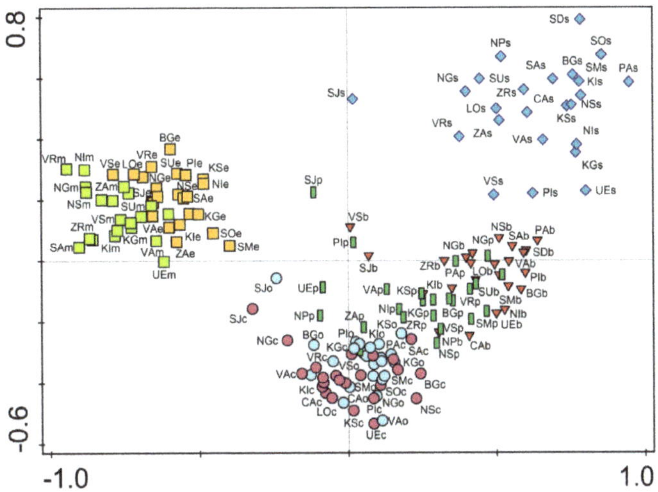

Figure 2. PCA ordination of plots according to plant species composition. Eigenvalues: axis 1, 0.2492; axis 2, 0.0925. Abbreviations: the first two uppercase letters represent the city code (see Table 2), the third lowercase letter represents the habitat type (see Material and Methods). Legend: red triangles—boulevards, orange squares—early successional sites, light green squares—mid-successional sites, dark green boxes—parks, purple circles—residential area (compact building pattern), light blue circles—residential area (open building pattern), blue diamonds—squares.

Table 2. Characterization of the studied cities in Serbia.

City (City Code)	N (°)	E (°)	Alt (m a.s.l.)	T (°C)	ΔT (°C)	P (mm)	Climate Type/Subtype
Niš (NI)	43.32083	21.89528	199	11.36	31.66	625	transitional submediterranean-Aegean subcontinental (IV6)
Vranje (VR)	42.55472	21.89778	481	9.80	30.65	625	
Beograd (BG)	44.81583	20.46000	113	11.9	29.74	672	transitional subcontinental-semiarid continental (VI3b/VII)
Novi Sad (NS)	45.25500	19.84528	85	11.34	30.67	611	
Pančevo (PA)	44.87083	20.64083	81	11.74	29.78	644	
Šabac (ŠA)	44.75694	19.69444	81	11.56	30.67	692	
Smederevo (SD)	44.66500	20.92694	78	11.36	29.99	650	
Sremska Mitrovica (SM)	44.96806	19.60694	84	11.52	30.79	649	
Vršac (VŠ)	45.12111	21.29555	92	11.30	30.5	666	
Čačak (ČA)	43.89111	20.35000	241	10.93	30.82	785	semi-arid temperate continental (subcontinental)—central-southeastern Balkan or Moesian (IV3)
Kragujevac (KG)	44.01000	20.91667	177	11.03	31.01	690	
Kruševac (KŠ)	43.58194	21.32639	162	11.16	31.55	652	
Loznica (LO)	44.53361	19.22389	126	11.14	29.73	844	
Novi Pazar (NP)	43.14028	20.51722	495	9.97	31.28	794	
Pirot (PI)	43.15611	22.58528	371	10.26	31.83	617	
Valjevo (VA)	44.26861	19.88417	188	10.96	30.70	803	
Kikinda (KI)	45.83000	20.46500	83	11.31	31.61	557	semi-arid continental Pannonian (VII)
Sombor (SO)	45.77278	19.11500	90	11.16	30.85	602	
Subotica (SU)	46.10000	19.66500	116	10.98	31.13	555	
Zrenjanin (ZR)	45.38028	20.39083	84	11.55	31.22	572	

Table 2. *Cont.*

City (City Code)	N (°)	E (°)	Alt (m a.s.l.)	T (°C)	ΔT (°C)	P (mm)	Climate Type/Subtype
Negotin (NG)	44.22806	22.53056	47	11.33	32.13	603	semi-humid continental Danubian (VII)
Zaječar (ZA)	43.90333	22.27833	132	10.88	31.92	623	
Užice (UE)	43.85667	19.84028	414	9.52	29.16	899	humid temperate-continental-west Balkan or Illyrian (IV2b)
Sjenica (SJ)	43.27306	20.00028	1006	6.48	30.91	755	humid mountain alpine (XI)

N—latitude, E—longitude, Alt—elevation, T—mean annual temperature, ΔT—difference between mean temperature in July and January, P—precipitation.

Table 3. Total explained variation, the influence of habitat type, climate, and their shared effect on the plant species composition in the analyzed cities.

	Habitat Type	Climate	Shared Effect
Unadjusted R2 (%)	12.7	3.8	<0.1
Adjusted R2 (%)	9.7	2.3	0

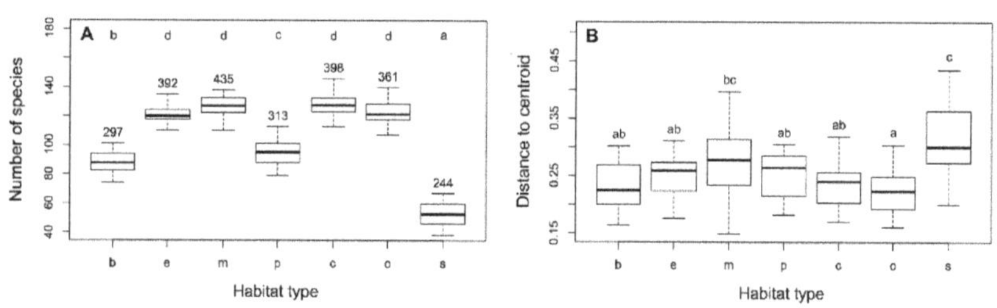

Figure 3. (**A**) Alpha (box plots) and gamma (numbers above) diversity in particular urban habitat types. (**B**) Beta diversity of plant taxa within urban habitat types. X-axis abbreviations: b—boulevard, e—early successional sites, m—mid-successional sites, p—city park, c—residential area (compact building pattern), o—residential area (open building pattern), s—historical city square. Homogeneous groups of urban habitat types are denoted by the same letters ($p < 0.01$).

A strong positive statistically significant correlation was found between alpha and gamma diversity ($r = 0.93$, $p = 0.002$). Alpha and beta diversity were strongly negatively and statistically significantly correlated ($r = -0.82$, $p = 0.02$). Gamma and beta diversity were also negatively correlated, but this correlation was statistically insignificant ($r = -0.62$, $p > 0.1$).

3. Discussion

Significant differences in species composition were found between individual types of urban habitats. As observed from the results of ordination analysis and the plots grouping, species composition is primarily influenced by urban habitat type and intensity of human influence, much more than by climatic features. This is consistent with the findings of Lososová et al. [26], who analyzed the urban flora of Central European cities along a gradient of distinctly different biogeographical regions with contrasting climatic characteristics, and with the results of Rebele [1] and Savard et al. [2], which confirm the hypothesis of uniformity of the urban environment.

Based on the ordination analysis of species composition, plots were grouped primarily based on habitat type affiliation. Four groups of plots can be observed: plots representing squares (1), plots representing boulevards and parks (2), plots representing residential areas (3), and plots representing successional sites on the urban peripheries (4). This suggests that certain habitat types have greater similarities in species composition than others. These similarities are the result of a similar character of anthropogenic influence and the location of the plots in the city (center or periphery; Figure 2).

The results of alpha and gamma diversity analysis in the urban habitats of the studied area showed that the lowest number of species was found in city square, which is consistent with the results from Central European cities [6,27]. Sealed and paved surfaces, which dominate in city squares are hostile habitats, and a relatively small number of plant species are adapted to such extreme conditions (trampling, high insolation, drought, etc.). Hence, it is not surprising that therophytes, as a disturbance tolerant life form, have the largest share in city squares (Figure 1). However, although well-adapted to highly influenced urban habitats, therophytes are thought to be more prone to extinction compared to other life forms [33,34]. Intensive human influence and vegetation limited to small patches strongly support the presence of cosmopolitan and even alien species, which negatively affect the native flora and consequently leads to their local extinction [34,35]. The richness of native flora is additionally affected by environmental filters that have to be overcome in order for plants to arrive from the surrounding natural habitats [12]. Furthermore, these habitat types have a significant share of alien and ornamental plants that have been spread by human activities. Due to all this, squares are unique in their species composition compared to the other urban habitat types, as indicated by their clear separation from them in the ordination diagram (Figure 2).

As opposed to a small number of species in city squares in Serbia and in Central European cities [27], the high plant species richness has been observed in the old centers of the Italian Mediterranean cities [30], due to a number of plants growing undisturbed on the ancient walls. However, such microhabitats beyond intense anthropogenic influences are rare in the city squares in Serbia, and chasmophytic flora were represented by only a few species (e.g., *Cymbalaria muralis* and *Sedum* spp.). Simultaneously with the lowest alpha and gamma diversity, urban squares in Serbia had the highest beta diversity, indicating that the differences between specific habitats of this type were greater than in other habitat types. The same was found in Central European cities as a result of a very low number of species [26].

A somewhat higher floristic richness was observed in boulevards. Bearing in mind that boulevards are usually spatially connected to city squares, the human influence is similar to that for squares but of lower intensity, which resulted in somewhat higher species richness. A major contributor to species richness in boulevards is tree lines, especially those with unsealed surfaces around the trees, considering that these microhabitats harbored a wide range of plant species. Due to the presence of planted trees, a significant number of their seedlings were observed in boulevards. This is one of the reasons for the marked similarities in the species composition of boulevards and parks, whose plots are grouped in the ordination diagram (Figure 2). The species detected in this habitat type are generally common in the cities and almost half of them were detected in all the habitat types analyzed, whereas only nine were recorded only in the boulevards, confirming that the species found in the boulevards are those that generally survive very successfully in urban conditions [12]. Similar to the city squares, the boulevard's flora were characterized by a significant participation of therophytes, particularly compared to the other habitat types. However, in addition to therophytes, boulevards were also characterized by perennials adapted to growing in small cracks in concrete (e.g., *Sagina procumbens*) and seedlings of tree species, primarily those deliberately planted (e.g., *Acer pseudoplatanus*). In contrast, in city squares and boulevards, the least represented life form was geophytes (Figure 1). Although bulbs and rhizomes enable them to survive in the hostile environment [36], their numbers were found to be negatively correlated with the number of inhabitants and traffic density [32].

Parks are very specific urban habitats. Although they resemble natural forest habitats (e.g., similar mild-mesoclimate, similar light regime, etc.), with a number of typical forest plants that can be found in them (e.g., *Brachypodium sylvaticum*, *Clematis vitalba*, seedlings of *Quercus robur*, etc.), parks are artificial habitats created by deliberate planting of trees, often of non-native origin and heavily impacted by human activity (e.g., trampling, regular mowing, planting horticultural annuals and perennials, etc.). In addition to the planted tree species, city parks in Serbia are also characterized by significant participation of spontaneously growing ornamental alien plants, especially typical ruderal ones (e.g., *Amaranthus deflexus* and *Lactuca serriola*, etc.), but also shrubs and tree seedlings (e.g., *Philadelphus coronarius*, *Symphoricarpos albus*, and *Broussonetia papyrifera*). Hence, in addition to phanerophytes as the predominant life form, a significant number of hemicryptophytic species occur in this habitat type (Figure 1). Contrary to almost half of the species detected in the urban parks that were found in all seven groups of the analyzed urban habitat types, a total of 19 species, mainly tree and shrub species, were detected exclusively in this habitat type. However, alpha and gamma diversity in city parks in Serbia was not high, especially compared to residential areas and early and mid-successional sites, primarily because their homogeneity and the lack of specific microhabitats that could harbor different species types. Regardless of this, city parks are recognized as very important habitats for conservation of local biodiversity [37–39].

Residential areas of cities in Serbia, especially those with compact building patterns, harbor a wide range of species, with the highest alpha and significant gamma diversity observed in this type of urban habitat. This high diversity is a consequence of the marked heterogeneity of habitats, considering that residential areas with densely distributed individual housing units represent a mosaic of diverse habitat types: sidewalks and other paved and sealed surfaces, lawns with different mowing regimes, ornamental gardens, urban gardens with various cultivated plants and accompanying weed flora, tree lines, etc., which contribute significantly to the high species richness, especially when compared to squares, boulevards, and urban parks [40]. However, it should be considered that the shape of the sampling plot may also have an influence on increasing species richness and that elongated plots, such as those used to study of this type of habitat, tend to have more species than compact plots of the same size [41–43]. Although our analyses revealed that the highest alpha diversity was found in residential areas with compact building patterns, Godefroid and Koedam [44] indicated a contrasting pattern, i.e., that half open and open areas in Brussels promote species richness, whereas areas with compact structures lead to its reduction. Due to the complexity of these habitat types, residential areas in analyzed cities in Serbia were characterized by different groups of plants, including species commonly found in squares or boulevards (e.g., *Arenaria serpyllifolia* and *Eragrostis minor*), a wide range of grasses (e.g., *Poa* spp. and *Lolium perenne*), escaped ornamental plants (e.g., *Kerria japonica* and *Antirrhinum majus*), juveniles of crop plants (e.g., *Zea mays* and *Solanum lycopersicum*), crop weeds (e.g., *Cynodon dactylon* and *Elymus repens*), and spontaneously growing native and alien trees and shrubs (e.g., *Campsis radicans*, *Syringa vulgaris*, and *Rhus typhina*).

Urban habitat types that stood out in terms of species composition and plant diversity were successional sites, particularly mid-successional ones, abandoned long enough to form species-rich grasslands. Similar findings have been made in several other urban studies [26,44]. Considering that these habitats usually develop on the urban periphery, they are characterized by the absence of strong human impact, with the main difference between them being the duration of the disturbance-free period. Due to recent disturbance, early successional sites are characterized by a vegetation cover dominated by annual plants with ruderal life strategies, capable of rapid colonization of bare ground (e.g., *Bromus tectorum* and *Petrorhagia prolifera*). Additionally, the results of previous studies [35,45] show that compared to the older succession stages, the younger ones are more susceptible to alien species invasion. Mid-successional sites hosted a higher number of plant species than early successional ones and are characterized by the highest alpha diversity compared to the urban habitat types studied. Because of a longer period of non-disturbance, these sites are suitable for inhabiting species with different life strategies. Hence, a lower proportion of therophytes was observed at the mid-successional sites compared to the early successional ones, whereas both habitat types were characterized by a greater participation of hemicryptophytes and therophytes in relation to the other types analyzed (Figure 1). Additionally, proximity to natural vegetation and openness to the urban surroundings facilitated the influx of native species into these habitats, and various shrub and tree species are common in these habitat types (e.g., *Prunus* spp., *Ulmus* spp., and *Juglans regia*). In general, mid-successional sites are characterized by a higher share of native flora than other urban habitat types, particularly city squares and boulevards [5,35]. Despite the very different physiognomy, early and mid-successional sites have a very similar species composition. Many of the same species of shrubs and trees are found in both types of successional sites, but in the early successional sites they appear as seedlings and much smaller individuals. Additionally, their similarity is also contributed by the fact that due to the complex human influence of varying intensity, some patches of older vegetation were found in the early successional sites analyzed because it was hard to find a completely uniform 1 ha area.

Although the results of this study indicated that local site conditions are the predominant factors determining plant species composition, the results of similar studies of urban flora in Italy contradict this hypothesis because climate was found to be the main factor determining species composition, most likely due to the strong climatic gradient from the north to the south of the Apennine Peninsula [30]. Despite the fact that the climate in Serbia is very diverse, ranging from continental to sub-Mediterranean and mountainous, the differences are not pronounced enough to point to climate as the main factor determining the diversification of flora in urban habitats, which is also true for the Central European urban flora [26]. However, the effect of the climate on the composition of plant species in urban habitats in Serbia should not be neglected. This applies particularly to Sjenica, which is located at an altitude of 1026 m and is characterized by a humid mountain alpine climate (Table 2), as the ordination analysis indicated a grouping of several plots of this city. The climatic differences between the other cities analyzed are less noticeable, resulting in a greater similarity in the ordination diagram (Figure 2). Apart from the climate, the composition of species in urban habitats may be influenced by other factors, such as the geographical location, and the structure of the city and its historical features. However, as their effects can be significant, especially in large-scale studies [27], and almost negligible in studies focusing on cities from a smaller geographical area, these factors are usually neglected in the analyses.

4. Materials and Methods

4.1. Data Sampling

Investigation of urban flora was carried out in 24 cities in Serbia (Table 2). Cities were selected to represent all major climate types and subtypes in Serbia. An additional condition that cities had to meet was the existence of the preselected typical urban habitat types. According to Stevanović and Stevanović [46], the following climate types and subtypes are represented in the territory of Serbia (climate type and subtype designations, according to Walter and Leith [47] and Horvat et al. [48], are given in parenthesis): transitional submediterranean-Aegean subcontinental (IV6), semi-arid continental Pannonian (VII), semi-humid continental Danubian (VII), transitional subcontinental-semiarid continental (VI3b/VII), semi-arid temperate continental (subcontinental)—central-southeastern Balkan or Moesian (IV3), humid temperate-continental—west Balkan or Illyrian (IV2b), and humid mountain alpine (XI).

To make comparable samples, floristic data were collected using a standardized protocol, which has already been used in similar studies for Central European cities [5,6,12,26–29]. In each of 24 selected cities in Serbia, we recorded the plant composition at seven specific sites of 1 ha in size. Each of the seven sites represents a different type of urban habitat:

1. historical city square (s), mostly with the buildings constructed before 19th century; sealed or paved more than 90% of the total area.
2. boulevard (b), with the buildings from 19th century, tree lines and small-size lawns; sealed or paved more than 70% of the total area.

3. residential area, with compact building pattern (c), represented by family houses (at least 50 years old) and private yards.
4. residential area, with open building pattern (o), represented by apartment blocks (40-60 years old), and lawns with sparse shrubs and trees.
5. city park (p), with coverage of old deciduous trees from 10 to 50% and regularly mowed lawns.
6. early successional sites (e), severely disturbed in the last 1–3 years, dominated by bare soil and scarce vegetation cover.
7. mid-successional sites (m), abandoned 5 to 15 years ago, with predominance of perennial grassland, and sparse young trees and shrubs.

Adequate sites (habitats) were selected using maps and satellite images in Google Earth. The 1 ha plots were selected within each habitat type and all vascular plants were recorded within them. This included seedlings from spontaneously grown planted trees and garden plants. However, intentionally planted individuals were omitted. In residential areas with a compact building pattern (c), a different approach was applied due to limited access to private gardens. In these cases, instead of an area of 1 ha, transects along a street 500 m long were analyzed, recording all species found in accessible areas, in addition to all those that could be seen from the street inside private yards. The research was conducted in 2015–2019 in the period from June to the end of August, to avoid spring and autumn, i.e., plants with significant variations in phenology. The nomenclature of species corresponding to the diagnostic species of classes of plant communities dominated by vascular plants follows Electronic Appendix S6 (EVC1) of Vegetation of Europe [49], and for other species follows the nomenclature of Flora Europaea [50].

4.2. Data Analysis

The composition of plant species recorded within individual habitat types was shown in synoptic tables. To determine diagnostic species for particular habitat types, the phi coefficient of association (Φ) was used as a statistical measure of the concentration of occurrence of species in particular habitat types [51]. Diagnostic species for a particular habitat type were defined as species that preferentially occur in that habitat type. Fisher's exact test ($p < 0.05$) was used to assess the statistical significance of the species-habitat association, quantified by Φ, as shown in the following equation [52]:

$$\Phi = \frac{N \times n_p - n \times N_p}{\sqrt{n \times N_p \times (N - n) \times (N - N_p)}}$$

where N is the number of all sites in the data set, N_p is the number of sites in the particular habitat type, n is the number of occurrences of the species in the data set, and n_p is the number of occurrences of the species in the particular habitat type. Diagnostic species were considered to be those that had a statistically significant species–habitat association and $\Phi > 0.30$. Synoptic tables and calculations of phi coefficient were carried out in the JUICE program [53].

All recorded species were categorized into four different groups according to their life form: geophytes, phanerophytes and chamaephytes (trees and shrubs), therophytes, and hemicryptophytes [54]. To compare the differences in the frequency of individual life forms in relation to urban habitat types, ANOVA [55] was used.

According to detrended correspondence analysis (DCA), the length of the gradient in species composition was 2.27 SD units. Therefore, to assess the general variation patterns in composition of plant species among urban habitat types, unconstrained and constrained linear ordination methods were used (PCA and RDA, respectively). Analysis and visualization of the PCA and RDA diagrams were performed using CANOCO 5.12 program [56].

To distinguish the influences of climatic variables and urban habitat types on plant composition, redundancy variation partitioning for RDA was carried out [57]. Two groups of variables were collected for each site: (1) habitat type, which was given as a categorical variable with seven expressions, and (2) three climatic variables: mean annual temperature, annual temperature range, i.e., the difference between mean temperatures in July and January, and annual precipitation total. The climate variables were taken from the WorldClim dataset. The significance of the influence of climatic variables and urban habitat types was tested by Monte Carlo permutations (999 permutations). These calculations were performed in the CANOCO 5.12 program [56].

Alpha and gamma diversity were used to indicate differences in plant species richness between habitat types. Alpha diversity was defined as the average number of taxa recorded per plot in each habitat type, and gamma diversity was estimated as the total number of taxa observed in all plots belonging to a particular habitat type. To determine beta diversity, we calculated an index of beta diversity: S/a-1, where S is the total number of taxa, and a is the average number of taxa per plot. The calculations of alpha, gamma, and beta diversity and the visualization of their differences between habitat types were carried out in the "vegan" package of the R programming language [58].

5. Conclusions

The results of the study of urban flora of Serbia indicate significant floristic richness of the investigated areas. Pronounced differences were found between the analyzed habitat types, both in terms of diversity, species composition, and dominant life forms. It was shown that the species composition and dominance of particular life forms are directly related to the intensity of anthropogenic influence. The lowest alpha and gamma diversity was found in city squares and boulevards, with a dominance of therophytes, a life form related to disturbed habitats, whereas habitats in the urban periphery under less pronounced human influence and heterogeneous residential urban areas, are characterized by significant floristic richness and much more uniform distribution of life forms. Comparison of the influence of different factors on the composition of urban flora shows that urban flora in the analyzed cities is more influenced by the type of habitat than by climatic features. Our results indicated that the most important ways to increase plant diversity in cities are the following: (1) allowing natural succession and reducing the intensity of the anthropogenic factor in certain parts of the city; (2) providing greater heterogeneity of urban areas, and forming specific urban microhabitats that allow the survival of those species that are not typical urbanophiles.

Author Contributions: Conceptualization, M.G., K.J., J.Š.-S., D.L. and S.J.; writing—original draft preparation, M.G., K.J., J.Š.-S. and S.J.; methodology, M.G., J.Š.-S., D.L. and K.J.; software, M.G. and J.Š.-S.; investigation, M.G. and M.T.; validation, M.G., D.L. and S.V. All authors have read and agreed to the published version of the manuscript.

Funding: This research was funded by Ministry of Education, Science and Technological Development of the Republic of Serbia, grant number 451-03-9/2021-14/ 200178.

Institutional Review Board Statement: Not applicable.

Informed Consent Statement: Not applicable.

Data Availability Statement: The data are not publicly available due to the database is part of the Ph.D. thesis. The complete database will be available only after the publication of the dissertation in 2022. Also, we will use the same database for other articles we plan to publish. For these reasons, we believe that the complete database should not be publicly available at this moment.

Conflicts of Interest: The authors declare no conflict of interest.

References

1. Rebele, F. Urban ecology and special features of urban ecosystems. *Glob. Ecol. Biogeogr. Lett.* **1994**, *4*, 173–187. [CrossRef]
2. Savard, J.P.L.; Clergeau, P.; Mennechez, G. Biodiversity concepts and urban ecosystems. *Landsc. Urban Plan.* **2000**, *4*, 131–142. [CrossRef]
3. McKinney, M.L.; Lockwood, J.L. Biotic homogenization: A few winners replacing many losers in the next mass extinction. *Trends Ecol. Evol.* **1999**, *14*, 450–453. [CrossRef]

4. Olden, J.D.; Poff, N.L.; Douglas, M.R.; Douglas, M.E.; Fausch, K.D. Ecological and evolutionary consequences of biotic homogenization. *Trends Ecol. Evol.* **2004**, *19*, 18–24. [CrossRef] [PubMed]
5. Lososová, Z.; Chytrý, M.; Tichý, L.; Danihelka, J.; Fajmon, K.; Hájek, O.; Kintrová, K.; Kühn, I.; Laniková, D.; Otypková, Z.; et al. Native and alien floras in urban habitats: A comparison across 32 cities of Central Europe. *Glob. Ecol. Biogeogr.* **2012**, *21*, 545–555. [CrossRef]
6. Čeplová, N.; Kalusová, V.; Lososová, Z. Effects of urban heat island, settlement size and habitat type on urban plant biodiversity. *Landsc. Urban Plan.* **2017**, *159*, 15–22. [CrossRef]
7. Pyšek, P.; Bacher, S.; Chytrý, M.; Jarošík, V.; Wild, J.; Celesti-Grapow, L.; Gassó, N.; Kenis, M.; Lambdon, P.W.; Nentwig, W.; et al. Contrasting patterns in the invasions of European terrestrial and freshwater habitats by alien plants, insects and vertebrates. *Glob. Ecol. Biogeogr.* **2010**, *19*, 317–331. [CrossRef]
8. Pyšek, P. Alien and native species in Central European urban floras: A quantitative comparison. *J. Biogeogr.* **1998**, *25*, 155–163. [CrossRef]
9. Gong, C.; Chen, J.; Yu, S. Biotic homogenization and differentiation of the flora in artificial and near-natural habitats across urban green spaces. *Landsc. Urban Plan.* **2013**, *120*, 158–169. [CrossRef]
10. Kowarik, I. On the role of alien species in urban flora and vegetation. In *Urban Ecology*; Marzluff, J.M., Shulenberger, E., Endlicher, W., Alberti, M., Bradley, G., Ryan, C., Simon, U., Brunnen, C.Z., Eds.; Springer: Boston, MA, USA, 1995; pp. 321–338. [CrossRef]
11. Kühn, I.; Brandl, R.; Klotz, S. The flora of German cities is naturally species rich. *Evol. Ecol. Res.* **2004**, *6*, 749–764.
12. Kalusová, V.; Čeplová, N.; Lososová, Z. Which traits influence the frequency of plant species occurrence in urban habitat types? *Urban Ecosyst.* **2017**, *20*, 65–75. [CrossRef]
13. Knapp, S.; Dinsmore, L.; Fissore, C.; Hobbie, S.E.; Jakobsdottir, I.; Kattge, J.; King, J.Y.; Klotz, S.; McFadden, J.P.; Cavender-Bares, J. Phylogenetic and functional characteristics of household yard floras and their changes along an urbanization gradient. *Ecology* **2012**, *93*, 83–98. [CrossRef]
14. Simberloff, D.; Souza, L.; Nuñez, M.A.; Barrios-Garcia, M.N.; Bunn, W. The natives are restless, but not often and mostly when disturbed. *Ecology* **2012**, *93*, 598–607. [CrossRef] [PubMed]
15. Aronson, M.F.; Lepczyk, C.A.; Evans, K.L.; Goddard, M.A.; Lerman, S.B.; MacIvor, J.S.; Nilon, C.; Vargo, T. Biodiversity in the city: Key challenges for urban green space management. *Front. Ecol. Environ.* **2017**, *15*, 189–196. [CrossRef]
16. Ajmone-Marsan, F.; Biasioli, M. Trace elements in soils of urban areas. *Water Air Soil Pollut.* **2010**, *213*, 121–143. [CrossRef]
17. Samoli, A.; Stergiopoulou, A.; Santana, P.; Rodopoulou, S.; Mitsakou, C.; Dimitroulopoulou, C.; Bauwelinck, M.; de Hoogh, K.; Costa, C.; Marí-Dell'Olmo, M.; et al. Spatial variability in air pollution exposure in relation to socioeconomic indicators in nine European metropolitan areas: A study on environmental inequality. *Environ. Pollut.* **2019**, *249*, 345–353. [CrossRef]
18. Godefroid, S. Temporal analysis of the Brussels flora as indicator for changing environmental quality. *Landsc. Urban Plan.* **2001**, *52*, 203–224. [CrossRef]
19. Unger, J.; Savić, S.; Gal, T. Modelling of the annual mean urban heat island pattern for planning of representative urban climate station network. *Adv. Meteorol.* **2011**, *2011*, 398613. [CrossRef]
20. Schmidt, K.J.; Poppendieck, H.-H.; Jensen, K. Effects of urban structure on plant species richness in a large European city. *Urban Ecosyst.* **2014**, *17*, 427–444. [CrossRef]
21. Luck, G.W.; Smallbone, L.T. The impact of urbanization on taxonomic and functional similarity among bird communities. *J. Biogeogr.* **2011**, *38*, 894–906. [CrossRef]
22. Statista. Degree of Urbanization (Percentage of Urban Population in Total Population) by Continent in 2020. Available online: http://www.statista.com/statistics/270860/urbanization-by-continent/ (accessed on 15 August 2021).
23. Breuste, J.; Haase, D.; Elmquist, T. Urban landscapes and ecosystem services. In *Ecosystem Services in Agricultural and Urban Landscapes*; Wratten, S., Sandhu, H., Cullen, R., Costanza, R., Eds.; John Wiley & Sons Ltd.: Hoboken, NJ, USA, 2013; pp. 83–102. [CrossRef]
24. Hulme, P.E.; Bacher, S.; Kenis, M.; Klotz, S.; Kühn, I.; Minchin, D.; Nentwig, W.; Olenin, S.; Panov, V.; Pergl, J.; et al. Grasping at the routes of biological invasions: A framework for integrating pathways into policy. *J. Appl. Ecol.* **2008**, *45*, 403–414. [CrossRef]
25. Jovanović, S.; Glišić, M. An analysis of research into urban flora and vegetation in Southeast Europe. *Acta Bot. Croat.* **2021**, *80*, 74–81. [CrossRef]
26. Lososová, Z.; Horsak, M.; Chytrý, M.; Čejka, T.; Danihelka, J.; Fajmon, K.; Hájek, O.; Juřičková, L.; Kintrová, K.; Laniková, D.; et al. Diversity of Central European urban biota: Effect of human-made habitat types on plants and snails. *J. Biogeogr.* **2011**, *38*, 1152–1163. [CrossRef]
27. Lososová, Z.; Chytrý, M.; Tichý, L.; Danihelka, J.; Fajmon, K.; Hájek, O.; Kintrová, K.; Laniková, D.; Otypková, Z.; Řehořek, V. Biotic homogenization of Central European urban floras depends on residence time of alien species and habitat types. *Conserv. Biol.* **2012**, *145*, 179–184. [CrossRef]
28. Lososová, Z.; Chytrý, M.; Danihelka, J.; Tichý, L.; Ricotta, C. Biotic homogenization of urban floras by alien species: The role of species turnover and richness differences. *J. Veg. Sci.* **2016**, *27*, 452–459. [CrossRef]
29. Lososová, Z.; Tichý, L.; Divíšek, J.; Čeplová, N.; Danihelka, J.; Dřevojan, P.; Fajmon, K.; Kalníková, V.; Kalusová, V.; Novák, P.; et al. Projecting potential future shifts in species composition of European urban plant communities. *Divers. Distrib.* **2018**, *24*, 765–775. [CrossRef]

30. Celesti-Grapow, L.; Blasi, C. A comparison of the urban flora of different phytoclimatic regions in Italy. *Glob. Ecol. Biogeogr.* **1998**, *7*, 367–378. [CrossRef]
31. Salinitro, M.; Alessandrini, A.; Zappi, A.; Tassoni, A. Impact of climate change and urban development on the flora of a southern European city: Analysis of biodiversity change over a 120-year period. *Sci. Rep.* **2019**, *9*, 9464. [CrossRef] [PubMed]
32. Rat, M.; Gavrilović, M.; Radak, B.; Bokić, B.; Jovanović, S.; Božin, B.; Boža, P.; Anačkov, G. Urban flora in the Southeast Europe and its correlation with urbanization. *Urban Ecosyst.* **2017**, *20*, 811–822. [CrossRef]
33. Stehlik, I.; Caspersen, J.P.; Wirth, L.E.A.; Holderegger, R. Floral free fall in the Swiss lowlands: Environmental determinants of local plant extinction in a peri-urban landscape. *J. Ecol.* **2007**, *95*, 734–744. [CrossRef]
34. Williams, N.S.G.; Hahs, A.K.; Vesk, P.A. Urbanisation, plant traits and the composition of urban floras. *Perspect. Plant Ecol. Evol. Syst.* **2015**, *17*, 78–86. [CrossRef]
35. Deák, B.; Hüse, B.; Tóthmérész, B. Grassland vegetation in urban habitats—Testing ecological theories. *Tuexenia* **2016**, *36*, 379–393. [CrossRef]
36. Fascetti, S.; Potenza, G.; Castronuovo, D.; Candido, V. Wild geophytes of ornamental interest in the native flora of southern Italy. *Ital. J. Agron.* **2014**, *9*, 99–106. [CrossRef]
37. Celesti-Grapow, L.; Pyšek, P.; Jarošík, V.; Blasi, C. Determinants of native and alien species richness in the urban flora of Rome. *Divers. Distrib.* **2006**, *12*, 460–501. [CrossRef]
38. Sulaiman, S.; Mohamad, N.; Idilfitri, S. Contribution of vegetation in urban parks as habitat for selective bird community. *Procedia Soc. Behav. Sci.* **2013**, *85*, 267–281. [CrossRef]
39. Banaszak-Cibicka, B.; Twerd, L.; Fliszkiewicz, M.; Giejdasz, K.; Langowska, A. City parks vs. natural areas—Is it possible to preserve a natural level of bee richness and abundance in a city park? *Urban Ecosyst.* **2018**, *21*, 599–613. [CrossRef]
40. Goddard, M.A.; Dougill, A.J.; Benton, T.G. Scaling up from gardens: Biodiversity conservation in urban environments. *Trends Ecol. Evol.* **2010**, *25*, 90–98. [CrossRef] [PubMed]
41. Bossuyt, B.; Hermy, M. Species turnover at small scales in dune slack plant communities. *Basic Appl. Ecol.* **2004**, *5*, 321–329. [CrossRef]
42. Heegaard, E.; Økland, R.H.; Bratli, H.; Dramstad, W.E.; Engan, G.; Pedersen, O.; Solstad, H. Regularity of species richness relationships to patch size and shape. *Ecography* **2007**, *30*, 589–597. [CrossRef]
43. Stohlgren, T.J. *Measuring Plant Diversity—Lessons from the Field*; Oxford University Press: Oxford, UK, 2007.
44. Godefroid, S.; Koedam, N. Urban plant species patterns are highly driven by density and function of built-up areas. *Landsc. Ecol.* **2007**, *22*, 1227–1239. [CrossRef]
45. Bastl, M.; Kočár, P.; Prach, K.; Pyšek, P. The effect of successional age and disturbance on the establishment of alien plants in man-made sites: An experimental approach. In *Plant Invasions: Studies from North America and Europe*; Brock, J.H., Wade, M., Pyšek, P., Green, D., Eds.; Backhuys Publishers: Leiden, The Netherlands, 1997; pp. 191–201.
46. Stevanović, V.; Stevanović, B. Osnovni klimatski, geološki i pedološki činioci biodiverziteta kopnenih ekosistema Jugoslavije. In *Biodiverzitet Jugoslavije sa Pregledom Vrsta od Međunarodnog Značaja*; Stevanović, V., Vasić, V., Eds.; Ecolibri & Faculty of Biology: Belgrade, Serbia, 1995; pp. 75–95.
47. Walter, H.; Lieth, H. *Klimadiagramm–Weltatlas*; Gustav Fisher Verlag: Jena, Germany, 1967.
48. Horvat, I.; Glavač, V.; Ellenberg, H. *Vegetation of Southeast Europe*; Gustav Fisher Verlag: Jena, Germany, 1974.
49. Mucina, L.; Bültmann, H.; Dierßen, K.; Theurillat, J.P.; Raus, T.; Čarni, A.; Šumberová, K.; Willner, W.; Dengler, J.; García, R.G.; et al. Vegetation of Europe: Hierarchical floristic classification system of vascular plant, bryophyte, lichen, and algal communities. *Appl. Veg. Sci.* **2016**, *19*, 3–264. [CrossRef]
50. Tutin, T.G.; Heywood, V.H.; Burges, N.A.; Moore, D.M.; Valentine, D.W.; Walters, S.M.; Webb, D.A. *Flora Europaea*; Cambridge University Press: Cambridge, UK, 1968–1980; Volumes 2–5.
51. Chytrý, M.; Tichý, L.; Holt, J.; Botta-Dukát, Z. Determination of diagnostic species with statistical fidelity measures. *J. Veg. Sci.* **2002**, *13*, 79–90. [CrossRef]
52. Tichý, L.; Chytrý, M. Statistical determination of diagnostic species for site groups of unequal size. *J. Veg. Sci.* **2006**, *17*, 809–818. [CrossRef]
53. Tichý, L. JUICE, software for vegetation classification. *J. Veg. Sci.* **2002**, *13*, 451–453. [CrossRef]
54. Raunkiaer, C. *The Life Forms of Plants and Statistical Plant Geography; Being the Collected Papers of C. Raunkiaer*; Clarendon: London, UK, 1934.
55. StatSoft. *Statistica for Windows, Version 9.0*; StatSoft Inc.: Tulsa, OK, USA, 2009.
56. ter Braak, C.J.F.; Šmilauer, P. *CANOCO Reference Manual and CanoDraw for Windows User's Guide: Software for Canonical Community Ordination, Version 4.5*; Microcomputer Power: Ithaca, NY, USA, 2002.
57. Peres-Neto, P.R.; Legendre, P.; Dray, S.; Borcard, D. Variation partitioning of species data matrices: Estimation and comparison of fractions. *Ecology* **2006**, *87*, 2614–2625. [CrossRef]
58. Oksanen, J.; Blanchet, F.G.; Friendly, M.; Kindt, R.; Legendre, P.; McGlinn, D.; Minchin, P.R.; O'Hara, R.B.; Simpson, G.L.; Solymos, P.; et al. Vegan: Community Ecology Package. R Package Version 2.5–6. Available online: https://CRAN.R-project.org/package=vegan (accessed on 10 August 2021).

plants

MDPI

Article

Orchids of Azerbaijani Cemeteries

Attila Molnár V. [1,*], Viktor Löki [2], Marc Verbeeck [3] and Kristóf Süveges [1]

[1] Department of Botany, University of Debrecen, Egyetem tér 1, H-4032 Debrecen, Hungary; eska1994@gmail.com
[2] Wetland Ecology Research Group, Centre for Ecological Research, Bem tér 18/C, H-4026 Debrecen, Hungary; loki.viktor@ecolres.hu
[3] Arbeitskreis Heimische Orchideen Baden-Württemberg, D-69469 Weinheim, Belgium; mv2know@hotmail.com
* Correspondence: mva@science.unideb.hu

Abstract: In order to explore their orchid flora, we performed surveys of 96 Azerbaijani burial places in 2018 and 2019. Altogether, 28 orchid taxa were found in 37 visited cemeteries. In the orchid diversity a remarkable pattern was observed: geographic latitude was significantly and positively related to the number of taxa and number of individuals. The most widespread and abundant orchids in Azerbaijani graveyards were *Anacamptis pyramidalis* and *A. papilionacea* (found in 23 and 8 cemeteries, respectively). Azerbaijani cemeteries can be important refuges for rare and threatened orchids, e.g., *Himantoglossum formosum* (three cemeteries), *Ophrys sphegodes* subsp. *mammosa* (eight), *Orchis adenocheila* (two), *O. punctulata* (three), *O. stevenii* (one) and *Steveniella satyrioides* (one). *Epipactis turcica*, detected in a single locality, was previously unknown to the flora of Azerbaijan. Additionally, we documented orchid tuber (salep) collection in two cemeteries.

Keywords: anthropogenic habitats; Caucasus; *Himantoglossum formosum*; human-made habitats; Orchidaceae; salep harvesting; Transcaucasia

Citation: Molnár V., A.; Löki, V.; Verbeeck, M.; Süveges, K. Orchids of Azerbaijani Cemeteries. *Plants* **2021**, *10*, 2779. https://doi.org/10.3390/plants10122779

Academic Editor: Robert Philipp Wagensommer

Received: 30 November 2021
Accepted: 11 December 2021
Published: 16 December 2021

1. Introduction

The Earth's surface has changed dramatically in recent centuries, with human activities serve as a leading cause of the drastic reduction in the area of natural habitats [1,2]. In parallel with the degradation and fragmentation of natural environments throughout the world, isolated natural habitat patches as remnants of the original wildlife have been revalued [3]. Anthropogenically influenced habitats now occupy a significant part of the Earth's surface and expand rapidly [4]. In order to conserve the remaining biodiversity, it is of the utmost importance to identify and protect the remaining habitats with a high conservation value, to develop a sustainable habitat management practice, and to plan future developments in the light of nature conservation priorities [5].

Recently, conservation professionals have recognized that some of the anthropogenically influenced or even human-made habitats, such as abandoned mines and industrial sites [6–8], road verges [9–11], tree plantations [12–14], river dikes [15], burial mounds [16], and urban habitats [17,18], play significant roles in conserving biodiversity. During the last decades, it has become increasingly evident that cemeteries also play an important role in maintaining biodiversity [19]. Although the orchid flora of cemeteries is globally rather poorly known, occurrences of orchids were published from Australian, Asian, and European burial places [20]. Based on previous knowledge on the occurrence and diversity of orchids in Turkish [21–24], Albanian [25] and central European [26] burial grounds, we predicted potential conservational importance of traditional Caucasian cemeteries. One of the main goals of our study was to search for *Himantoglossum formosum*, the rarest and perhaps the least known orchid of the Caucasian region [27]. During the 180 years after its description [28], almost nothing was known about the species [29], and it was re-discovered in 1994 [30]. According to recent studies, this is an ancient, phylogenetically isolated [31] and morphologically well separated [32] bona fide species. It is listed as Vulnerable (Rare) in the IUCN Red List of Threatened Plants [33].

The aims of this paper were to survey Azerbaijani cemeteries as orchid habitats, and to test which geographic factors influence the prevalence of orchids in the surveyed cemeteries.

2. Materials and Methods

We studied burial grounds (Azerbaijani: məzarlıq, hereafter cemeteries) regardless of their spatial dimension, position within settlements, or presence of built facilities. We surveyed 96 Azerbaijani cemeteries (Figure 1, Table A1) during 2018 (17–30 May by Molnár V., Löki, Mizsei and Süveges, and 28 June–4 July by Molnár V. and Szabó) and 2019 (29 April–6 May by Verbeeck, Duijnhouwer, Segers and Bobocea) and (31 May–6 June by Verbeeck, Duijnhouwer and Bradeanu). Most cemeteries were visited only once (90 and 3 cemeteries in May 2018 and in April 2019, respectively), but three cemeteries were visited in both years. All orchid taxa and the number of individuals were counted or estimated in the whole area of each visited cemetery. Species were identified based on the comprehensive book of Kuehn et al. [34]. Authors of plant names were listed in Table 1. The geocoordinates and the elevation of the visited cemeteries were determined using a Garmin eTrex Legend handheld GPS device and recorded in WGS84 format. During field trips, particular attention was devoted to documenting salep collection activity in cemeteries.

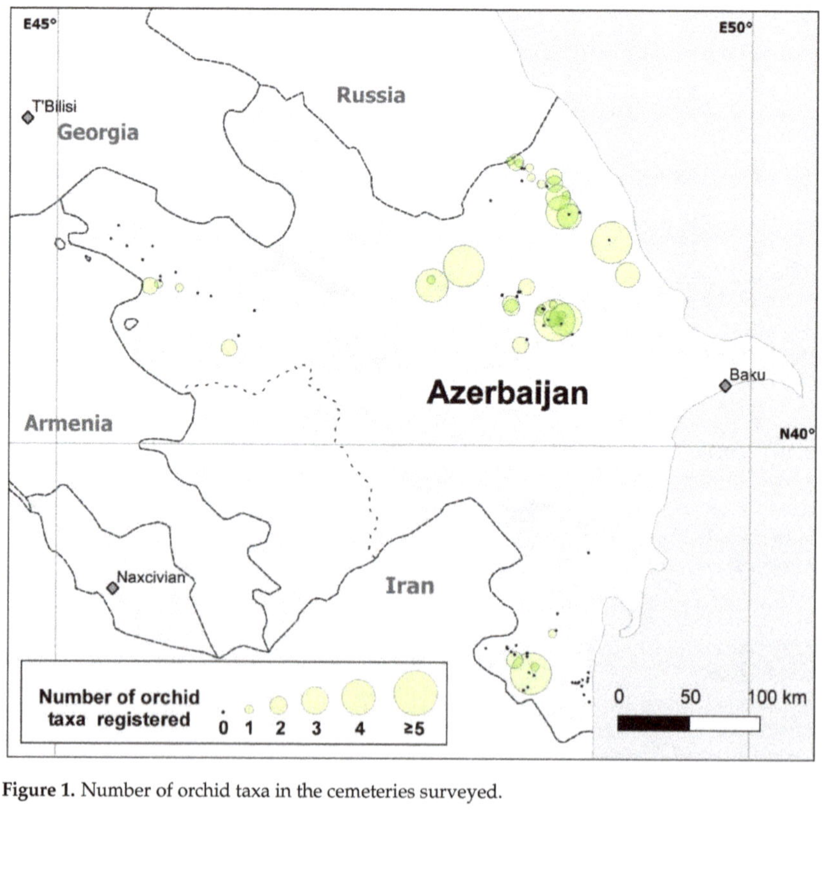

Figure 1. Number of orchid taxa in the cemeteries surveyed.

Table 1. Orchid taxa recorded in Azerbaijani cemeteries.

Taxon	No. of Cemeteries	Total Number of Individuals
Anacamptis pyramidalis (L.) Rich.	23	1901
Anacamptis papilionacea L.	8	567
Ophrys sphegodes subsp. *mammosa* (Desf.) Soó ex Nelson	8	227
Orchis simia Lam.	7	492
Limodorum abortivum (L.) Sw.	4	23
Anacamptis morio (L.) Bateman et al.	3	514
Anacamptis collina (Banks and Sol. ex Russell) Bateman et al.	3	86
Himantoglossum formosum (Steven) K. Koch	3	27
Orchis punctulata Steven ex Lindl.	3	14
Cephalanthera rubra (L.) Rich.	3	9
Ophrys oestrifera M. Bieb.	3	3
Orchis adenocheila Czerniak.	2	239
Orchis mascula subsp. *longicalcarata* Akhalk. et al.	2	9
Anacamptis coriophora (L.) Bateman et al.	2	6
Ophrys apifera Huds.	2	3
Orchis caucasica Regel	1	50
Orchis ×*chabalensis* B. Baumann et al. (*O. punctulata* × *O. stevenii*)	1	30
Ophrys sp.	1	24
Dactylorhiza romana (Sebast.) Soó	1	20
Orchis stevenii Rchb. F.	1	20
Epipactis turcica Kreutz	1	7
Anacamptis sp.	1	6
Ophrys caucasica Woronow ex Grossh.	1	5
Epipactis microphylla (Ehrh.) Sw.	1	3
Neotinea tridentata (Scop.) Bateman et al.	1	3
Epipactis sp.	1	2
Cephalanthera damasonium (Mill.) Druce	1	1
Steveniella satyrioides (Spreng.) Schltr.	1	1

To understand the role of geographic factors in determining variation in taxon richness and abundance of orchids across Azerbaijan, we built statistical models with either of these variables as dependent variables, and latitude, longitude and altitude as explanatory variables. Both the number of individuals and the number of taxa had Poisson distributions, but due to the overdispersion in these variables, we used generalized linear model (GLMs) with quasi-Poisson distribution. All models were built in the R statistical environment [35].

3. Results

Numbering (ID), geographic location, and altitude above see level of the cemeteries visited, together with lists of the orchid taxa found in each one, are given in Table A1. In total, 28 orchid taxa were found, and considerable differences can be observed in the number of individuals and frequency of each taxon (Table 1), as well as in orchid species richness and abundance of each cemetery (Table 2).

Table 2. Descriptive statistics orchid flora of Azerbaijani cemeteries.

Number of cemeteries studied	96
Number of cemeteries hosting orchids	37
Rate of cemeteries hosting orchids	38.5%
Mean (±SD) number of orchid taxa/cemeteries	0.92 (±1.73)
Maximum number of orchid taxa/cemeteries	9
Mean (±SD) number of orchid individuals/cemeteries	44.4 (±173.8)
Maximum number of orchid individuals/cemeteries	1050

Each taxon was found total in 1–24 cemeteries (mean ± SD = 3.2 ± 4.5), with the number of individuals varying from 1 to 1902 (mean ± SD = 150 ± 374). The most widespread and abundant species was *Anacamptis pyramidalis* (Figure 2A). The number of taxa detected in only one graveyard was 15, whereas four species were found in more than five cemeteries. The highest number of taxa in a given cemetery was 9. In most cases only one taxon (18 cemeteries (15%)) or two taxa (11 cemeteries (9.4%)) occurred. Cemeteries that serve as habitats for five or more taxa were extremely rare (4 (3.4%)). The most orchid-rich cemeteries were found near Lerik (AZ-16, 9 species) Ağabəyli (AZ-52, 8 species), Nohurqishlaq (AZ-93, Figure 2B, 8 species), and DashliJalgan (AZ-90, 5 species).

Figure 2. Orchids in Azerbaijani cemeteries. (**A**) *Anacamptis pyramidalis* population in the cemetery of əngixaran (AZ–61). (**B**) Cemetery of Nohurqishlaq (AZ-93), habitat of *Orchis punctulata*, *O. stevenii* and their hybrids (*Orchis* ×*chabalensis*). (**C**) Viable population of *Himantoglossum formosum* was found on a few tens of square meter of refuge under some old oak trees in cemetery of Zizik (AZ-74). (**D**) Spurs of salep harvesting in the cemetery of Ağabəyli (AZ-52). (**E**) Effect of fencing around cemetery against grazing: plant cover is considerable lower outside (left) than inside (right, with flowering individuals of *Anacamptis pyramidalis*) of cemetery of Zurnabad (AZ-32). (**F**) Inflorescence of *Steveniella satyrioides*. (**G**) Occurrence of *Epipactis turcica* was formerly unknown from Azerbaijan (Tengealti, AZ-85). (**H**) A very localized and rare endemic species, *Himantoglossum formosum* in cemetery of Zizik (AZ-74). Photo credit: **A**, **C**, **D**, **G** and **H** by A. Molnár V.; **B** and **F** by M. Verbeeck; **E** by V. Löki.

The harvest of orchid tubers ("salep") was observed in two cemeteries during 2018. In Ağabəyli cemetery (AZ-52, Figure 2D) three species (*Anacamptis papilionacea, Orchis adenocheila, O. simia*), and in Dashli Jalgan cemetery (AZ-90) five species (*Anacamptis collina,*

A. papilionacea, *Ophrys sphegodes* subsp. *mammosa*, *Orchis simia*, *Neotinea tridentata*), were collected. Both of these localities host notable orchid populations with eight and five species, respectively.

The number of orchid taxa and individuals found in Azerbaijani cemeteries was significantly positively related to latitude (Tables 3 and 4, respectively), but not to longitude and altitude. When non-significant predictors were removed from the model in a stepwise manner (based on the largest *p*-values), only latitude remained in the final model as a significant predictor of orchid species richness and abundance.

Table 3. Effect of geographic location on number of orchid taxa per cemetery. Parameter estimates, their standard errors (SE), associated t-values (t) and significance levels (p) are presented.

	Full Model				Minimal Model			
	Estimate	SE	t	*p*	Estimate	SE	t	*p*
Intercept	−0.2317	0.2018	−1.148	0.251	−0.1916	0.2017	−0.95	0.342
Altitude	0.0850	0.1746	0.487	0.627				
Latitude	0.5750	0.1943	2.960	0.003	0.5471	0.1968	2.78	0.005
Longitude	0.3305	0.1916	1.725	0.084				

Table 4. Effect of geographic location on number of orchid individuals in Azerbaijani cemeteries.

	Full Model				Minimal Model			
	Estimate	SE	t	*p*	Estimate	SE	t	*p*
Intercept	3.8439	0.3783	10.161	<0.001	3.8765	0.3788	10.234	<0.001
Altitude	0.0784	0.1801	0.435	0.6635				
Latitude	0.2891	0.1930	1.498	0.1342	0.5770	0.2032	2.839	0.0045
Longitude	0.6063	0.2023	2.997	0.0027				

4. Discussion

During our work, it has been proved that Muslim Azerbaijani cemeteries host significant orchid populations. The key conservation importance of Azerbaijani cemeteries can be explained by two facts: (1) Religious privileges protected these sacred sites and their natural values, because they have largely been exempt from forest and agricultural utilization ever since; and (2) the mostly fenced area of cemeteries provide protection against excessive grazing (Figure 2E).

Azerbaijani cemeteries provide shelters for several valuable populations of rare and threatened orchids. From a conservation point of view, one of the most valuable species is the Eastern Caucasian endemic *Himantoglossum formosum* (Figure 2H), which was found in three of the visited cemeteries (Zizik, AZ-74, Figure 2C; Yasab, AZ-78; Piral, AZ-79). Viable populations of the rare *Orchis adenocheila* were found in two cemeteries (Lerik, AZ-16; Ağabəyli, AZ-52). The occurrence of *Steveniella satyrioides* was detected in cemetery of Lerik (Lerik, AZ-16, Figure 2F). The occurrence of *Epipactis turcica* (Figure 2G) was also found near Tengealti (AZ-85); this taxon was formerly unreported in Azerbaijan.

The long-term survival of these orchid populations in cemeteries strongly depends on long-established, sustainable management practices and traditional burial habits [22,36]. Establishment of graves (especially modern graves covered by marble or concrete tombstones) on the most valuable parts of these cemeteries is expressly undesirable from a conservation perspective, as well as the use of herbicides or electric trimmers. However, mowing or moderate grazing of grassy areas around the burial ground is preferred and encouraged for a more efficient conservation of the local biodiversity and valuable flora elements. Based on their diverse and abundant orchid community in some of the visited cemeteries, we strongly recommend the local councils and the nature protection authorities to protect certain burial places, especially near Lerik (AZ-16), Ağabəyli (AZ-52), DashliJalgan (AZ-90), Nohurqishlaq (AZ-93), and Nugadi (AZ-92).

A special threatening factor of tuberous orchids, namely the harvest of their tubers (making salep for culinary purposes [37]) was observed in Azerbaijani cemeteries. On the one hand, the right of local human communities to continue using traditional natural resources is unquestionable and seems also sustainable [38,39]. On the other hand, the effects of tuber collection on populations of frequent and widespread orchids is little known, while the sustainability of salep harvesting is at least controversial [40–47]. However, destroying the rarest taxa (*Himantoglossum formosum, Orchis adenocheila*) should definitely be avoided.

Author Contributions: Conceptualization: A.M.V.; formal analysis: K.S.; investigation: A.M.V., V.L., M.V. and K.S.; writing—original draft preparation: A.M.V.; writing—review and editing: V.L., M.V. and K.S.; visualization: A.M.V., V.L. and M.V. All authors have read and agreed to the published version of the manuscript.

Funding: This research was funded by the National Research, Development and Innovation Office of Hungary (grant number NKFI-OTKA K132573).

Institutional Review Board Statement: Not applicable.

Informed Consent Statement: Not applicable.

Data Availability Statement: All data analyzed in this study are available in Appendix A.

Acknowledgments: The authors are grateful to Edvárd Mizsei, Éva Szabó, Luc Segers, Roel Duijnhouwer, Alin Bradeanu and Mihai Bobocea for their assistance during the field work, to Attila Takács for editing the map of surveyed cemeteries and to the anonymous reviewers for their valuable suggestions and useful recommendations. We would like to express our gratitude to C. A. J. Kreutz (The Netherlands) for identification of *Epipactis turcica*, and to Orsolya Vincze for her linguistic corrections of the earlier version of our manuscript.

Conflicts of Interest: The authors declare no conflict of interest.

Appendix A

Table A1. Numbering (ID), geographic location, altitude, year of observation and orchid taxa of the 96 cemeteries studied in Azerbaijan. A dash "–" indicates that no orchid taxa were recorded. Generic name abbreviations: *A.—Anacamptis, C.—Cephalanthera, D.—Dactylorhiza, E.—Epipactis, H.—Himantoglossum, L.—Limodorum, O.—Orchis, Op.—Ophrys, S.—Steveniella.*

ID	Settlement	Latitude, Longitude	Alt. (m)	Year	Taxa (Number of Individuals)
01	Şorsulu	39.42429° N, 48.82938° E	26	2018	–
02	TəzəAlvadı	39.09155° N, 48.61027° E	2	2018	–
03	Lənkəran	38.77246° N, 48.83487° E	29	2018	–
04	Lənkəran	38.74200° N, 48.83201° E	19	2018	–
05	Velədi	38.72414°N, 48.82849°E	15	2018	–
06	Şürük	38.69889° N, 48.78815° E	21	2018	–
07	Telman	38.65184° N, 48.80355° E	1	2018	–
08	Kəkülus	38.61265° N, 48.84406° E	17	2018	–
09	Kərgəlan	38.73326° N, 48.79404° E	0	2018	–
10	Shaglakuche	38.72074° N, 48.76931° E	58	2018	–
11	Shaglakuche	38.71618° N, 48.76020° E	60	2018	–
12	Shaglakuche	38.71618° N, 48.74096° E	80	2018	–
13	Shaglakuche	38.71839° N, 48.72004° E	97	2018	–
14	Lerik	38.80257° N, 48.45105° E	803	2018	*O. mascula* (4)
15	Lerik	38.77126° N, 48.41079° E	1108	2018	–
16	Lerik	38.76576° N, 48.42419° E	1062	2018, 2019	*A. papilionacea* (2019: 3), *A. pyramidalis* (2018: 500, 2019: 300), *D. romana* (2019: 20), *Op. caucasica* (2019: 5), *Op. sphegodes* subsp. *mammosa* (2018: 50, 2019: 100), *O. adenocheila* (2018: 5, 2019: 200), *O. mascula* (2019: 5), *O. simia* (2018: 50, 2019: 200), *S. satyrioides* (2019: 1)

Table A1. *Cont.*

ID	Settlement	Latitude, Longitude	Alt. (m)	Year	Taxa (Number of Individuals)
17	Ambu	38.75581° N, 48.44571° E	1238	2018	–
18	Gosmalijion	38.69194° N, 48.40264° E	1308	2018	–
19	Laman	38.85799° N, 48.39724° E	800	2018	–
20	Aran	38.87474° N, 48.39668° E	780	2018	–
21	Aran	38.87868° N, 48.39648° E	780	2018	–
22	Bülüdül	38.83971° N, 48.30756° E	805	2018	*A. pyramidalis* (7), *L. abortivum* (4)
23	Züvüç	38.86909° N, 48.30655° E	885	2018	–
24	Yardımlı	38.88330° N, 48.28216° E	1020	2018	–
25	Yardımlı	38.90051° N, 48.25522° E	827	2018	–
26	Yardımlı	38.90921° N, 48.24933° E	720	2018	–
27	Perimbel	38.90218° N, 48.09793° E	1366	2018	–
28	Yardımlı	38.91879° N, 48.31857° E	580	2018	–
29	Yeyənkənd	38.98335° N, 48.57487° E	102	2018	*L. abortivum* (8)
30	Allahyarlı	39.00025° N, 48.60194° E	66	2018	–
31	Göygöl	40.57610° N, 46.30989° E	720	2018	–
32	Zurnabad	40.51304° N, 46.24282° E	958	2018	*A. pyramidalis* (100), *Op. sphegodes* subsp. *mammosa* (8)
33	Dağ Kəsəmən	41.08915° N, 45.38852° E	401	2018	–
34	Poylu	41.15871° N, 45.44410° E	328	2018	–
35	Köçəsgər	41.05043° N, 45.50680° E	456	2018	–
36	Qaraxanlı	41.05042° N, 45.68841° E	346	2018	–
37	Tovuz	40.97800° N, 45.62000° E	440	2018	–
38	İbrahimhacılı	40.89159° N, 45.74460° E	548	2018	–
39	Ağaməmmədli	40.85088° N, 45.73310° E	581	2018	*A. morio* (1)
40	Yanıqlı	40.84016° N, 45.67080° E	669	2018	*A. pyramidalis* (100), *O. simia* (20)
41	Məşədilər	40.87028° N, 45.74312° E	548	2018	–
42	Düyərli	40.91173° N, 45.85246° E	369	2018	–
43	Gəncə	40.71074° N, 46.42177° E	370	2018	–
44	Şiştəpə	40.83121° N, 45.87962° E	592	2018	*A. morio* (8)
45	Şəmkir	40.80390° N, 46.01184° E	559	2018	–
46	Çinarlı	40.78807° N, 46.10889° E	431	2018	–
47	Ağsu	40.53386° N, 48.33778° E	152	2018	*A. pyramidalis* (25), *O. simia* (20)
48	Ağsu	40.56429° N, 48.38220° E	152	2018	–
49	Muğanlı	40.63904° N, 48.50003° E	732	2018	–
50	Muğanlı	40.66912° N, 48.52951° E	878	2018	–
51	Böyük Xınıslı	40.65781° N, 48.61163° E	870	2018	*A. papilionacea* (1)
52	Ağabəyli	40.66747° N, 48.57887° E	927	2018	*A. collina* (1), *A. papilionacea* (465), *A. pyramidalis* (150), *Op. apifera* (1), *Op. sphegodes* subsp. *mammosa* (3), *O. adenocheila* (39), *O. punctulata* (1), *O. simia* (26)
53	Şamaxı	40.64936° N, 48.62496° E	783	2018	–
54	Muğanlı	40.67730° N, 48.55807° E	971	2018	*A. pyramidalis* (9), *Op. apifera* (2)
55	Suraxanı	40.71495° N, 48.47015° E	900	2018	*A. pyramidalis* (1)
56	Kalva	40.72985° N, 48.48152° E	907	2018	*A. pyramidalis* (1)
57	Xatman	40.72947° N, 48.49023° E	862	2018	–
58	Dilman	40.72596° N, 48.49876° E	831	2018	–
59	Məlhəm	40.69225° N, 48.62865° E	1115	2018	*A. papilionacea* (1)
60	Qəleybuğurd	40.75093° N, 48.56726° E	912	2018	*L. abortivum* (10)

Table A1. *Cont.*

ID	Settlement	Latitude, Longitude	Alt. (m)	Year	Taxa (Number of Individuals)
61	əngixaran	40.67143° N, 48.65670° E	978	2018	*A. papilionacea* (2), *A. pyramidalis* (140), *Op. sphegodes* subsp. *mammosa* (8), *Op. oestrifera* (1)
62	Gandov	40.81714° N, 48.31921° E	952	2018	–
63	Müşkəmir	40.81919° N, 48.33202° E	1100	2018	–
64	Lahıc	40.84393° N, 48.37823° E	1214	2018	*A. coriophora* (1), *A. pyramidalis* (7)
65	Qaraqaya	40.79417° N, 48.30885° E	1111	2018	–
66	Talıstan	40.79954° N, 48.20033° E	827	2018	–
67	Talıstan	40.80186° N, 48.20287° E	850	2018	–
68	İkinci Yeniyol	40.75425° N, 48.26252° E	850	2018	*A. pyramidalis* (12), *Op. oestrifera* (1)
69	Təzəkənd	40.73435° N, 48.27164° E	671	2018	*A. papilionacea* (10), *A. pyramidalis* (2)
70	Sabir	40.59207° N, 48.70527° E	567	2018	–
71	Digah	41.38517° N, 48.47876° E	662	2018	*A. pyramidalis* (3)
72	Qirmizi Qəsəbə	41.37358° N, 48.51741° E	607	2018	–
73	Ağbil	41.42650° N, 48.56662° E	410	2018	*C. damasonium* (1), *E.* sp. (2)
74	Zizik	41.38543° N, 48.57021° E	482	2018, 2019	*H. formosum* (2018: 21, 2019: 7), *Op. sphegodes* subsp. *mammosa* (2018: 1)
75	Avadjuk	41.47244° N, 48.39413° E	670	2018	*A. pyramidalis* (1)
76	Hil	41.46756° N, 48.35726° E	770	2018	–
77	Hil	41.46891° N, 48.33767° E	768	2018	-
78	Yasab	41.49661° N, 48.31670° E	787	2018, 2019	*H. formosum* (2018: 4, 2019: 0)
79	Piral	41.50062° N, 48.29514° E	838	2018	*A. pyramidalis* (13), *H. formosum* (2)
80	Hazra	41.50588° N, 48.25472° E	725	2018	*A. pyramidalis* (2)
81	Laza	41.29840° N, 48.11429° E	1703	2018	–
82	Urva	41.40196° N, 48.34058° E	1046	2018	–
83	Qusar	41.41850° N, 48.40676° E	768	2018	*C. rubra* (2)
84	Pirvahid	41.32669° N, 48.65619° E	370	2018	*A. pyramidalis* (6)
85	Tengealti	41.23764° N, 48.62676° E	701	2018	*A. pyramidalis* (10), *C. rubra* (5), *E. turcica* (7), *L. abortivum* (1)
86	Sirt-Chichi	41.22650° N, 48.67541° E	672	2018	–
87	Chichi	41.21673° N, 48.67737° E	538	2018	*A. pyramidalis* (2), *C. rubra* (2), *E. microphylla* (3)
88	Gilanov	41.23790° N, 48.75276° E	325	2018	–
89	Mashrif	41.09305° N, 48.96764° E	420	2018	–
90	Dashli Jalgan	41.08023° N, 48.98348° E	180	2018	*A. collina* (77), *A. papilonacea* (35), *Neotinea tridentata* (3), *Op. sphegodes* subsp. *mammosa* (96), *O. simia* (11)
91	Tıxlı	40.90958° N, 49.10128° E	593	2018	*A. collina* (8), *A.* sp. (6), *Op.* sp. (24)
92	Nugadi	41.31504° N, 48.59641° E	506	2018	*A. pyramidalis* (800), *O. cf. caucasica* (50), *O. simia* (200)
93	Nohurqishlaq	40.95280° N, 47.92485° E	749	2019	*A. coriophora* (5), *A. pyramidalis* (5), *A. morio* (505), *Op. sphegodes* subsp. *mammosa* (10), *Op. oestrifera* (1), *O. ×chabalensis* (*O. punctulata* × *O. stevenii*) (30), *O. punctulata* (10), *O. stevenii* (20)
94	Chukhur Gabala	40.87934° N, 47.69153° E	404	2019	*A. pyramidalis* (5)
95	Şəfili	40.84899° N, 47.69877° E	354	2019	*A. papilionacea* (50), *Op. sphegodes* subsp. *mammosa* (1), *O. punctulata* (3), *O. simia* (15)
96	Gosmalijion	38.67424° N, 48.37322° E	1450	2018	–

References

1. Le Roux, J.J.; Hui, C.; Castillo, M.L.; Iriondo, J.M.; Keet, J.H.; Khapugin, A.A.; Médail, F.; Rejmánek, M.; Theron, G.; Yanelli, F.A.; et al. Recent anthropogenic plant extinctions differ in biodiversity hotspots and cold spots. *Curr. Biol.* **2019**, *29*, 2912–2918. [CrossRef]
2. Fahrig, L. Effects of habitat fragmentation on biodiversity. *Annu. Rev. Ecol. Evol. Syst.* **2003**, *34*, 487–515. [CrossRef]
3. Saunders, D.A.; Hobbs, R.J.; Margules, C.R. Biological consequences of ecosystem fragmentation: A review. *Conserv. Biol.* **1991**, *5*, 18–32. [CrossRef]
4. Ellis, E.C. Evolution: Biodiversity in the Anthropocene. *Curr. Biol.* **2019**, *29*, R831–R833. [CrossRef]
5. Areendran, G.; Sahana, M.; Raj, K.; Kumar, R.; Sivadas, A.; Kumar, A.; Deb, S.; Gupta, V.D. A systematic review on high conservation value assessment (HCVs): Challenges and framework for future research on conservation strategy. *Sci. Total Environ.* **2020**, *709*, 135425. [CrossRef] [PubMed]
6. Ratcliffe, D.A. Ecological effects of mineral exploitation in the United Kingdom and their significance to nature conservation. *Proc. R. Soc. Lond. A* **1974**, *339*, 355–372.
7. Esfeld, K.; Hensen, I.; Wesche, K.; Jakob, S.S.; Tischew, S.; Blattner, F.R. Molecular data indicate multiple independent colonizations of former lignite mining areas in Eastern Germany by *Epipactis palustris* (Orchidaceae). *Biodivers. Conserv.* **2008**, *17*, 2441. [CrossRef]
8. Shefferson, R.P.; Kull, T.; Tali, K. Mycorrhizal interactions of orchids colonizing Estonian mine tailings hills. *Am. J. Bot.* **2008**, *95*, 156–164. [CrossRef]
9. Way, J.M. Roadside verges and conservation in Britain: A review. *Biol. Conserv.* **1977**, *12*, 65–74. [CrossRef]
10. Fekete, R.; Nagy, T.; Bódis, J.; Biró, É.; Löki, V.; Süveges, K.; Takács, A.; Tökölyi, J.; Molnár, V.A. Roadside verges as habitats of rare lizard-orchids (*Himantoglossum* spp.): Ecological traps or refuges? *Sci. Tot. Environ.* **2017**, *607–608*, 1001–1008. [CrossRef] [PubMed]
11. Fekete, R.; Bódis, J.; Fülöp, B.; Süveges, K.; Urgyán, R.; Malkócs, T.; Vincze, O.; Silva, L.; Molnár, V.A. Roadsides provide refuge for orchids characteristic of the surrounding landscape. *Ecol. Evol.* **2020**, *10*, 13236–13247. [CrossRef] [PubMed]
12. Adamowski, W.; Conti, F. Mass occurrence of orchids in poplar plantations near Czeremcha village as an example of apophytism. *Phytocoenosis* **1991**, *3*, 259–267.
13. Adamowski, W. Population expansion of native orchids in anthropogenous habitats. *Neobiota* **2004**, *3*, 86.
14. Süveges, K.; Löki, V.; Lovas-Kiss, Á.; Ljubka, T.; Fekete, R.; Takács, A.; Vincze, O.; Lukács, B.A.; Molnár, V.A. From European priority species to characteristic apophyte: *Epipactis tallosii* (Orchidaceae). *Willdenowia* **2019**, *49*, 401–409. [CrossRef]
15. Bátori, Z.; Körmöczi, L.; Zalatnai, M.; Erdős, L.; Ódor, P.; Tölgyesi, C.; Margóczi, K.; Torma, A.; Gallé, R.; Cseh, V.; et al. River dikes in agricultural landscapes: The importance of secondary habitats in maintaining landscape-scale diversity. *Wetlands* **2016**, *36*, 251–264. [CrossRef]
16. Deák, B.; Tóthmérész, B.; Valkó, O.; Sudnik-Wojcikowska, B.; Moysiyenko, I.I.; Bragina, T.M.; Apostolova, I.; Dembicz, I.; Bykov, N.I.; Török, P. Cultural monuments and nature conservation: A review of the role of kurgans in the conservation and restoration of steppe vegetation. *Biodivers. Conserv.* **2016**, *25*, 2473–2490. [CrossRef]
17. Kantsa, A.; Tscheulin, T.; Junker, R.R.; Petanidou, T.; Kokkini, S. Urban biodiversity hotspots wait to get discovered: The example of the city of Ioannina, NW Greece. *Landsc. Urban Plan.* **2013**, *120*, 129–137. [CrossRef]
18. Molnár, V.A.; Siffer, S.; Molnár, H.A.; Fekete, R. Occurrence of the rare plant *Sternbergia colchiciflora* in urban environment. *Biol. Futur.* **2020**, *71*, 93–98. [CrossRef] [PubMed]
19. Barrett, G.W.; Barrett, T.L. Cemeteries as repositories of natural and cultural diversity. *Conserv. Biol.* **2001**, *15*, 1820–1824. [CrossRef]
20. Löki, V.; Deák, B.; Lukács, B.A.; Molnár, V.A. Biodiversity potential of burial places—A review on the flora and fauna of cemeteries and churchyards. *Glob. Ecol. Conserv.* **2019**, *18*, e00614. [CrossRef]
21. Löki, V.; Tökölyi, J.; Süveges, K.; Lovas-Kiss, Á.; Hürkan, K.; Sramkó, G.; Molnár, V.A. The orchid flora of Turkish graveyards: A comprehensive field survey. *Willdenowia* **2015**, *45*, 231–243. [CrossRef]
22. Löki, V.; Molnár, V.A.; Süveges, K.; Heimeier, H.; Takács, A.; Nagy, T.; Fekete, R.; Lovas-Kiss, Á.; Kreutz, C.A.J.; Sramkó, G.; et al. Predictors of conservation value of Turkish cemeteries: A case study using orchids. *Landsc. Urban Plan.* **2019**, *186*, 36–44. [CrossRef]
23. Molnár, V.A.; Nagy, T.; Löki, V.; Süveges, K.; Takács, A.; Bódis, J.; Tökölyi, J. Turkish graveyards as refuges for orchids against tuber harvest. *Ecol. Evol.* **2017**, *7*, 11257–11264. [CrossRef] [PubMed]
24. Molnár, V.A.; Süveges, K.; Molnár, Z.S.; Löki, V. Using localpeople's traditional ecological knowledge in discovery of rare plants: A case study from Turkey. *Acta Soc. Bot. Pol.* **2017**, *86*, 3541. [CrossRef]
25. Molnár, V.A.; Takács, A.; Mizsei, E.; Löki, V.; Barina, Z.; Sramkó, G.; Tökölyi, J. Religious differences affect orchid diversity of Albanian graveyards. *Pak. J. Bot.* **2017**, *49*, 289–303.
26. Löki, V.; Schmotzer, A.; Takács, A.; Süveges, K.; Lovas-Kiss, Á.; Lukács, B.A.; Tökölyi, J.; Molnár, V.A. The protected flora of long-established cemeteries in Hungary: Using historical maps in biodiversity conservation. *Ecol. Evol.* **2020**, *10*, 7497–7508. [CrossRef] [PubMed]
27. Vakhrameeva, M.G.; Tatarenko, I.V.; Varlygina, T.I.; Torosyan, G.K.; Zagulskii, M.N. *Orchids of Russia and Adjacent Countries*; ARG GantnerVerlag KG: Ruggel, Lichtenstein, 2008; 690p.
28. Steven, C. Stirpes rariores in itinere Caucasico A 1810 lectae a C. Steven. *Mémoires Société Impériale Nat. Moscou* **1813**, *4*, 89–112.

29. Baumann, H.; Kunkele, S. *Die Wildwachsenden Orchideen Europas*; Kosmos: Stuttgart, Germany, 1982; 432p.
30. Rückbrodt, D.; Rückbrodt, H.; Gügel, E.; Zaiss, W.-W. Bemerkungen zu Orchideen des Kaukasusgebietes. *Ber. Arbeitskr. Heim. Orch.* **1994**, *11*, 46–56.
31. Sramkó, G.; Molnár, V.A.; Hawkins, J.A.; Bateman, R.M. Molecular phylogeny and evolutionary history of the Eurasiatic orchid genus *Himantoglossum* s. l. (Orchidaceae). *Ann. Bot.* **2014**, *114*, 1609–1626. [CrossRef] [PubMed]
32. Bateman, R.M.; Molnár, V.A.; Sramkó, G. In situ morphometric survey elucidates the evolutionary systematic of the Eurasian *Himantoglossum* clade (Orchidaceae: Orchidinae). *PeerJ* **2017**, *5*, e2893. [CrossRef] [PubMed]
33. Walter, K.S.; Gillett, H.J. *1997 IUCN Red List of ThreatenedPlants;* IUCN—The World Conservation Union: Gland, Switzerland, 1998; 932p.
34. Kuehn, R.; Pedersen, H.A.; Cribb, P. *Field Guide to the Orchids of European and the Mediterranean*; Kew Publishing: Kew, UK, 2019; 430p.
35. R: A Language and Environment for Statistical Computing. Available online: http://www.rproject.org/ (accessed on 30 November 2021).
36. Tamer, E.C.; Karaman, B.; Copur, U.O. A traditional Turkish beverage: Salep. *Food Rev. Int.* **2006**, *22*, 43–50. [CrossRef]
37. Molnár, V.A.; Löki, V. Disappearing botanical and cultural heritage of wooden headboard-graveyards in Eastern-Hungary and Transylvania (Romania). *Kitaibelia* **2018**, *23*, 51–64.
38. Berkes, F.; Folke, C.; Gadgil, M. Traditional ecological knowledge, biodiversity, resilience and sustainability. In *Biodiversity Conservation;* Perrings, C., Ed.; Kluwer Academic Publishers: Dordrecht, The Netherlands, 2005; pp. 269–287.
39. Turner, N.J.; Turner, K.L. Traditional food systems, erosion and renewal in Northwestern North America. *Indian J. Tradit. Knowl.* **2007**, *6*, 57–68.
40. Kasparek, M.; Grimm, U. European trade in Turkish salep with special reference to Germany. *Econ. Bot.* **1999**, *53*, 396–406. [CrossRef]
41. Pieroni, A. Reader ponders trade in salep (ground orchid bulbs). *Econ. Bot.* **2000**, *54*, 138–139. [CrossRef]
42. Sezik, E. Destruction and conservation of Turkish orchids. In *Biodiversity—Biomolecular Aspects of Biodiversity and Innovative Utilization;* Springer: Boston, MA, USA, 2002; pp. 391–400.
43. Sezik, E. Turkish orchids and salep. *Acta Pharm. Turc.* **2002**, *44*, 151–157.
44. Ghorbani, A.; Gravendeel, B.; Naghibi, F.; de Boer, H. Wild orchid tuber collection inIran: A wake-up call for conservation. *Biodivers. Conserv.* **2014**, *23*, 2749–2760. [CrossRef]
45. Kreziou, A.; de Boer, H.; Gravendeel, B. Harvesting of salep orchids in north-western Greece continues to threaten natural populations. *Oryx* **2016**, *50*, 393–396. [CrossRef]
46. Mincheva, I.; Petrova, A.; Yordanova, M.; Kozuharova, E. Is the traditional use of "salep" in the Bulgarian Rhodopes hazardous for the wild populations of terrestrial orchids? *Flora Mediterr.* **2018**, *28*, 399–418.
47. Charitonidou, M.; Stara, K.; Kougioumoutzis, K.; Halley, J.M. Implications of salep collection for the conservation of the Elder-flowered orchid (*Dactylorhiza sambucina*) in Epirus, Greece. *J. Biol. Res. Thessalon.* **2019**, *26*, 18. [CrossRef] [PubMed]

plants

MDPI

Article

Vegetation Composition of the Halophytic Grass *Aeluropus lagopoides* Communities within Coastal and Inland Sabkhas of Saudi Arabia

Basharat A. Dar [1], Abdulaziz M. Assaeed [1], Saud L. Al-Rowaily [1], Abdullah A. Al-Doss [1] and Ahmed M. Abd-ElGawad [1,2,*]

1 Plant Production Department, College of Food & Agriculture Sciences, King Saud University, P.O. Box 2460, Riyadh 11451, Saudi Arabia; bdar@ksu.edu.sa (B.A.D.); assaeed@ksu.edu.sa (A.M.A.); srowaily@ksu.edu.sa (S.L.A.-R.); aaldoss@ksu.edu.sa (A.A.A.-D.)

2 Department of Botany, Faculty of Science, Mansoura University, Mansoura 35516, Egypt

* Correspondence: aibrahim2@ksu.edu.sa; Tel.: +966-59-945-8396

Abstract: Sabkhas are unique, highly saline ecosystems, where specially adapted plants can grow. *Aeluropus lagopoides* (L.) Thwaites is a halophytic forage plant growing in salt marsh habitats of inland and coastal sabkhas of Saudi Arabia. The present study provides an analysis of vegetation composition and distribution of the *A. lagopoides* community in five different regions within Saudi Arabia, emphasizing the environmental factors that affect species distribution. The floristic survey revealed the presence of 48 species, belonging to 26 families. Poaceae, Chenopodiaceae, Mimosaceae, Zygophyllaceae, and Asteraceae are the largest families (50% of total species). Phanerophyte, followed by chamaephytes, are the most frequent forms, indicating a typical saline desert life-form spectrum. The vegetation analysis revealed the dominance of *A. lagopoides* in all locations, where it was the most dominant species in Qareenah, Qaseem, and Salwa locations, and the second most dominant species in Jouf and Jizan locations. The flourishment of this halophytic grass within a wide soil range in sabkhas revealed its adaptability to the harsh environment, which could be ascribed to its structural adaptations and modifications, as well as the phenotypic plasticity. The Qareenah and Qaseem locations attained the highest species richness and evenness, while the Jizan location was the least diverse. Within the studied locations, other highly salt-tolerant species were determined with high abundances, such as *Suaeda aegyptiaca* (Hasselq.) Zohary, *Zygophyllum album* L.f., *Tamarix nilotica* (Ehrenb.) Bunge, *Cressa cretica* L., and *Salicornia europaea* L. The soil analysis showed a significant variation for all parameters among the studied locations, except for pH, chloride, and clay content. The Qaseem location revealed the highest values of most soil parameters, while the Jizan location showed the lowest. The canonical correspondence analysis (CCA) showed that the community structure and diversity are mainly affected by the soil salinity and moisture. Due to the economic potentialities of *A. lagopoides* as a forage plant and sand stabilizer, the conservation of its habitats is of vital importance. In addition, this grass could be integrated as a promising forage candidate that can be planted in saline-affected areas, even in the summer dry season.

Keywords: phenotypic plasticity; mangrove grass; salt marshes; coastal habitat; biodiversity

Citation: Dar, B.A.; Assaeed, A.M.; Al-Rowaily, S.L.; Al-Doss, A.A.; Abd-ElGawad, A.M. Vegetation Composition of the Halophytic Grass *Aeluropus lagopoides* Communities within Coastal and Inland Sabkhas of Saudi Arabia. *Plants* **2022**, *11*, 666. https://doi.org/10.3390/plants11050666

Academic Editors: Robert Philipp Wagensommer and Panayiotis Dimitrakopoulos

Received: 9 January 2022
Accepted: 24 February 2022
Published: 28 February 2022

Publisher's Note: MDPI stays neutral with regard to jurisdictional claims in published maps and institutional affiliations.

1. Introduction

Sabkhas are geological phenomena formed in an arid or semi-arid climate, as broad plains or salt flats, containing evaporates dictated by the local water table [1]. Geographically, it has a large habitat range, distributed worldwide through Southeast Europe, the siliciclastic coast of California, Mexico, North Africa from Morocco to Somalia, the Middle East and the Arabian Peninsula, Australia, and Asia [2–4]. Sabkhas are grouped into two major landform types [5], low-lying coastal salt marshes [6,7] or inland interdune areas as salt-crusted depressions [8]. Nearly all key species in these saline habitats are perennial halophytes forming different plant communities [4,9].

The Arabian Peninsula is also characterized by high salinity, high annual temperature variations, and shallow waters. Saudi Arabia possesses harsh natural desert environments without rivers or lakes. It is distinguished by various ecosystems, including mountains, wadis, meadows, rocky mountains, sandy deserts, and saltpans with distinct plant communities [10,11]. The salt-affected areas of Saudi Arabia are classified into the coastal plain, inland zone, and littoral salt marshes [12]. Coastal and inland saline habitats, called sabkhas, are highly stressful environments, as they are highly saline and wet unique ecosystems, where specially adapted halophytic plants can grow [13]. These sabkhas are mostly saturated with brine, and the soil surface is often encrusted with thick salt crust [14]. Natural saline habitats vary in salinity levels due to differences in topography, soil properties, and micro-climate, both spatially and temporally [15]. Vegetation composition in these ecosystems is influenced by complex heterogonous environmental factors, including duration and degree of inundation by seawater and both overground and underground freshwater input [16], coastal and inland geomorphology, microtopography, soil moisture content, and soil type [17].

Specific plant species can dominate sabkha habitats, forming monospecific stands. The vegetation of these monospecific stands forms zones with distinct plant communities [18], forming a variety of specialized habitats of distinct vegetation mosaics [19]. Most of the key species in the saline habitats are perennial halophytes, which constitute about 2% of the world's flora [18,20], predominantly belonging to the families Chenopodiaceae, Zygophyllaceae, Plumbaginaceae, Poaceae, and Juncaceae [21]. The distribution of some halophytic species is best correlated along a gradient of soil variables, such as salinity, moisture content, soil texture, organic matter, and calcium carbonate [22].

Among the Poaceae family, *Aeluropus lagopoides* (L.) Thwaites is one important key species of saline habitats of Saudi Arabia [23]. Geographically, it has a wide habitat range, distributed through Southeast Europe, North Africa, the Middle East, Arabian Peninsula, and Central Asia [3]. In Saudi Arabia, *A. lagopoides* grows in various coastal and inland sabkhas [23,24]. It is restricted in the form of specialized vegetation patches in the Wadi Hargan, Riyadh, salt marsh sabkha of the Qaseem and Jouf, and coastal zones of the Salwa and Jizan regions [25,26]. It is of economic importance, where it is utilized as fodder in arid areas, used to stabilize sand dunes [27], and can be used for landscaping of the urban areas [28].

Knowledge of the ecological distribution of key species is the primary characteristic of its conservation strategy in its ecosystem [29]. Apart from the establishment and maintenance of the protected area, baseline information about the key species, vis a vis the associated species, edaphic factors of the habitats, environmental variation, anthropogenic activities of the regions, are crucial [25,30]. This ecogeographical survey is considered central to all the conservation issues and a key requisite in the development of the conservation strategy [31,32]. According to our field observations, *A. lagopoides* has various phenotypic characteristics and forms distinct vegetative patches within various regions of Saudi Arabia. To the best of our knowledge, no study has dealt with the vegetation composition of the key forage halophyte *A. lagopoides* in Saudi Arabia. Therefore, the present study aims to assess the vegetation structure of *A. lagopoides* communities in the various regions/habitats around Saudi Arabia, considering regional heterogeneity, edaphic factors, and variation of climatic gradients. This study will help in understanding the distribution of this important forage plant that flourished in one of Earth's harsh environments and, in consequence, shows the potential of conserving this plant, as well as integrating it in a foraging system.

2. Results

2.1. Floristic Composition of the Studied Regions

As expected, the plant diversity of the studied sabkha regions is low, in which the species had to withstand harsh environmental conditions, i.e., the high salinity content

(Table S1). The floristic analysis revealed the presence of 48 species of vascular plants, which are mainly perennials (75%). The highest number of species (24 species: 22 perennials and 2 annual) were recorded in the inland Sabkha of Qareenah, Riyadh region, which is represented by about 34% of the total recorded species, while the sabkha of the northern Al-Jouf region recorded the lowest number of species (all eight perennial species) which is about 10% of the total recorded species (Figure S1). However, the species' evenness was highest in the inland sabkhas of the Qaseem region, representing about 42% of the recorded species, and the lowest was in the coastal sabkha of Jizan, representing about 9% (Figure S1). Summing up, the coastal sabkhas (Salwa and Jizan) recorded the highest number of species (43 species: 36 perennials and 6 annuals), which is represented by 60% of total recorded species, compared to the inland sabkhas (Qareenah, Qassem, and Jouf) that represented 40% of the total recorded species (29 species: 21 perennials and 8 annuals).

The identified plant species belonged to 26 families, where Poaceae, Chenopodiaceae, Mimosaceae, Zygophyllaceae, and Asteraceae were the major families that represented 50% of the total species (Figure 1A and Table S1).

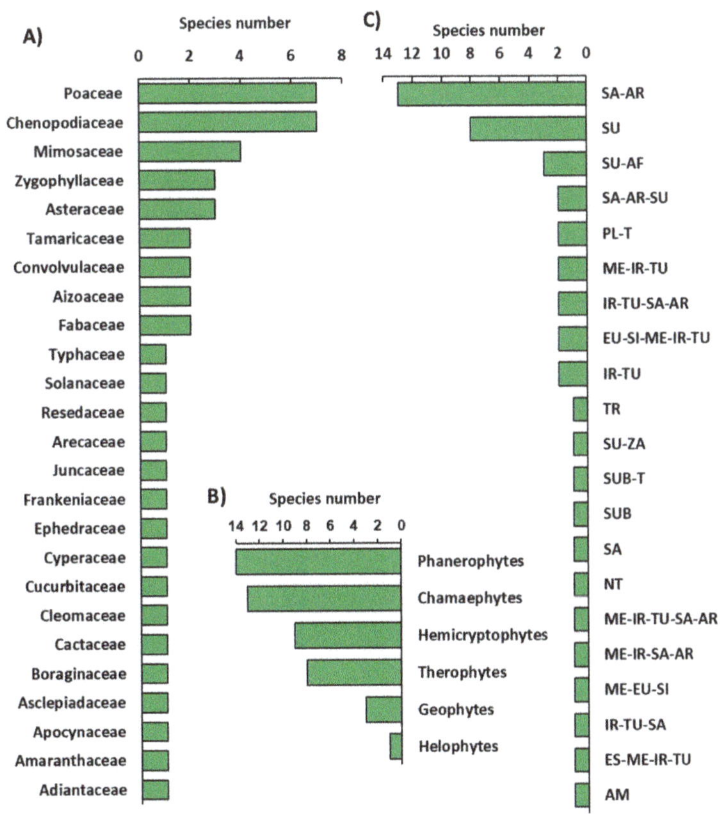

Figure 1. Floristic composition of the surveyed region. (**A**) Represented families, (**B**) life forms, and (**C**) chorotype spectra. SA-AR: Saharo-Arabian, SU: Sudian, AF: African, ME: Mediterranean, IR-TU: Irano-Turanian, EU-SI: Euro-Siberian, PL-T: Plurireginalbor-trop, SU-ZA: Sudano-Zambezian, SUB-T: Subtropical-Tropical, SA-South-American, NT-Neotropical, AM-American.

On the one hand, The recorded plant species were classified into six life forms, according to Raunkiaer's system, of which 29% were phanerophytes and 27% chamaephytes, while hemicryptophytes, geophytes, and helophytes were represented by 19%, 17%, and 2%, respectively (Figure 1B). On the other hand, the chorological analysis of the identified species revealed that 62.50% of those species were monoregional, and the Saharo-Arabian element was the most present chorotype (27.08%) (Figure 1C). However, 22.92% of the identified species were classified as biregional plants, whereas Sudanian-African was the most represented element (6.25%). The pluriregional chorotype was also represented, with 14.58% of the total recorded species, and the most represented ones were the Euro-Siberian-Mediterranean-Irano-Turanian and Saharo-Arabian-Mediterranean-Irano-Turanian, which represented 4.17% each.

2.2. Vegetation Analysis of the Studied Regions

The analysis of importance values of each species, based on the relative cover and density, led to the recognition of the dominant and important species within each location (Table 1). The details of all species are presented in Table S2. The Qareenah region was the most diversified, with 24 species. This group attained the highest richness (Simpson diversity index = 0.95). In this location, *A. lagopoides* was the first dominant species (importance value = 44.41), while *Zygophyllum coccineum* L. was the second most dominant (importance value = 25.64). The other important species recorded were *Juncus rigidus* Desf., *Tamarix nilotica* (Ehrenb.) Bunge, *Rhazya stricta* Decne., *Acacia gerrardii* Benth., and *Phragmites australis* (Cav.) Trin. ex Steud. (Table 1).

Table 1. Plant species richness, evenness, and dominance of the studied sabkha locations of Saudi Arabia.

Location	Richness	Evenness	1st Dominant	2nd Dominant	Important Species
Qareenah	2.53	0.90	*Aeluropus lagopoides* (L.) Thwaites (44.41) *	*Zygophyllum coccineum* L. (25.64)	*Juncus rigidus* Desf. (24.14) *Tamarix nilotica* (Ehrenb.) Bunge (12.88) *Rhazya stricta* Decne. (10.79) *Acacia gerrardii* Benth. (10.77) *Phragmites australis* (Cav.) Trin. ex Steud. (9.72)
Qaseem	1.6	0.74	*Aeluropus lagopoides* (L.) Thwaites (94.20)	*Suaeda aegyptiaca* (Hasselq.) Zohary (28.61)	*Cressa cretica* L. (28.23) *Juncus rigidus* Desf. (15.45) *Lycium shawii* Roem. & Schult. (8.97) *Salicornia europaea* L. (6.87)
Salwa	1.04	0.81	*Aeluropus lagopoides* (L.) Thwaites (66.62)	*Zygophyllum album* L.f. (41.05)	*Phragmites australis* (Cav.) Trin. ex Steud. (22.40) *Juncus rigidus* Desf. (15.39) *Suaeda aegyptiaca* (Hasselq.) Zohary (14.95) *Phoenix dactylifera* L. (6.48)
Jouf	1.53	0.75	*Tamarix nilotica* (Ehrenb.) Bunge (74.68)	*Aeluropus lagopoides* (L.) Thwaites (45.61)	*Zygophyllum album L.f.* (33.48) *Cressa cretica* L. (28.80) *Suaeda aegyptiaca* (Hasselq.) Zohary (8.46)
Jizan	1.84	0.75	*Suaeda aegyptiaca* (Hasselq.) Zohary (80.76)	*Aeluropus lagopoides* (L.) Thwaites (54.52)	*Panicum repens* L. (11.23) *Cyperus conglomeratus* Rottb. (11.12) *Aerva javanica* (Burm.f.) Juss. ex Schult. (8.29) *Zygophyllum simplex* L. (8.20)

* represents the importance value based on the relative plant density and cover.

The location of Qaseem is dominated by *A. lagopoides* (importance value = 94.20). In this location, *Suaeda aegyptiaca* was determined as the second most dominant species (importance value = 28.61). The other important species that attained high importance values were *Cressa cretica* L., *J. rigidus*, *Lycium shawii* Roem. and Schult., and *Salicornia europaea* L. This location attained a Simpson diversity index of 1.6 and Shannon evenness of 0.74. On the other hand, the Salwa location showed the lowest Simpson diversity index (1.04), and it was dominated by *A. lagopoides* (importance value = 66.62), followed by *Zygophyllum album* L.f. (importance value = 41.05). The other important species within this location were *P. australis*, *J. rigidus*, *S. aegyptiaca* (Hasselq.) Zohary, and *Phoenix dactylifera* L. (Table 1).

The Jouf location was the least diversified (eight species) among the recognized groups, and it is dominated by *T. nilotica* as the most dominant and *A. lagopoides* as second most dominant species. This group attained a Simpson diversity index of 1.53 and Shannon evenness of 0.75. The other important species recorded in this group were *Z. album*, *C. cretica*, and *S. aegyptiaca* (Table 1). Lastly, the Jizan location comprised 15 recorded species. This community attained a Simpson diversity index of 1.81 and Shannon evenness of 0.75. The most dominant in this location was *S. aegyptiaca* (importance value = 80.76), while *A. lagopoides* was the second most dominant species. The other important species of this group were *Panicum repens*, *Cyperus conglomeratus* Rottb., *Aerva javanica* (Burm.f.) Juss. ex Schult., and *Zygophyllum simplex* L. (Table 1). *A. lagopoides* dominated in all studied locations, either as most dominant, as in the inland sabkhas of Qareenah, Qaseem regions, and Coastal sabkhas of Salwa, or second most dominant species, in the inland sabkhas of Jouf and coastal sabkha of the Jizan region (Table 1).

The application of detrended correspondence analysis (DCA) on the vegetation data showed the separation of the Jizan location, on the right side of the DCA diagram (Figure 2). However, the other location showed quite significant overlapping, with a close correlation between the Qassem and Salwa locations.

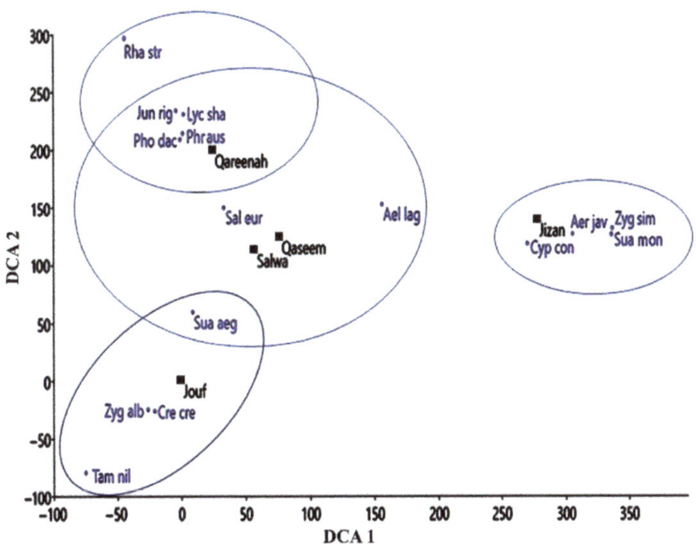

Figure 2. Detrended correspondence analysis (DCA) ordination of the studied locations (■) based on the importance value of dominant, co-dominant, and important species (•) recorded from each location. Ael lag: *Aeluropus lagopoides*, Tam nil: *Tamarix nilotica*, Jun rig: *Juncus rigidus*, Lyc sha: *Lycium shawii*, Pho dac: *Phoenix dactlifera*, Phr aus: *Phragmites australis*, Sal eur: *Salicornia europaea*, Sua aeg: *Suaeda aegyptiaca*, Zyg alb: *Zygophyllum album*, Cre cre: *Cressa cretica*, Cyp con: *Cyperus conglomeratus*, Zyg sim: *Zygophyllum simplex*, Aer jav: *Aerva javanica*, Pan rep: *Panicum repens*, Sau mon: *Suaeda monoica*, Rha str: *Rhazya stricta*.

Moreover, the cluster analysis of the vegetation data of all recorded species confirmed the data of DCA, where it revealed that the Jizan location is different than other locations (Figure 3). Salwa and Qaseem locations showed similar vegetation composition, while Jouf and Qareenah were different in the vegetation structure.

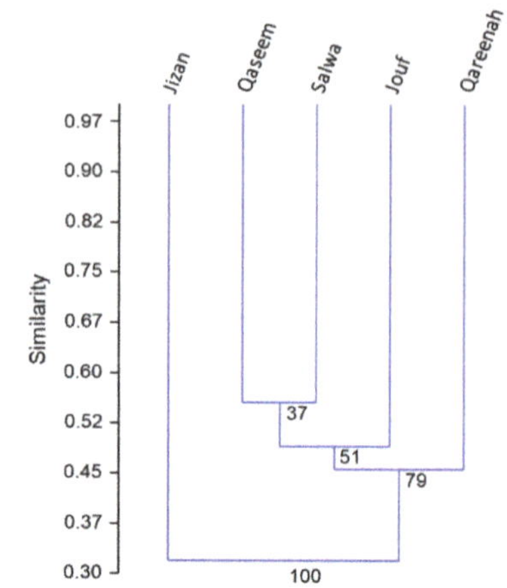

Figure 3. Hierarchical clustering of different studied locations based on the importance values of the recorded plant species (n = 48).

2.3. Vegetation-Soil Relationship

The soil analysis of the five studied locations showed significant variations regarding all measured parameters, except for pH, Cl^-, and clay (Table 2). The Qaseem location attained the highest moisture, clay, silt, salinity, Ca, Mg, Na, K, Cl, SO_4, HCO_3, and organic matter (Table 2). The Salwa location is characterized by high sand content, while the soil of the Qareenah location revealed the highest content of calcium carbonate. The Jizan location had the lowest moisture content, pH, organic matter, K, sulphate, bicarbonate, and calcium carbonate, while the Jouf location attained the lowest salinity content.

Table 2. Soil chemical and physical properties of the studied locations dominated by *Aeluropus lagopoides* community.

Parameter	Location					*p*-Value
	Qareenah	Qaseem	Salwa	Jouf	Jizan	
Moisture %	8.90 ± 1.896 [b,#]	21.56 ± 3.456 [a]	5.86 ± 2.439 [b,c]	4.97 ± 1.005 [b,c]	1.82 ± 0.424 [c]	0.0001 ***
pH	8.38 ± 0.18 [a]	8.39 ± 0.14 [a]	8.42 ± 0.12 [a]	8.12 ± 0.11 [a]	8.07 ± 0.21 [a]	0.474
Clay %	12.48 ± 0.65 [b]	16.80 ± 1.26 [a]	14.20 ± 1.99 [a,b]	13.24 ± 0.95 [a,b]	13.87 ± 0.994 [a,b]	0.188
Silt %	19.28 ± 3.22 [b,c]	48.40 ± 2.76 [a]	8.60 ± 1.60 [d]	26.00 ± 10.77 [b,c]	42.01 ± 5.272 [a,b]	0.0005 ***
Sand %	68.24 ± 3.77 [a]	34.80 ± 2.52 [c]	77.66 ± 3.23 [a]	60.76 ± 10.90 [a,b]	44.11 ± 5.667 [b,c]	0.0004 ***
EC ($dS{\cdot}m^{-1}$)	13.02 ± 2.83 [b,c]	26.30 ± 3.87 [a]	22.17 ± 7.65 [a,b]	8.63 ± 3.02 [c]	9.69 ± 6.76 [c]	0.003 **
Ca (meq/L)	19.10 ± 2.33 [b]	39.86 ± 3.74 [a]	40.07 ± 4.54 [a]	22.50 ± 2.93 [b]	24.31 ± 16.96 [c]	0.003 **
Mg (meq/L)	41.50 ± 10.83 [b]	69.13 ± 14.74 [a]	31.18 ± 12.79 [a]	15.00 ± 3.23 [b]	25.76 ± 18.78 [c]	<0.0001 ***
Na (meq/L)	59.53 ± 17.32 [a,b]	134.81 ± 33.36 [a]	132.43 ± 52.52 [b,c]	37.92 ± 17.24 [b,c]	42.56 ± 28.836 [c]	0.001 **
K (meq/L)	10.49 ± 4.46 [a,b]	20.22 ± 4.97 [a]	18.25 ± 7.55 [a]	11.01 ± 8.95 [b]	4.72 ± 3.366 [b]	0.02 *
Cl (meq/L)	111.40 ± 25.40 [a,b]	241.50 ± 40.92 [a]	211.80 ± 75.15 [a]	78.80 ± 26.48 [a,b]	92.62 ± 65.560 [b]	0.193
SO_4 (meq/L)	15.64 ± 4.63 [a,b]	17.78 ± 4.47 [a]	6.70 ± 1.67 [b,c]	4.73 ± 3.42 [c]	0.73 ± 0.048 [c]	0.007 **
HCO_3 (%)	3.14 ± 0.20 [a]	3.58 ± 0.54 [a]	2.89 ± 0.39 [a,b]	2.04 ± 0.19 [b]	0.83 ± 0.229 [c]	0.0001 ***
OM %	1.63 ± 0.19 [a]	1.78 ± 0.37 [a]	0.70 ± 0.09 [b]	0.67 ± 0.09 [b]	0.31 ± 0.041 [b]	0.0001 ***
$CaCO_3$%	34.84 ± 2.10 [a]	15.54 ± 3.38 [b]	9.60 ± 4.92 [b,c]	4.16 ± 1.47 [c,d]	0.55 ± 0.152 [d]	0.0001***

Values are mean ± standard errors. EC: electrical conductivity, OM: organic matter. Superscript letters within each row showed significant variation at $p < 0.05$ (Duncan's test). * $p < 0.05$, ** $p < 0.01$, *** $p < 0.001$ at degree of freedom (*df*) for region (n − 1) = 4 and replications (n − 1) = 9.

The correlation between the vegetation composition and soil properties was assessed by canonical correspondence analysis (CCA). The CCA showed that inland sabkhas of the Qareenah region were separated on the upper left side of the CCA biplot and showed a close correlation to $CaCO_3$, Mg, organic matter content, pH, and sulphate (Figure 4). In contrast, the inland sabkha of the Jouf region is segregated in the CCA biplot's lower left side, where they are affected by Ca and Clay. On the other hand, the coastal sabkha of the Jizan region was segregated on the lower right side of the CCA biplot, where it showed a correlation to silt contents. Finally, the inland sabkha of Qaseem and coastal sabkha of Salwa were separated on the central part of the CCA biplot, where they showed a positive correlation with salinity, moisture, K, bicarbonates, Na, and sand (Figure 4).

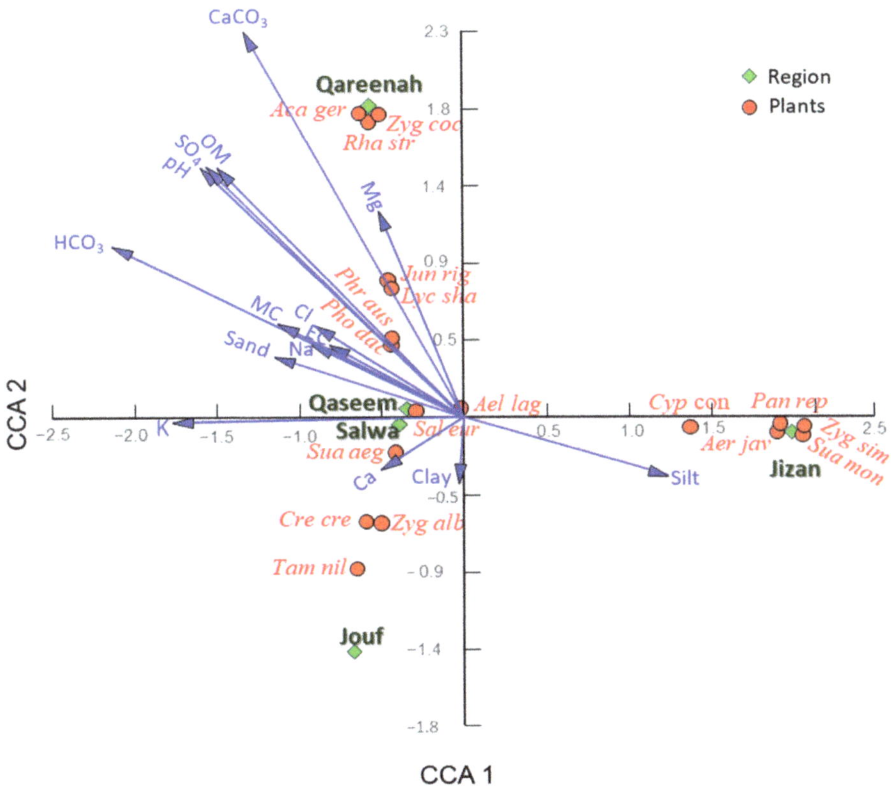

Figure 4. Canonical correspondence analysis (CCA) showing the correlation between the soil variables and dominant and important species representing the studied locations. Aca ger: *Acacia gerrardii*, Ael lag: *Aeluropus lagopoides*, Zyg coc: *Zygophyllum coccineum*, Sua aeg: *Suaeda aegyptiaca*, Zyg alb: *Zygophyllum album*, Tam nil: *T. nilotica*, Jun rig: *Juncus rigidus*, Rha str: *Rhazya stricta*, Phr aus: *Phragmites australis*, Cre cre: *Cressa cretica*, Lyc sha: *Lycium shawii*, Sal eur: *Salicornia europaea*, Pho dac: *Phoenix dactlifera*, Pan rep: *Panicum repens*, Cyp con: *Cyperus conglomeratus*, Aer jav: *Aerva javanica*, Zyg sim: *Zygophyllum simplex*. OM: organic matter, EC: electrical conductivity.

The Pearson's correlation analysis between soil variables and dominant, co-dominant and important species is shown in Figure 5. *A. lagopoides*, the most dominant species of the Qareenah, Qaseem, and Salwa regions, and second most dominant species of the Jouf and Jizan regions, showed a strong positive correlation to all tested soil parameters, except for $CaCO_3$ (r = −0.07) and sand (r = −0.54). Similarly, *S. aegyptiaca* (the most dominant species of the Jizan region and the second most dominant species of the Qaseem region), *J. rigidus*, *L. shawii*, *P. dactylifera*, *P. australis*, and *S. europaea*, revealed a positive correlation for all tested characteristics, except for $CaCO_3$ and sand contents.

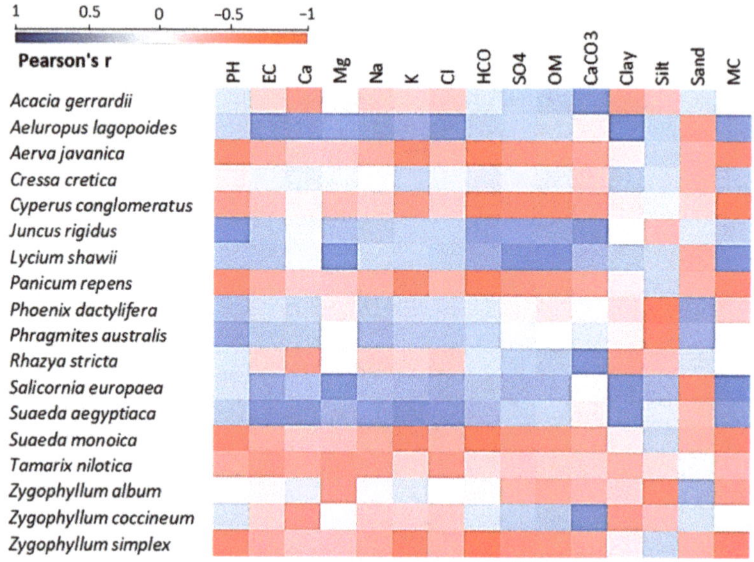

Figure 5. Pearson's correlation heatmap between the soil variables and the dominant, co-dominant, and important associated plant species within the studied locations. EC: electrical conductivity, OM: organic matter, and MC: moisture content.

However, *T. nilotica*, the most dominant species of the Jouf Region, was negatively correlated with all the soil parameters, except for sand (r = 0.20). *Z. coccineum*, the second most dominant species of the Qareenah region, showed a negative correlation with most of the soil parameters, except for pH (r = 0.35), anions, and sand (r = 0.36). Among the important associated species, *J. rigidus*, *L. shawii*, *P. dactylifera*, *P. australis*, and *S. europaea* showed a strong positive correlation with almost all the soil parameters (Figure 5).

3. Discussion

Saudi Arabia is located within an arid and semi-arid zone. It is distinguished by its different ecosystems, including mountains, meadows, valleys, rocky and sandy deserts, and salt marshes [11,25]. Among salt marsh ecosystems, there are inland salt marshes and saline coastal habitats, called sabkhas. These sabkhas have scarce vegetation due to severe environmental conditions (wind exposure, high temperature, and high salinity). However, sabkha edges (transition towards sand) are characterized by well-defined zones, and each zone occupies a particular plant community. The salt-tolerant or halophytic plants, which constitute about 2% of the world's flora [20], grow in these habitats. The spatial distribution of plants in these vegetation zones is affected by soil salinity and soil composition. This explained the low plant diversity in the studied sabkha locations.

The floristic analysis of this study revealed the predominance of phanerophytes and chamaephytes, reflecting the domination of perennial halophytes over annuals and ephemerals in saline conditions. The perennial halophytes can tolerate high salt content in soil [33]. The predominance of Poaceae, Chenopodiaceae, and Mimosaceae in the studied habitats was in harmony with other previous studies in different saline habitats [25,34]. On the other hand, most of the recorded species belong to the Saharo-Arabian element. This finding is in harmony with other studies on salt marsh habitats [10].

Based on the data of diversity indexes, the Qareenah location showed the highest richness. This could be ascribed to the water factor, as this location is a wadi system and has a relative amount of water. In our previous work, a total of 111 plant species were recorded within this wadi, even some hydrophytes and ferns are flourished [25]. The diversification of the Qareenah location could be ascribed to the fact that it forms a woodland community by colonizing various acacia species (*A. gerardii* Benth., *Acacia ehrenbergiana* Hayne, and *Acacia tortilis* (Forssk.) Hayne), along with *T. nilotica* and other xerophytes, such as *L. shawii* and *R. stricta*. Since this habitat is interconnected with wadis, these plants are the common plants in various wadis in desert habitats of Saudi Arabia [35–37].

On the other hand, the Qaseem location showed higher evenness and low richness. This location is a wetland, with a water content of 21.56%, that enables specific species (species that flourished in wetlands, i.e., with high water content) to colonize this habitat with a higher number of individuals. The soil analysis showed that the Qaseem location accomplished the highest moisture, clay, silt, salinity, Ca, Mg, Na, K, Cl, SO_4, HCO_3, and organic matter. The harsh soil conditions, such as high salinity, can be the factor of low species diversification due to the non-survival of annuals. The inland sabkha of Qaseem forms a brine on the soil, making it difficult for less salt-tolerant plants to grow. Usually, in saline environments, soil factors control plant species' growth and survival rates, thus, affecting vegetation patterns [38]. Soil factors also reduce plants' fecundity and germination ability, thus, shaping plant competition and population fitness [39,40]. Most of the dominant and associated species in this location are typical halophytes, such as *C. cretica*, *J. rigidus*, *L. shawii*, and *S. europaea*, which grow in high salt and wet habitats [39,41–43], thus, determining the vegetation zonation pattern of salt marshes.

The cluster analysis of the studied locations showed that the northern inland plain of Jouf is dissimilar to the other locations. This could be attributed to the environmental factors, where the soils of the Jouf location showed the lowest salinity. In this location, *T. nilotica* was determined as the dominant species. *T. nilotica* usually colonizes *A. lagopoides* patches, where the altitude is greater than 500 m ASL but grows in soil with low salinity [44].

Among all the regions, the southern coastal sabkha of Jizan has the least diverse vegetation. The soils in this location revealed the lowest moisture content, pH, organic matter, K, sulphate, bicarbonate, calcium carbonate. This could explain the low diversity of this location, as well as the presence of xerophytic plants, such as *C. conglomeratus*, *A. javanica*, and *Z. simplex*.

Overall, the vegetation analysis of the studied location revealed the dominance of *A. lagopoides* in all locations, where it was the most dominant species in the Qareenah, Qaseem, and Salwa locations, and second most dominant species in Jouf and Jizan locations. *A. lagopoides* is one of the most important halophytic grasses in Saudi Arabia. It is a salt excretive grass that grows in the form of patches or mats in highly saline and moistened soil, where it is characterized by structural adaptations and modifications [27,45]. It can tolerate the harsh and saline habitat by expelling the salts it gains, and the plant itself has a very low salt content, making it a palatable forage grass [46]. In addition, this grass has small and waxy leaves, as well as a network of roots and underground rhizomes that help the plant to survive in high salty conditions, even in the summer season, where the salinity becomes even higher [47]. According to our field observation, the *A. lagopoides* has phenotypic plasticity, where its morphology is changed from one location to another. This observation could be a way of adapting to the harsh conditions.

Within the studied locations, other halophytic plants (highly salt-tolerant species) were determined with high abundances, such as *S. aegyptiaca*, *Z. album*, *T. nilotica*, *S. aegyptiaca*, *C. cretica*, and *S. europaea*. These species flourish in saline habitats, where moisture and salinity shaped the community structure [25,48].

4. Materials and Methods

4.1. Study Area

The study was carried out from 2020 through 2021 around the entirety of Saudi Arabia to explore the vegetation zones of the halophyte *A. lagopoides* (Figure 6). *A. lagopoides* mosaic vegetation has been shaped in different eco-regions of Saudi Arabia based on soil properties of the habitat and the morphological adaptation of the indicator plant via phenotypic plasticity.

Figure 6. The populations of *Aeluropus lagopoides* (L.) Thwaites in different locations of Saudi Arabia (**a–c**), and different morphological growth forms (**d–i**). AP: *Aeluropus* patch, FMA: full mature *Aeluropus* patch, Cu: culm, GI: globose inflorescence, St: stolon.

We monitored this grass in five sabkha regions (Figure 7) of Saudi Arabia which represents both coastal and inland sabkhas as follows:

(1) Salwa; coastal sabkha as lowland on the coast of the Arabian Gulf,
(2) Jizan; coastal sabkha on the Southern Coastal Region of Jizan on the Red Sea coast,
(3) Qareenah; inland sabkha in wadi Hargan, Riyadh Region,
(4) Qaseem, inland sabkha of the Al-Aushazia location, and
(5) Jouf; inland sabkha in Domat Aljandal.

Figure 7. Map of Saudi Arabia showing the different locations of sampled *Aeluropus lagopoides* communities. Qareenah (Qar), Qaseem (Qas), Salwa (Sal), Jouf (Jou), and Jizan (Jiz).

The climate of Saudi Arabia is dry-hot and is classified as an arid region occupying about 5% of the world's arid zones [49]. It has low relative humidity except along the coastal zones, where it sometimes reaches 100%. The mean annual temperature is 33 °C during summer and 14 °C in winter, with a wide seasonal and diurnal variation [50]. The mean solar radiation was recorded highest for July and August and lowest for December and January except in Jizan. However, the rate of pan-evaporation was minimal along the coastal and high mountainous terrain and maximal in the interior due to the maximum presence of desert conditions. High rainfall variations and long drought periods have been recorded between the years without any rain. The climate data were collected between 1999 and 2019 from https://en.climate-data.org/asia/saudi-arabia-29/ (accessed on 15 May 2021) (Figure S2).

4.2. Vegetative Sampling

The vegetation sampling for each location was conducted from September to March when *A. lagopoides* was fully growing in each location and was in the full maturity stage. For each location and depending on species growth and form of vegetation composition of the studied area, 10 quadrats were selected randomly from each region (Table 3). The nested quadrat's area was 10 m × 10 m for shrubs and 5 m × 5 m for small shrubs and herbaceous species. These quadrats were chosen within the sabkha area dominated by patches of *A. lagopoides* population. The species in each quadrat were identified and named according to Chaudhary [35] and Miller et al. [51], as well as following the website http://www.powo.org (accessed on 1 February 2022). The plant density was determined according to Bonham [52], while the plant cover was estimated based on the scale of Braun-Blanquet [53]. To assess species dominance in each location, species importance value was calculated by the summation of relative density and relative cover of each species. The species life forms were identified according to Raunkiaer [54], while the chorotypes of all the species were made to assess the recorded species to World Geographical Groups.

Table 3. The coordinates and elevation of the different sample quadrats of different locations, Saudi Arabia.

Location	Quadrat No.	Coordinates	Elevation m a.s.l
Qareenah	1	25°03′59.7″ N 46°10′47.7″ E	833
	2	25°03′58.5″ N 46°10′48.1″ E	824
	3	25°03′56.6″ N 46°10′49.4″ E	816
	4	25°03′55.4″ N 46°10′51.1″ E	812
	5	25°03′53.4″ N 46°10′53.8″ E	810
	6	25°03′50.5″ N 46°10′52.9″ E	811
	7	25°03′53.0″ N 46°10′59.6″ E	806
Qaseem	1	26°03′17.7″ N 44°08′10.1″ E	590
	2	26°03′18.5″ N 44°08′15.2″ E	654
	3	26°03′14.2″ N 44°08′13.2″ E	621
	4	26°03′46.2″ N 44°08′16.3″ E	603
Salwa	1	24°45′23.5″ N 50°45′13.5″ E	−10
	2	25°43′39.8″ N 50°08′16.4″ E	−9
	3	25°43′45.5″ N 50°08′02.7″ E	−8
	4	24°45′04.3″ N 50°45′20.9″ E	−11
Jouf	1	29°49′12.0″ N 39°58′23.6″ E	565
	2	29°49′16.2″ N 39°58′27.4″ E	563
	3	29°49′51.0″ N 39°58′55.9″ E	558
	4	29°49′05.1″ N 39°58′08.9″ E	519
Jizan	1	16°58′06.1″ N 42°33′50.9″ E	15
	2	16°58′07.3″ N 42°33′42.4″ E	9
	3	16°58′08.2″ N 42°33′40.0″ E	6
	4	16°58′08.6″ N 42°34′04.9″ E	4
	5	16°58′06.8″ N 42°34′01.0″ E	4

4.3. Soil Sampling and Analysis

From each quadrat (n = 10) where vegetative sampling was conducted, three soil samples (0–30 cm depth) were collected from three random positions in plastic bags and pooled as a composite sample. All the soil samples were duly labeled and transferred to Range Science Lab, College of Food Science and Agriculture, King Saud University, Riyadh,

Saudi Arabia for further analyses. In addition, a portion of each sample was collected in moisture tins for the determination of soil moisture content by the weight-loss method. The soil samples were spread over separate plastic sheets, air-dried at room temperature, filtered through a 2 mm sieve to remove any debris, if present, and stored in a plastic bag until further analyses. Soil texture for sand, silt, and clay fractions were analyzed by the hydrometer method [55]. Soil organic matter (OM) was determined by wet combustion with dichromate at 450 °C [56]. Soil water extracts (1:5) were prepared for the estimation of soil electrical conductivity (EC) and pH [56]. Soluble inions (Cl and SO_4) were determined by titration method, while the determination of soluble cations (Ca, Mg, Na, and K), using a flame photometer according to Rhoades [57].

4.4. Data Analysis

In this study, multivariate analysis was applied viz. classification and ordination. Based on the data of relative density and cover of all species inside the quadrats (n = 10) of each region, a matrix of species importance values (relative density + relative cover) was constructed and subjected to hierarchical cluster analysis for classification and detrended correspondence analysis (DCA) for ordination using PAST 4.03 software [58]. Species rarefaction for base species richness and abundance in all the studied regions was analyzed using PAST 3X. The soil variables for the studied regions were subjected to one-way ANOVA and the mean values were separated based on Duncan's test at 0.05 probability level to examine the significant difference among studied regions. In order to detect the relationship between plants (dominant and important plant species with high importance values) of the studied area on one hand and soil variable data, on the other hand, canonical correspondence analysis (CCA) was conducted using MultiVariate Statistical Package (MVSP Version 3.2, Kovach Computing Services, Pentraeth, Wales, UK) according to Ter Braak and Smilauer [59]. In the CCA analysis, two datasets were constructed; one regarding the importance values of the dominant and important species (performed like that for DCA) and the second of the soil parameters of the quadrats (n = 10) of each region. Also, Pearson's correlation heatmap between the soil variables and the dominant and important species was performed using the XLSTAT software program (version 2018, Addinsoft, NY, USA).

5. Conclusions

The present study revealed variance among the community structure of *A. lagopoides*, within different sabkhas in Saudi Arabia. The community of inland sabkhas (Qareenah and Qaseem locations) showed higher plant diversity compared to the coastal sabkhas. The plant diversity of the *A. lagopoides* communities is mainly shaped by the salinity and water content. Moreover, the survival and flourishment of the halophytic grass *A. lagopoides* within a wide soil range in sabkhas revealed the adaptability of this plant to the harsh environment, which could be ascribed to its structural adaptations and modifications, as well as the phenotypic plasticity. Since *A. lagopoides* has many economic potentialities, where it is utilized as fodder, stabilizes sand dunes, used for landscaping of urban areas, the conservation of these natural vegetation zone habitats is of vital importance. Also, this valuable plant could be integrated as a promising forage candidate in saline-affected areas, even in the summer dry season.

Supplementary Materials: The following supporting information can be downloaded at: https://www.mdpi.com/article/10.3390/plants11050666/s1, Table S1: Floristic analysis of the recorded plant species in the studied sabkha locations of Saudi Arabia, Table S2: Vegetation composition of studied locations dominated with *A. lagopoides* in Saudi Arabia, Figure S1: Species richness and abundance based on the relative density of all the studies regions, Figure S2: Monthly climate data of the surveyed regions.

Author Contributions: Conceptualization, A.M.A. and A.A.A.-D.; software, A.M.A.-E.; formal analysis, A.M.A.-E., B.A.D. and A.M.A.; investigation, A.M.A., A.M.A.-E., S.L.A.-R., B.A.D. and A.A.A.-D.; writing—original draft preparation, A.M.A.-E., A.M.A. and B.A.D.; writing—review and editing, A.M.A., A.M.A.-E., S.L.A.-R., B.A.D. and A.A.A.-D. All authors have read and agreed to the published version of the manuscript.

Funding: Deanship of Scientific Research at King Saud University supported this work through research group no. RG-1441-302.

Institutional Review Board Statement: Not applicable.

Informed Consent Statement: Not applicable.

Data Availability Statement: Not applicable.

Acknowledgments: The authors extend their appreciation to the Deanship of Scientific Research at King Saud University for supporting this work through research group no. RG-1441-302.

Conflicts of Interest: The authors declare no conflict of interest.

References

1. Shearman, D. Origin of marine evaporites by diagenesis. *Trans. Inst. Min. Metall.* **1966**, *75*, 208–215.
2. Barakat, N.; El-Gawad, A.; Laudadio, V.; Kabiel, H.; Tufarelli, V.; Cazzato, E. A contribution to the ecology and floristic markers of plant associations in different habitats of Sinai Peninsula, Egypt. *Rend. Lincei* **2014**, *25*, 479–490. [CrossRef]
3. Cope, T. *Flora of Pakistan, No. 143: Poaceae*; Pakistan Agricultural Research Council Islamabad and University of Karachi: Karachi, Pakistan, 1982.
4. Sciandrello, S.; Musarella, C.; Puglisi, M.; Spampinato, G.; Tomaselli, V.; Minissale, P. Updated and new insights on the coastal halophilous vegetation of southeastern Sicily (Italy). *Plant Sociol.* **2019**, *56*, 81–98.
5. Al-Jaloud, A.A.; Hussain, G. Sabkha ecosystem and halophyte plant communities in Saudi Arabia. In *Sabkha Ecosystems*; Khan, M.A., Böer, B., Kust, G.S., Barth, H., Eds.; Society of Petroleum Engineers of Aime: Dhahran, Saudi Arabia, 2006; pp. 1–7.
6. Philby, H.S.J. Rub'Al Khali: An account of exploration in the Great South Desert of Arabia under the Auspices and Patronage of His Majesty'Abdul'Aziz ibn Sa'ud, King of the Hejaz and Nejd and its dependencies. *Geogr. J.* **1933**, *81*, 1–21. [CrossRef]
7. Thesiger, W. A new journey in southern Arabia. *Geogr. J.* **1946**, *108*, 129–145. [CrossRef]
8. Thomas, B. The South-Eastern Borderlands of Rub'al Khali. *Geogr. J.* **1929**, *73*, 193–212. [CrossRef]
9. Joshi, A.; Kanthaliya, B.; Arora, J. Halophytic plant existence in Indian salt flats: Biodiversity, biology, and uses. In *Handbook of Halophytes: From Molecules to Ecosystems towards Biosaline Agriculture*; Springer International Publishing: New York, NY, USA, 2020; pp. 1–22.
10. Abdel Khalik, K.; El-Sheikh, M.; El-Aidarous, A. Floristic diversity and vegetation analysis of wadi Al-Noman, Mecca, Saudi Arabia. *Turk. J. Bot.* **2013**, *37*, 894–907. [CrossRef]
11. AlNafie, A.H. Phytogeography of Saudi Arabia. *Saudi J. Biol. Sci.* **2008**, *15*, 159–176.
12. Shaltout, K.; Sheded, M.; El-Kady, H.; Al-Sodany, Y. Phytosociology and size structure of Nitraria retusa along the Egyptian Red Sea coast. *J. Arid Environ.* **2003**, *53*, 331–345. [CrossRef]
13. Doody, J.P. *Saltmarsh Conservation, Management and Restoration*; Springer Science & Business Media: New York, NY, USA, 2008; Volume 12.
14. Aljaloud, A.A. *Effect of Reclamation on Al-Hassa Saline Soil in Saudi Arabia*; California State University: Fresno, CA, USA, 1983.
15. Bazihizina, N.; Barrett-Lennard, E.G.; Colmer, T.D. Plant responses to heterogeneous salinity: Growth of the halophyte Atriplex nummularia is determined by the root-weighted mean salinity of the root zone. *J. Exp. Bot.* **2012**, *63*, 6347–6358. [CrossRef]
16. Burrows, C.J. (Ed.) Influences of strong environmental pressures. In *Processes of Vegetation Change*; Springer: Dordrecht, The Netherlands, 1990; pp. 207–268.
17. Álvarez-Rogel, J.; Carrasco, L.; Marín, C.; Martínez-Sánchez, J. Soils of a dune coastal salt marsh system in relation to groundwater level, micro-topography and vegetation under a semiarid Mediterranean climate in SE Spain. *Catena* **2007**, *69*, 111–121. [CrossRef]
18. Ghazanfar, S.A.; Böer, B.; Al Khulaidi, A.W.; El-Keblawy, A.; Alateeqi, S. Plants of Sabkha ecosystems of the Arabian Peninsula. In *Sabkha Ecosystems*; Gul, B., Böer, B., Khan, M.A., Clüsener-Godt, M., Hameed, A., Eds.; Springer: Cham, Switzerland, 2019; pp. 55–80.
19. Alharthi, A.; El-Sheikh, M.A.; Elhag, M.; Alatar, A.A.; Abbadi, G.A.; Abdel-Salam, E.M.; Arif, I.A.; Baeshen, A.A.; Eid, E.M. Remote sensing of 10 years changes in the vegetation cover of the northwestern coastal land of Red Sea, Saudi Arabia. *Saudi J. Biol. Sci.* **2020**, *27*, 3169–3179. [CrossRef] [PubMed]
20. Flowers, T.J.; Colmer, T.D. Salinity tolerance in halophytes. *New Phytol.* **2008**, *179*, 945–963. [CrossRef]
21. Almahasheer, H. Assessment of coastal salt marsh plants on the Arabian Gulf region. *Saudi J. Biol. Sci.* **2021**, *28*, 5640–5646. [CrossRef] [PubMed]
22. Zahran, M.A.; Murphy, K.J.; Mashaly, I.A.; Khedr, A.A. On the ecology of some halophytes and psammophytes in the Mediterranean coast of Egypt. *Verh. Ges. Okol.* **1996**, *25*, 133–146.
23. Basahi, R.A. Plant diversity of the coastal regions of Gulf of Aqaba, Saudi Arabia. *Annu. Res. Rev. Biol.* **2018**, *26*, 1–11. [CrossRef]
24. Barth, H.-J. Sabkha edge vegetation of coastal and inland sabkhat in Saudi Arabia. In *Ecophysiology of High Salinity Tolerant Plants*; Khan, M.A., Weber, D.J., Eds.; Springer: Dordrecht, The Netherlands, 2008; pp. 215–224.

25. Abd-ElGawad, A.M.; Assaeed, A.M.; Al-Rowaily, S.L.; Dar, B.M.; Malik, J.A. Moisture and salinity drive the vegetation composition of Wadi Hargan, Riyadh, Saudi Arabia. *Diversity* **2021**, *13*, 587. [CrossRef]
26. Mandaville, J.P. *Flora of Eastern Saudi Arabia*; Routledge: New York, NY, USA, 2013.
27. Gulzar, S.; Khan, M.A.; Ungar, I.A. Effects of salinity on growth, ionic content, and plant–water status of *Aeluropus lagopoides*. *Commun. Soil Sci. Plant Anal.* **2003**, *34*, 1657–1668. [CrossRef]
28. Phondani, P.C.; Bhatt, A.; Elsarrag, E.; Alhorr, Y.M.; El-Keblawy, A. Criteria and indicator approach of global sustainability assessment system for sustainable landscaping using native plants in Qatar. *Ecol. Indic.* **2016**, *69*, 381–389. [CrossRef]
29. Brussard, P.F.; Reed, J.M.; Tracy, C.R. Ecosystem management: What is it really? *Landsc. Urban Plan.* **1998**, *40*, 9–20. [CrossRef]
30. Hulshof, C.M.; Spasojevic, M.J. The edaphic control of plant diversity. *Glob. Ecol. Biogeogr.* **2020**, *29*, 1634–1650. [CrossRef]
31. Ouédraogo, A.S. Conservation and use of forestmgenetic resources. In Proceedings of the XI World Forestry Congress, Antalya, Turkey, 13–22 October 1997; pp. 173–188.
32. Al Beyrouthy, J.; Karam, N.; Al-Zein, M.S.; Yazbek, M. Ecogeographic survey and gap analysis for *Medicago* L.: Recommendations for in situ and ex situ conservation of Lebanese species. *Genet. Resour. Crop Evol.* **2019**, *66*, 1009–1026. [CrossRef]
33. Breckle, S.-W. Salinity, halophytes and salt affected natural ecosystems. In *Salinity: Environment-Plants-Molecules*; Läuchli, A., Lüttge, U., Eds.; Kluwer Academic Publishers: New York, NY, USA, 2002; pp. 53–77.
34. Qasim, M.; Gulzar, S.; Shinwari, Z.K.; Aziz, I.; Khan, M.A. Traditional ethnobotanical uses of halophytes from Hub, Balochistan. *Pak. J. Bot.* **2010**, *42*, 1543–1551.
35. Chaudhary, S.A. *Flora of the Kingdom of Saudi Arabia*; Ministry of Agriculture and Water: Riyadh, Saudi Arabia, 2001; Volume 3.
36. Aref, I.M.; El-Juhany, L.; Hegazy, S.S. Comparison of the growth and biomass production of six acacia species in Riyadh, Saudi Arabia after 4 years of irrigated cultivation. *J. Arid Environ.* **2003**, *54*, 783–792. [CrossRef]
37. Shaltout, K.; Mady, M. Analysis of raudhas vegetation in central Saudi Arabia. *J. Arid Environ.* **1996**, *34*, 441–454. [CrossRef]
38. Ji, Y.; Zhou, G.; New, T. Abiotic factors influencing the distribution of vegetation in coastal estuary of the Liaohe Delta, Northeast China. *Estuaries Coasts* **2009**, *32*, 937–942. [CrossRef]
39. Gul, B.; Hameed, A.; Weber, D.J.; Khan, M.A. Assessing seed germination responses of great basin halophytes to various exogenous chemical treatments under saline conditions. In *Sabkha Ecosystems*; Gul, B., Böer, B., Khan, M.A., Clüsener-Godt, M., Hameed, A., Eds.; Springer: Cham, Switzerland, 2016; pp. 85–104.
40. Wang, C.-H.; Li, B. Salinity and disturbance mediate direct and indirect plant–plant interactions in an assembled marsh community. *Oecologia* **2016**, *182*, 139–152. [CrossRef]
41. Mercado, F.G.; Del Moral Torres, F.; Luque, E.G.; Carrique, E.L.; Delgado Fernández, I.C.; De Haro Lozano, S. Soil requirements of four salt tolerant species in two saline habitats. *Arid Land Res. Manag.* **2014**, *28*, 395–409. [CrossRef]
42. Di Bella, C.E.; Jacobo, E.; Golluscio, R.A.; Rodríguez, A.M. Effect of cattle grazing on soil salinity and vegetation composition along an elevation gradient in a temperate coastal salt marsh of Samborombón Bay (Argentina). *Wetl. Ecol. Manag.* **2014**, *22*, 1–13. [CrossRef]
43. Crawford, J.T.; Stone, A.G. Relationships between soil composition and Spartina alterniflora dieback in an Atlantic salt marsh. *Wetlands* **2015**, *35*, 13–20. [CrossRef]
44. Koull, N.; Chehma, A. Soil-vegetation relationships of saline wetlands in north east of Algerian Sahara. *Arid Land Res. Manag.* **2015**, *29*, 72–84. [CrossRef]
45. Kumar, A.; Kumar, A.; Lata, C.; Kumar, S. Eco-physiological responses of *Aeluropus lagopoides* (grass halophyte) and *Suaeda nudiflora* (non-grass halophyte) under individual and interactive sodic and salt stress. *S. Afr. J. Bot.* **2016**, *105*, 36–44. [CrossRef]
46. Gulzar, S.; Khan, M.A. Seed germination of a halophytic grass *Aeluropus lagopoides*. *Ann. Bot.* **2001**, *87*, 319–324. [CrossRef]
47. Mohsenzadeh, S.; Malboobi, M.; Razavi, K.; Farrahi-Aschtiani, S. Physiological and molecular responses of *Aeluropus lagopoides* (Poaceae) to water deficit. *Environ. Exp. Bot.* **2006**, *56*, 314–322. [CrossRef]
48. Naz, N.; Hameed, M.; Ashraf, M.; Arshad, M.; Ahmad, M.S.A. Impact of salinity on species association and phytosociology of halophytic plant communities in the Cholistan Desert, Pakistan. *Pak. J. Bot.* **2010**, *42*, 2359–2367.
49. Bashour, I.; Al-Mashhady, A.; Prasad, J.D.; Miller, T.; Mazroa, M. Morphology and composition of some soils under cultivation in Saudi Arabia. *Geoderma* **1983**, *29*, 327–340. [CrossRef]
50. Vincent, P. *Saudi Arabia: An Environmental Overview*; Taylor & Francis: New York, NY, USA, 2008.
51. Miller, A.G.; Cope, T.A.; Nyberg, J. *Flora of the Arabian Peninsula And Socotra*; Edinburgh University Press in Association with Royal Botanic Garden, Edinburgh, Royal Botanic Gardens, Kew: Edinburgh, UK, 1996.
52. Bonham, C.D. *Measurements for Terrestrial Vegetation*; John Wiley & Sons: New York, NY, USA, 2013.
53. Mueller-Dombois, D.; Ellenberg, H. *Aims and Methods of Vegetation Ecology*; Wiley and Sons: New York, NY, USA, 1974.
54. Raunkiaer, C. *Plant Life Forms*; Clarendon Press: Oxford, UK, 1937.
55. Bowles, J.E. *Engineering Properties of Soils and Their Measurement*; McGraw-Hill, Inc.: New York, NY, USA, 1992.
56. Rowell, D. *Soil Science: Methods and Applications*; Longman Group: Harlow, UK, 1994.
57. Rhoades, J. Soluble salts. In *Methods of Soil Analysis. Part 2, Chemical and Microbiological Properties*; John Wiley & Sons: Madison, WI, USA, 1982; pp. 167–179.

58. Hammer, Ø.; Harper, D.A.; Ryan, P.D. PAST: Paleontological statistics software package for education and data analysis. *Palaeontol. Electron.* **2001**, *4*, 9.
59. Ter Braak, C.J.; Smilauer, P. *CANOCO Reference Manual and CanoDraw for Windows User's Guide: Software for Canonical Community Ordination (Version 4.5)*; Microcomputer Power: Ithaca, NY, USA, 2002.

plants

MDPI

Article

First Survey of the Vascular and Cryptogam Flora on Bulgaria's Ancient Mounds

Iva Apostolova [1,*], Desislava Sopotlieva [1], Magdalena Valcheva [1], Anna Ganeva [1], Veselin Shivarov [1], Nikolay Velev [1], Kiril Vassilev [1], Tsvetelina Terziyska [1] and Georgi Nekhrizov [2]

1 Institute of Biodiversity and Ecosystem Research, Bulgarian Academy of Sciences, 2 Gagarin Str., 1113 Sofia, Bulgaria; dsopotlieva@gmail.com (D.S.); magdalena.i.valcheva@gmail.com (M.V.); annaganeva8@gmail.com (A.G.); v.shivarov@abv.bg (V.S.); nikolay.velev@abv.bg (N.V.); kiril5914@abv.bg (K.V.); ts.terziyska@gmail.com (T.T.)

2 National Archaeological Institute with Museum, Bulgarian Academy of Sciences, 2 Saborna Str., 1000 Sofia, Bulgaria; nehrizov@gmail.com

* Correspondence: iva.apostolova@gmail.com

Abstract: This work represents the first study of the floristic diversity on Bulgaria's ancient mounds. The objective of this research was to assess the importance of the mounds for the preservation of the native vascular and cryptogam flora. Our sampling design included 111 ancient mounds distributed throughout the country. We recorded a total of 1059 vascular plants, 58 bryophytes and 61 lichen taxa. Despite their small area, the mounds were shown to preserve nearly a quarter of the Bulgarian flora. The vegetation cover on the mounds included 61% perennials indicating a long-term persistence and stability. The majority (98%) of the established vascular plants were native species. Although the conservation significance of the vascular plant species were not common, we recorded 2 critically endangered, 9 endangered and 14 Balkan endemics during the present study. The lichen *Arthopyrenia salicis* was recorded for the first time in Bulgaria and a new locality of the rare bryophyte *Ceratodon conicus* was discovered. The established compositional difference between plots from the northern and southern slopes of the mounds (88.95%) is a testament to the high local habitat diversity. The prevalence of species characteristic for *Festuco-Brometea* suggests that the mounds preserve fragments of native grasslands and steppes. The variation in cover of agricultural and other human modified areas in the mounds' immediate surroundings did not substantially affect their species richness. We argue that the ancient mounds should be taken into consideration in future green space planning.

Keywords: bryophytes; generalist plants; grassland specialists; historical monuments; invasive alien plants; kurgans; lichens; native plants

Citation: Apostolova, I.; Sopotlieva, D.; Valcheva, M.; Ganeva, A.; Shivarov, V.; Velev, N.; Vassilev, K.; Terziyska, T.; Nekhrizov, G. First Survey of the Vascular and Cryptogam Flora on Bulgaria's Ancient Mounds. *Plants* **2022**, *11*, 705. https://doi.org/10.3390/plants11050705

Academic Editor: Robert Philipp Wagensommer

Received: 11 February 2022
Accepted: 4 March 2022
Published: 6 March 2022

1. Introduction

Ancient mounds (also called tumuli or more commonly kurgans) were constructed in temperate Eurasia between 4th millennium BC and 4th century AD and used primarily for burial purposes. A remarkable number of these mounds have been preserved due to their spiritual and cultural importance [1–4]. Bulgaria is exceptionally rich in ancient mounds with a known number of approximately 50,000 [5]; 11,000 of these mounds have been registered in the Archaeological Map of Bulgaria (http://www.naim-bas.com/akb/ accessed on 3 November 2021). Some of Bulgaria's ancient mounds are remarkable historical monuments, including massive underground stone buildings often decorated with wall paintings (e.g., Kazanlak and Aleksandrovo tombs). The most attractive of these structures are important tourist destinations open to the public. At present, most of Bulgaria's ancient mounds are surrounded by vast agricultural lands [4]. Similar to ancient mounds in other European countries, these structures are often standing as sole "islands" of semi-natural vegetation in an otherwise human-modified landscape [3,6,7]. Along with field margins, road verges and buffer strips adjacent to arable land, the ancient mounds preserve small semi-natural fragments and provide an opportunity for the long-term survival of indigenous flora. Moreover, ancient mounds consist of different microhabitats, which enrich the suitability for the development of an ecologically diverse flora [8,9]. The long-term persistence of the mounds within agricultural lands, primarily due to sacred and religious respect, naturally makes them a part of the Green and Blue Infrastructure defined at the European level as a "strategically planned network of natural and semi-natural areas with other environmental features designed and managed to deliver a wide range of ecosystem services" [10].

Recently, there has been an increased interest in burial mounds as biodiversity hotspots situated in an otherwise homogenous agricultural landscape [2,3,7,11–15]. Plants, and especially flowering plants (Angiosperms), are one of four groups of living organisms (along with Heteroptera, Symphyta and aculeate Hymenoptera) that have been shown to be best served for the biodiversity evaluation of cultivated areas [16]. Recent research has shown that ancient mounds preserve a remarkable plant diversity [3,11,13,15,17,18]. To date, no studies on the natural value of Bulgaria's ancient mounds have been conducted and no records of their floristic diversity are known to exist. At the beginning of the current research, we assumed that the cultural significance and principal sacrosanct nature of Bulgaria's ancient mounds, akin to other countries, provided long-term repository conditions for natural communities and that they served as refugia for indigenous flora in anthropogenically transformed areas. Considering cryptograms' signal for increased degree of community stability and naturalness [19], and in order to enrich the current biodiversity assessment, we included cryptograms (bryophytes and lichens) along with vascular plants in this survey.

The objectives of this study were (1) to collect completely novel information about the floristic diversity of Bulgarian ancient mounds and (2) to assess the potential of ancient mounds to preserve native vascular and cryptogam flora, despite being largely isolated.

2. Materials and Methods

2.1. Study Objects and Study Area

We used the Archaeological Map of Bulgaria (http://www.naim-bas.com/akb/ accessed on 3 December 2021) to select the study objects. Our selection criteria included mounds that were (1) undisturbed by archaeological investigation, (2) were clearly recognizable, and (3) were higher than 1 m and more than 9 m in diameter (Table 1). We visited a total of 111 ancient mounds spread out across the territory of Bulgaria (Figures 1 and 2). The mounds were located in the lowlands and hilly plains of the country between 60 and 900 m a.s.l. In our study, a larger mound base usually corresponded to a larger height. The correlation of height to diameter was $r = 0.57518$ ($p < 0.05$) and the correlation of height to 2D area (calculated as $\pi d^2/4$, where "d" is diameter) was $r = 0.61815$ ($p < 0.05$). Therefore, we used the 2D area of the mounds as a representation of the relative mound size.

Table 1. Basic topographic parameters of the studied mounds (n = 111).

Parameter	Min	Max	Mean	SD
Altitude (m, a.s.l.)	61.0	920.0	354.4	238.8
Height (m)	1.2	24.5	5.3	3.1
Diameter (m)	9.5	88.4	40.7	15.8
Coverage of herbaceous vegetation (%)	0	100	78.3	30.5
Coverage of shrub/forest vegetation (%)	0	100	21.7	30.5
Surrounding of semi-natural vegetation in 200 m buffer (%)	0	100	23.1	26.4
Surrounding of agricultural and other anthropogenic lands in 200 m buffer (%)	0	100	76.9	26.4

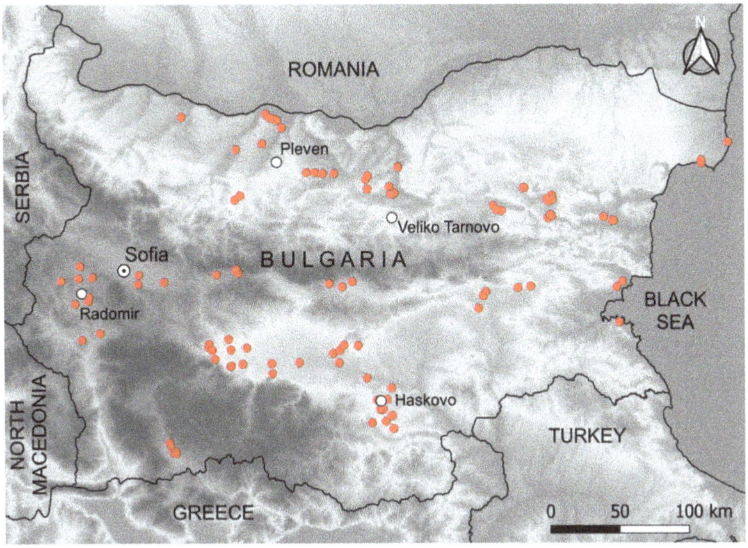

Figure 1. Map of Bulgaria with marked locations of the sampled mounds.

Figure 2. An example of a typical mound view (photo credit: I. Apostolova).

We visually estimated the percentage of grassland and woody vegetation cover of the mounds using Google Earth images. We chose a 10% threshold to facilitate the rough estimation in vegetation cover of the mounds. There were 86 mounds with more than 70% herbaceous cover, 16 with more than 70% forest vegetation cover and 9 with mixed vegetation cover.

In order to estimate the degree of mound isolation, we created a buffer area with a radius of 200 m around the base of each mound and calculated the land cover of natural vs. non-natural habitats within. The land cover types in the buffer area were obtained from the Land Parcel Identification System (LPIS) database maintained by the Ministry of Agriculture, Food and Forestry of Bulgaria and generalized as semi-natural vegetation and agricultural and other anthropogenic lands (for details see [4]). More than half of the studied mounds were highly isolated—80 mounds were surrounded by more than 70% of agricultural and other anthropogenic lands.

The investigated mounds fall within the temperate and the continental–Mediterranean climatic zones. The temperate zone, which incorporates the northern parts of the country, has an average mean annual temperature of 11.9 °C and annual precipitation of 573 mm (town of Pleven, 1971–2000), while the continental–Mediterranean zone, which is more typical for the southern part of the country, has an average mean annual temperature of 12 °C and annual precipitation of 637 mm (town of Haskovo, 1971–2000) [20]. A major part of the study area falls within the broadleaved deciduous forests zone (Map of Natural Vegetation of Europe, [21]), while the southern areas include Mediterranean vegetation fragments with typical plant species [22]. Some small areas in the north-east are influenced by steppe vegetation [23].

2.2. Sampling Design, Data Collection and Data Analysis

The field work was conducted during the maximum period for vegetation development (June and July) in 2019 and 2020. Our sampling design was focused on the major ecological differences exhibited at the northern and southern slopes of the mounds. This design considered findings reported by previous studies on burial mounds [9] regarding the difference in floristic composition at different exposures due to habitat heterogeneity. In order to obtain more detailed floristic data, we first sampled all species within 5 × 5 m plots situated on the northern and southern slopes (two plots per mound). We then carefully explored the remaining mound area and recorded any additional species until the floristic variety was exhausted. We used presence/absence species data, both at plot and at mound level, for our analyses.

Despite the fact that some taxa were identified to subspecies, we set the final plant list to species level. Some closely related species were joined in species aggregates (*Achillea millefolium* aggr.–including *A. millefolium*, *A. pannonica* and *A. setacea*) or determined at aggregate level (*Rubus hirtus* aggr.). Plants that were in a phenological stage unsuitable for correct species determination, or that were difficult to identify (e.g., *Taraxacum* spp.), were determined to genus level. The vascular plant species nomenclature follows The Euro+Med PlantBase [24], with the exception of *Brassica juncea*, which follows the Plant List [25]. Bryophyte nomenclature follows Hill et al. [26], and lichen nomenclature follows Nimis et al. [27].

For each taxon, we attributed a set of characteristics regarding biological type, functional role, floristic element, conservation and native status (Supplementary ESM S1). Data regarding the biological type of vascular plants were extracted from national literature sources [28,29]. The biological types were grouped as follows: short lived (including annual and biannual plants), perennial (including biannual to perennial and perennial plants), dwarf-shrub, shrub and tree. The association of vascular plant species to higher rank syntaxa was defined following Mucina et al. [30]. In cases when more than a single phytosociological class was proposed for a certain diagnostic species, we selected the best representative for the country's vegetation based on our expertise. The diagnostic role of a species was used to assign each species to one of the following 3 functional groups: generalists, grassland specialists and forest specialists. We considered species to be generalists if they were diagnostic of synanthropic vegetation or if they had a broad distribution across different habitat types. The functional affiliation of species that were not assigned to a specific syntaxon was determined based on their most common habitat occurrence in the country. The determination of phytogeographical (floristic) elements for vascular plants follows Assyov and Petrova [31], for bryophytes—Ganeva and Düll [32] and for lichens—Wirth [33] and Nimis [34].

The native status of vascular plants was retrieved from Euro+Med PlantBase [24] because of the lack of such data for Bulgaria. Only species listed by Petrova et al. [35] were recognized as invasive alien plants. Vascular plants with conservation importance included Balkan and Bulgarian endemics ([31,36] complemented by The Euro+Med PlantBase [24]), Bulgarian red list species [37], species protected by the Bulgarian legislation (Appendix 3 of the Bulgarian Biological Diversity Act [38]) and other European and international documents (e.g., Council Directive 92/43/EEC [39], as well as the Convention on International Trade in Endangered Species of Wild Fauna and Flora (CITES) [40]). The above mentioned attributes were not applicable for the taxa determined to genus level (N/A). The records of these species were not included in the analyses based on functional groups.

We calculated basic descriptive statistics for all biological characteristics of the registered vascular plants. We used similarity percentages analysis (SIMPER) [41] in PRIMER 7 [42] to determine the species that contributed the most to the floristic resemblances between mounds. The difference in species composition between the mounds was assessed by using the beta diversity index in PAST [43]. We used correlation analyses with Pearson correlation coefficient in STATISTICA 13 [44] to test for correlation between floristic richness (total vascular plant richness, richness of generalists and of grassland and forest specialists) and two other variables: 2D area and proportion of anthropogenic land in the buffer areas. We used plot level data to graphically express the differences in species richness of the different species groups between plots with northern and plots with southern exposure (by their mean values and standard deviation), also carried out in STATISTICA 13 [44].

3. Results

3.1. Diversity and Species Characteristics

3.1.1. Vascular Plants

The list of registered vascular plants includes 1059 taxa (Supplementary ESM S1). The average number of species per mound was 69.9 ± 22.6 SD (min 27, max 152). We identified 971 plants to species level, accepted 3 taxa as aggregates or species groups, and identified 85 taxa to genus level. The floristic diversity was confined to 82 vascular plant families. Flowering plants (Angiosperms) made up the majority of the observed species diversity and only seven species belonged to other groups: one horsetail—*Equisetum hyemale*, two ferns—*Polystichum aculeatum*, *Pteridium aquilinum*, and four Gymnosperms—*Juniperus communis*, *J. oxycedrus*, *Pinus nigra*, *P. sylvestris*. There were 23 families represented by more than 10 taxa, the most species rich of these were Asteraceae—126 taxa (11.9% of the established taxa), Fabaceae—111 taxa (10.5%), Poaceae—106 taxa (10%), Lamiaceae—61 taxa (5.8%), Brassicaceae—58 taxa (5.5%), Caryophyllaceae—57 taxa (5.4%), Rosaceae—48 taxa (4.5%), Apiaceae—46 taxa (4.3%), Boraginaceae—31 taxa (2.9%) and Plantaginaceae—31 taxa (2.9%). Twenty-three other families (28.1%) were represented by a single species. The flora of the studied mounds was composed primarily of perennial herbaceous plants (61%), followed by short-lived plants, and a low number of shrubs and trees. Generalists and grassland specialists dominated the species composition of the mounds. A major part (98%) of the established vascular plants consisted of native species (Table 2). Only 21 plants belonged to other categories (alien (status unknown)—13, naturalized alien—5, in large-scale cultivation—2 and doubtfully native—1). Invasive alien plants included *Acer negundo*, *Ailanthus altissima*, *Amaranthus albus*, *Conyza canadensis*, *Cuscuta campestris*, *Datura stramonium*, *Erigeron annuus*, *Phytolacca americana*, *Robinia pseudoacacia*, *Sorghum halepense*, *Xanthium orientale* subsp. *italicum* and *X. strumarium*. These taxa represent 20% of all plants included in the list of invasive or potentially invasive alien plants in Bulgaria and three of them (*Acer negundo*, *Ailanthus altissima* and *Robinia pseudoacacia*) are among the "top 10" invasive alien plants in Bulgaria. The largest number of invasive plant species we registered on a single mound was four. The mound (ID 586) was situated in the central part of north Bulgaria (north of the town of Veliko Tarnovo). A substantial number of the studied mounds (40 or 35.4%) had at least one invasive alien plant.

Table 2. Biological characteristics of the registered vascular plants. Total number and descriptive statistics are given.

		On a Mound			
Parameter	Total	Min	Max	Mean	SD
Biological type					
Short-lived	378	4	64	24.77	10.90
Perennial	495	10	88	36.13	13.57
Dwarf-shrub	17	0	7	0.80	1.09
Shrub	49	0	12	4.02	2.80
Tree	35	0	7	2.55	1.87
N/A	85	0	10	2.20	2.26
Plants with conservation importance					
Critically Endangered (CR)	2	0	1	0.04	0.19
Endangered (EN)	9	0	2	0.22	0.43
Vulnerable (VU)	7	0	2	0.15	0.39
Near Threatened (NT)	6	0	1	0.07	0.26
Least concern (LC)	2	0	2	0.21	0.45
Bulgarian Biodiversity Act	10				
Habitat Directive	1				
CITES	11				
N/A	85	0	10	2.20	2.26
No conservation status	947	27	137	67.50	20.87
Native status					
Native (including archaeophytes)	948	27	141	66.76	21.30
Naturalized alien	5	0	2	0.15	0.39
Doubtfully native	1				
In large-scale cultivation	2				
Alien (status unknown)	13	0	5	0.83	0.97
N/A	85	0	10	2.20	2.26
No data	5	0	2	0.45	0.63
Invasive alien plants (for Bulgaria)					
Invasive alien plants (IAP)	12	0	4	0.50	0.80
Not classified as IAP	962	27	142	67.76	21.38
N/A	85	0	10	2.20	2.26

Critically endangered species present on the mounds included *Anchusa stylosa* and *Limonium asterotrichum*. Endangered plants present on the mounds included *Astragalus haarbachii, A. wilmottianus, Chamaecytisus frivaldszkyanus, C. kovacevii, Dianthus pallidiflorus, Erysimum cheiranthoides, Festuca thracica, Goniolimon besseranum* and *Jurinea ledebourii*. Fourteen of the registered taxa were Balkan endemics: *Achillea clypeolata, A. pseudopectinata, Armeria rumelica, Astragalus wilmottianus, Asyneuma anthericoides, Cytisus eriocarpus, Dianthus moesiacus, Festuca thracica, Heptaptera triquetra, Koeleria simonkaii, Dichoropetalum vittijugum, Polygala supina* subsp. *rhodopea* (syn. *P. rhodopaea* (Velen.) Janch.), *Scabiosa triniifolia* and *Thymus longidentatus*. Eleven plants were included in the CITES Convention. A mound near the town of Radomir (ID 185) contained the richest number of plants of conservation interest and maintained populations of 5 such species.

According to our SIMPER analyses, species composition similarity across plots with northern exposure was 13.25% and was 11.84% across plots with southern exposure. We found a remarkable dissimilarity between plots with northern and southern exposure—88.95%. We registered 142 species unique to the plots with southern exposure. Species registered in more than three plots on southern slopes were *Camelina sativa, Heliotropium europaeum, Sedum hispanicum, Senecio leucanthemifolius, Valerianella dentata, Verbascum ovalifolium, V. densiflorum, Tribulus terrestris, Crepis sancta, Haplophyllum suaveolens, Herniaria incana, Thymelaea passerina* and *Filago arvensis*. Another 152 species were confined to plots with northern exposure. Taxa that exhibited higher frequency on the plots with northern exposure included *Astragalus glycyphylloides, Ficaria verna, Carlina vulgaris, Helleborus odorus, Leucanthemum vulgare, Phlomis tuberosa, Ranunculus polyanthemos, Anthoxanthum odoratum,*

Helictochloa compressa, *Quercus frainetto* and *Luzula campestris*. The mean species richness of grassland and forest specialists and generalists was similar between plots with northern and plots with southern exposure (Figure 3).

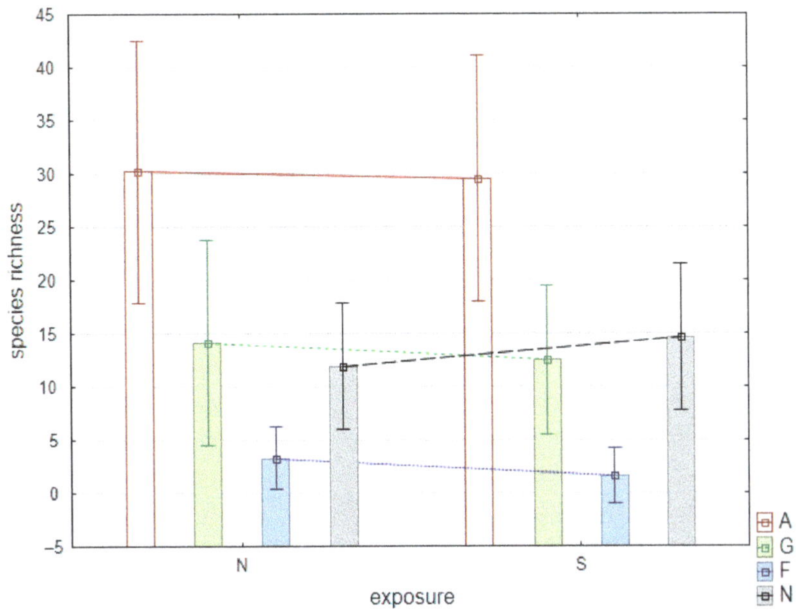

Figure 3. Differences in mean species richness (columns) and standard deviations (whiskers) of different species groups (A—all plant species, G—grassland specialists, F—forest specialists and N—generalists) between plots with northern (N) and plots with southern (S) exposure on the studied mounds.

Average dissimilarity between all pairs of mounds was 0.78 (Whittaker measure for beta diversity). Fifteen species were registered on more than 50% of the studied objects and included *Poa angustifolia* (77% of the studied mounds), *Achillea millefolium* aggr. (72.6%), *Eryngium campestre* (67.3%), *Galium verum* (64.6%), *Teucrium chamaedrys* (61.1%), *Tragopogon dubius* (60.2%), *Sanguisorba minor* (54%) and *Botriochloa ischaemum* (50.4%). According to our SIMPER analyses, these taxa along with *Prunus spinosa*, *Lactuca serriola*, *Dactylis glomerata*, *Falcaria vulgaris*, *Convolvulus arvensis*, *Crataegus monogyna*, *Elytrigia repens*, *Galium aparine* and *Potentilla recta* contributed the most to the floristic similarity between mounds. On the other hand, 480 species (45.3% of the total flora) were registered from one or two mounds. We found no relationship between the 2D area of the mounds and their floristic richness. With the enlargement of the 2D area of the mounds, the richness of forest specialists slightly increased ($r = 0.28$, $p < 0.05$). More than a half of the studied mounds (69) were highly isolated and surrounded by an agricultural matrix above the average in the buffer (Table 1). As the anthropogenically transformed lands around the mounds increased, the total number of registered species ($r = -0.28$, $p < 0.05$) as well as the total number of specialists ($r = -0.34$, $p < 0.05$), including grassland specialists ($r = -0.33$, $p < 0.05$), decreased.

The phytogeographical spectrum of the registered plants resembled that of the national flora (Figure 4). The number of species with Mediterranean distribution was the highest, followed by the plants with European and Eurasian distribution.

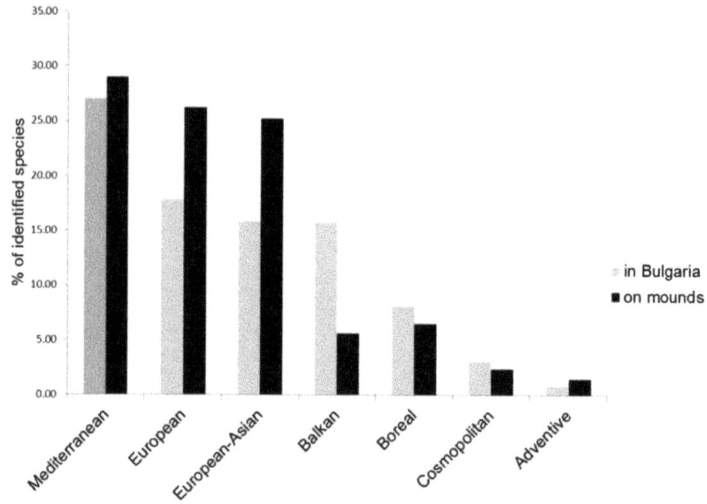

Figure 4. Phytogeographical spectrum of vascular plants on the mounds, compared to vascular plants in the Bulgarian flora. The national record was calculated by following [45].

Across all plants growing on the mounds, 853 were diagnostic of 40 different vegetation classes. Most numerous were the grassland specialists and generalists, and *Festuco-Brometea* diagnostic species constituted 44% of all identified plants (Figure 5). There were 193 species diagnostic of anthropogenic vegetation (classes *Artemisietea vulgaris, Papaveretea rhoeadis, Chenopodietea, Epilobietea angustifolii, Polygono-Poetea annuae* and *Sisymbrietea*) and their per mound abundance was relatively low (on average from 2.2 ± 1.8 SD for *Chenopodietea* to 7.87 ± 3.5 SD species for *Artemisietea vulgaris*). There were 100 species diagnostic for forest vegetation (classes *Alno glutinosae-Populetea albae, Carpino-Fagetea sylvaticae, Quercetea pubescentis, Quercetea robori-petraeae* and *Salicetea purpureae*). Plants diagnostic for *Quercetea pubescentis* prevailed with the highest average per mound presence (2.56 ± 2.5 SD). There were 206 species with wide ecological plasticity to which no diagnostic value toward a particular syntaxon was attributed.

3.1.2. Bryophytes

We registered a total of 58 bryophyte taxa; 54 were identified to species level (Supplementary ESM S1). They belong to 39 genera and 16 families. Pottiaceae included the highest number of species, all of which were confined to dry, skeletal and sandy substrates. The second most diverse family was Bryaceae, a group which includes species with diverse ecological preferences. On the mounds, this family was represented by species typical of dry eroded terrains. No bryophytes were registered on 40 of the studied mounds (36.04%). In cases where bryophytes were found, their number ranged from 1 to 10 species (average 3.1 ± 2.1SD). The only typical epiphytic moss recorded during this study was *Orthotrichum pumilum*, which was found on trees of the genus *Quercus*. The rest of the listed species usually occupied the soil substrate and were occasionally also found on woody stems. *Abietinella abietina, Barbula unguiculata, Pterygoneurum ovatum, Rhynchostegium megapolitanum* and *Thuidium assimile* were confined to herbaceous habitats. *Bryum pallescens* has broad ecological affiliation, while *Atrichum undulatum, Fissidens taxifolius* and *Plagiomnium affine* were observed only within the forested northern slopes. Nearly all of the bryophytes registered during the present study are taxa commonly found in Bulgaria where they are known to occur in a variety of different plant communities, usually in lowland areas. Phytogeographically, most of the bryophytes (35 species) belong to the temperate region. A significant share of the registered bryophytes were cosmopolites (*Polytrichum juniperinum*, *Polytrichum piliferum*, *Funaria hygrometrica*, *Schistidium apocarpum*, *Ceratodon purpureus*, *Weissia controversa*, *Syntrichia ruralis*, *Bryum argenteum*, *Pohlia nutans* and *Hypnum cupessiforme*). The *Ceratodon conicus* species, previously known from a single locality in Bulgaria (Vitosha Mt. [46]), deserves special attention. This taxon was registered on a mound in the Thracian lowland near the town of Plovdiv (ID249), a substantial distance from the previously reported location.

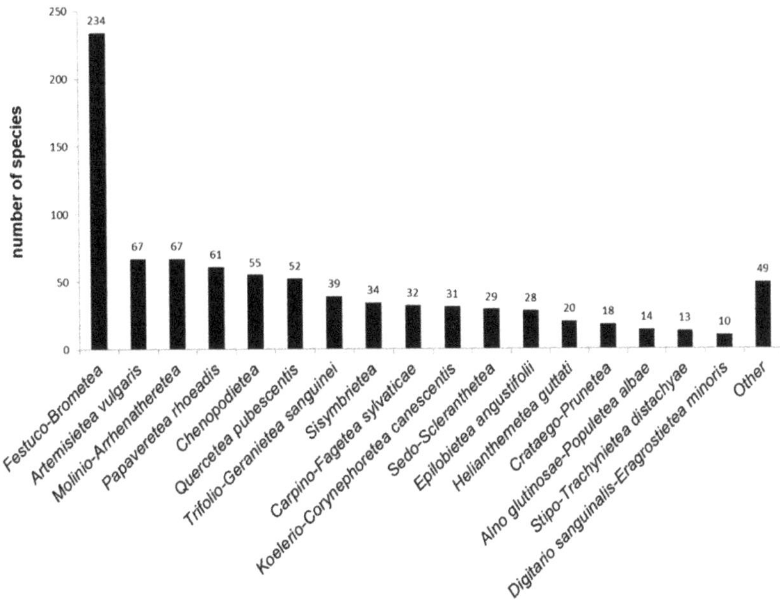

Figure 5. Vascular plants number associated with the phytosociological class level.

3.1.3. Lichens

A total of 61 lichen taxa, 56 identified to species level, were registered on 52% of the studied mounds (Supplementary ESM S1). The highest number of recorded lichens were epiphytes (35 taxa), followed by epigean taxa (16 taxa), and by species with wide ecological breadth occurring on all substrate types (2 taxa). We recorded eight epilithic lichen taxa. The number of species registered on a mound ranged between 1 and 18 taxa (average 4.0 ± 2.1SD). Lichens were not registered on 57 of the studied mounds. The most common family found on mounds was Parmeliaceae, followed by Cladoniaceae and Physciaceae. Frequently registered lichen species were *Xanthoria parietina*, *Parmelia sulcata* and *Physcia adscendens*. *Arthopyrenia salicis* is recorded for the first time for the country. Until present, *Caloplaca cerinella* and *Catillaria nigroclavata* were known from single Bulgarian localities and are reported here for a second time. Phytogeographically, many of the lichens (31 taxa) are typical for the temperate region. In general, there is a clear preponderance of warm-temperate lichens. Arctic–alpine lichens, common for the Bulgarian alpine and subalpine zones, were not encountered during the present study. It is interesting to note that the new country record and the two other poorly recorded taxa were found together on a single mound (ID 562) situated in the eastern part of the country, near the Black Sea.

4. Discussion

4.1. Species Diversity

This article presents new information regarding the vascular and cryptogam (bryophytes and lichens) flora of ancient mounds in Bulgaria. The species diversity established during this study is a testament to the importance of ancient mounds for preserving the native flora as previously emphasized for other parts of Eurasia [3,12,13,15]. According to the most recent report [47], the Bulgarian flora includes 4064 vascular plant species. The 1056 vascular plant taxa we recorded present nearly a quarter of the overall national plant diversity. These results emphasize the role of ancient mounds in the preservation of a high percentage of the national floristic diversity. Similar results have been previously reported in other European countries: a total of 346 plants were registered on 82 mounds in Hungary [18], which equals 15.5% of the national floristic diversity [48] and 721 species were registered on 106 mounds in Ukraine [12], which represents 11.6% of the national flora [49]. The flora of the studied ancient mounds mirrors the phytogeographical characteristics of the Bulgarian flora, i.e., the mounds contain a representative sample of the national plant species pool. The higher proportion of Mediterranean and European, as well as adventive, elements reflects the species pool of the lowlands. A similar finding, namely that ancient mounds reflect local biogeographical zones, was reported by Sudnik-Wójcikowska and Moysiyenko [50], in Ukraine.

We expected that the flora of the mounds in different countries will reflect local environmental conditions, species pools and human influence. Nevertheless, there was a surprising similarity regarding the number of species recorded per mound in different countries. Our results were close to the 72 species registered on the best documented mound, Csípő-halom (Hortobágy, Great Hungarian Plain) [51]. The range of species established per mound in Poland has been found to be from 44 to 81 [52], and the reported number of species per mound in Ukraine has been found to be between 82 and 125 [12]. Similarities among the mounds in Europe did not solely concern their species diversity. Another similarity is the weak correlation between mound area and species richness, originally established by Deák et al. [53]. Our results corroborate these findings and strengthen the idea that mound protection should be of high priority irrespective of their size. Studies of ancient mounds often refer to these structures as islands in a sea of anthropogenically modified areas [12,14,15]. This is a fair comparison, given the fact that the increase in agricultural and other modified areas around the mounds does not significantly affect their biodiversity, a finding also confirmed by our study.

The outstanding floristic diversity of Bulgarian mounds, especially given the fact that most of them are located in areas characterized by intensive large-scale agriculture, supports our assumption that ancient mounds are valuable refugia for indigenous flora and that they play an important role in its conservation. The number of species we registered on one or two mounds was high (45.3% of the established flora) and it reflected the local species pool, a finding similar to that of Sudnik-Wójcikowska and Moysiyenko [8]. The high rate of dissimilarity between the sampled mounds is a reflection of the diversity of biogeographical zones in the country. The difference in species composition between plots with northern and those with southern exposure corresponds to the well-established effect of slope aspect on vegetation [54]. The abundance of generalists on the south facing slopes and the abundance of grassland specialists on the northern slopes support the findings of Deák et al. [9]. Forest specialists are more abundant on north facing slopes due to the ecological requirements of trees to milder climatic conditions and especially to higher air and soil moisture.

According to our results, 6.4% of the bryophytes native to the country can be found on the mounds. The new *Ceratodon conicus* locality, established nearly a century later and on a substantial distance away from the original locality, gives us a reason to believe in the potential of ancient mounds for the long-term preservation of biodiversity. The cover and diversity of lichen species depends primarily on suitable substrates and environmental factors. Although the saxicolous lichens represent one of the largest ecological groups of lichens, their preferred bare and stable rock substrates were not found on the studied mounds. Most of the saxicolous lichens found during this study appeared on artificial substrates of contemporary anthropogenic structures, such as geodetic points. Only a few species were found on pebbles that are not a suitable substrate for lichens because of their instability. The predominance of epiphytic lichens is not surprising as most of the mounds had trees and shrubs. The terricolous taxa were rarely present due to their inability to compete with the well-established vegetation. The first record of the epiphytic lichen *Arthopyrenia salicis* from Bulgaria is noteworthy. To our knowledge, the closest *A. salicis* locality is in the Slovenian Julian Alps [55]. This species is especially frequent in northern and western Europe [56], where it occurs on the smooth bark of the trees, as experienced in our study. Lichens frequently established on the mounds are among the most common species characteristic of the lowland epiphytic communities on broadleaved trees in Bulgaria. However, the abundance of *Xanthoria parietina* and *Physcia adscendens* is indicative of a higher load of nitrogenous compounds [57]. Typical nitrophilous lichen community members, such as *Phaeophyscia orbicularis* and *Polycauliona polycarpa* [58], were also relatively frequently found on the studied mounds. We assume that these nitrophilous species might have been positively affected by the surrounding agricultural lands and treated with fertilizers.

4.2. Species Characteristics

The ecological characteristics of the established flowering plants strongly resemble the peculiarities of the national flora with a dominant presence of species typical for dry areas. These species use evolutionary advantages to survive via vegetative propagules (e.g., *Agropyron repens*, *Poa angustifolia*, *Botryochloa ischaemum*, *Achillea millefolium*, *Bryum dichotomum*, *B. klinggraeffii*, *B. moravicum*, *B. rubens* and *B. ruderale*), underground storage organs (e.g., *Arum maculatum*, *Crocus flavus* and *Ornithogalum* spp.) or via annual life cycles (e.g., *Trifolium striatum*, *Astragalus spruneri*, *Apera spica-venti*, *Arenaria serpyllifolia* and *Coronilla scorpioides*). Patches of bare ground and pits left after continuous illegal treasure hunting become suitable places where fruits or seeds of common trees and shrubs (*Pyrus pyraster*, *Prunus cerasifera*, *P. spinosa*) accidentally fall and develop free from competition. Open spaces also favor the establishment and continued persistence of certain bryophytes confined to eroded habitats (*Ceratodon purpureus*, *Polytrichim piliferum* and *Bryum argenteum*).

The role of ancient mounds in steppe preservation is continuously being reiterated in the published literature [3,6,7,17,59]. Our study confirms the prevalence of species characteristic of the *Festuco-Brometea* class associated with the mounds, similar to those reported for Ukraine [6]. This vegetation type includes dry grasslands and steppes from the sub-Mediterranean, nemoral and hemiboreal areas of Europe [30]. *Festuco-Brometea* is widespread in the lowlands and hilly plains of Bulgaria and unites mainly secondary herbaceous communities. Some exceptions occur in the north-eastern parts of the country where small fragments of true steppe communities are present. Given the anthropogenic origin of the mounds, a large number of anthropophytes are logically expected to be associated with these structures. We identified 19.9% of all species as related to anthropogenic vegetation of the classes *Artemisietea vulgaris*, *Papaveretea rhoeadis* and *Sisymbrietea*. Species characteristic of *Artemisietea vulgaris* provide a signal for ruderalization associated with dry habitats. Species diagnostic of *Papaveretea rhoeadis* and *Sisymbrietea* reveal the apparent influence of segetal vegetation in the mounds' immediate vicinity. The diagnostic features of the plant taxa registered on the mounds gives us a reason to characterize most of their vegetation as ruderalized steppes. Although the ancient mounds are considered to be important areas for the protection of steppe flora and vegetation [3], in Bulgaria they also preserve fragments of forest communities [4] characteristic of temperate lowlands. Most of the established trees and shrubs belong to *Quercetea pubescentis* and *Carpino-Fagetea sylvaticae*, two classes widely distributed on the territory of the country. The development of trees, which were visibly old in some places, is a result of the natural succession directed toward the potential vegetation of the Bulgarian lowlands under temperate climate [21]. This successional trend is also indicative of the apparent lack of management practices and activities as associated with the ancient mounds. The diversity of tree species and forest specialists was low compared to the overall floristic diversity. However, the increasing tree coverage leads to a decreased biodiversity under the canopy [60].

4.3. Significance for Nature Conservation

Preserved for 2–3 millennia, ancient mounds retain the natural flora as evidenced by the predominant presence of native species associated with these structures. The prevalence of perennials is a testament to the long-lasting stability of the local vegetation and creates conditions for the establishment of cryptogams. The lack of serious disturbances over prolonged periods, excluding archaeological investigations and damaging activities by treasure hunters, has favored the establishment of perennial species, respective of closed vegetation, which prevents the penetration of many neophytes and especially that of invasive alien plants. The common development of communities in vascular plants and cryptogams is considered an indication of sustainability [61].

Vascular plants of conservation importance were not common on the ancient mounds. The observed endemic species presented only 5.2% of the Balkan endemics reported for Bulgaria [36]. The sampled mounds preserve two critically endangered and nine endangered plants, which represent 1.8% of both categories evaluated at the national level [37].

Our results corroborate the notion that the undisturbed closed vegetation prevents establishment of alien invasive plants [62]. The presence of low neophytes and alien plant numbers has been previously reported for other archaeological sites [63]. Nevertheless, our results hint towards an impending threat given the proportion of mounds affected by the presence of even sole individuals of alien plant species. Although the presence of large agricultural fields isolates the mounds from immediate human influence, we found no evident relationship between the surrounding land use and the number of anthropophytes present on the mounds. Similar results were previously reported by Sudnik-Wójcikowska and Moysiyenko [64].

The existence of relatively well-preserved floristic diversity indicates that the mounds have the potential to provide not only cultural and spiritual but also other valuable ecosystem services (e.g., provisioning of biomass, maintenance of native plant populations and maintenance of physical, chemical and biological conditions of the locality) [65], which, in the face of accelerated fragmentation and land degradation, will become even more important in the future. Therefore, there is a clear need for further research on the topic. The high rate of isolation on the mounds along with their floristic richness, emphasize their remarkable role in the preservation of natural communities, and in providing connectivity when serving as stepping stones for species dispersal from and to other fragments of natural and semi-natural environments, as previously mentioned by others [7].

The Natura 2000 and the network of protected areas are naturally considered as part of the EU's strategy for green and blue infrastructure, but new areas are expected to be added in the future [10]. Within the growing body of literature regarding the concept of green infrastructure, we perceive green infrastructure as green space planning [66]. Situated in anthropogenically transformed lands, the mounds present ideal features for greenspace preservation. Some of Bulgaria's ancient mounds are situated within urban areas, but we did not include them in our sampling. The presence of semi-natural (green) space within arable fields will certainly support an increase in environmental benefits. Ancient mounds are existing structures that only need maintenance and protection. Protection from further treasure hunting is required because such disturbances could facilitate the penetration of alien and woody plants. Reduction of shrubs (especially *Prunus spinosa*) and non-native trees (*Robinia pseudoacacia*, *Ailanthus altissima*) will help to maintain species-rich grasslands. Although species of conservation importance were rarely observed during our study, the mounds should not be neglected as important areas for plant protection, especially in the cases of highly fragmented semi-natural areas where different subpopulations are well separated from each other.

5. Conclusions

Millennia-old ancient mounds are a characteristic feature of the Bulgarian landscape and play an important role for the conservation of indigenous flora. Agricultural practices in the surrounding areas have little effect on the floristic richness of the mounds. In this study, we show that Bulgaria's ancient mounds preserve a considerable proportion of the national vascular and cryptogam flora. Further research will certainly offer new knowledge about the natural significance of the ancient mounds. Current legal preservation of the ancient mounds as archaeological monuments does not guarantee their proper management in terms of nature conservation. With the enlargement of the green infrastructure in the cultural landscape at European level, the significance of ancient mounds will most certainly increase.

Supplementary Materials: The following supporting information can be downloaded at: https://www.mdpi.com/article/10.3390/plants11050705/s1. ESM S1: Original floristic data from 111 ancient mounds from Bulgaria.

Author Contributions: Conceptualization I.A.; methodology I.A. and D.S.; data collection I.A., D.S., M.V., N.V., K.V., A.G., T.T. and V.S.; formal analysis I.A., M.V. and D.S.; original draft preparation I.A., A.G. and V.S.; review and editing M.V., D.S. and G.N. All authors have read and agreed to the published version of the manuscript.

Funding: This research was funded by the Bulgarian National Science Fund (contract КП-06-H21/2, 2018).

Informed Consent Statement: Not applicable.

Data Availability Statement: All data analyzed in this study are available in Supplementary ESM S1.

Acknowledgments: We wish to thank Balazs Deák for his inspiration to begin studying mounds in Bulgaria. We also thank Salza Palpurina for her help during the field work. We would like to acknowledge Kaloyan Ivanov, who kindly polished the English in this paper.

Conflicts of Interest: The authors declare no conflict of interest. The funders had no role in the design of the study; in the collection, analyses, or interpretation of data; in the writing of the manuscript, or in the decision to publish the results.

References

1. Sudnik-Wójcikowska, B.; Moysiyenko, I.I. *Kurhany na "Dzikich Polach"—Dziedzictwo Kultury i Ostoja Ukraińskiego Stepu [Kurgans in the "Wild Field"—A Cultural Heritage and Refugium of the Ukrainian Steppe]*; Wydawnictwa Uniwersytetu Warszawskiego: Warsaw, Poland, 2012.
2. Deák, B. *Nature and Culture: The Role of Ancient Burial Mounds in the Conservation of Eurasian Steppe Vegetation*; Centre for Ecological Research: Tihany, Hungary, 2020; p. 172.
3. Deák, B.; Tóthmérész, B.; Valkó, O.; Sudnik-Wójcikowska, B.; Moysiyenko, I.; Bragina, T.; Apostolova, I.; Dembicz, I.; Bykov, N.; Török, P. Cultural monuments and nature conservation: A review of the role of kurgans in the conservation and restoration of steppe vegetation. *Biodivers. Conserv.* **2016**, *25*, 2473–2490. [CrossRef]
4. Apostolova, I.; Palpurina, S.; Sopotlieva, D.; Terziyska, T.; Velev, N.; Vassilev, K.; Nekhrizov, G.; Tsvetkova, N. Ancient Burial Mounds-Stepping Stones for Semi-Natural Habitats in Agricultural Landscape. *Ecol. Balk.* **2020**, *12*, 43–52.
5. Kitov, G. The Thracian tumuli. *Thracia* **1993**, *10*, 39–80.
6. Sudnik-Wójcikowska, B.; Moysiyenko, I.I. The Flora of Kurgans in the West Pontic Grass Steppe Zone of Southern Ukraine. *Chornomorski Bot. J.* **2006**, *2*, 14–44. Available online: http://eKhSUIR.kspu.edu/handle/123456789/2868 (accessed on 5 February 2022). [CrossRef]
7. Dembicz, I.; Moysiyenko, I.; Shaposhnikova, A.; Vynokurov, D.; Kozub, L.; Sudnik-Wójcikowska, B. Isolation and patch size drive specialist plant species density within steppe islands: A case study of kurgans in southern Ukraine. *Biodivers. Conserv.* **2016**, *25*, 2289–2307. [CrossRef]
8. Sudnik-Wójcikowska, B.; Moysiyenko, I.I. The floristic differentiation of microhabitats within kurgans in the desert steppe zone of southern Ukraine. *Acta Soc. Bot. Pol.* **2008**, *77*, 139–147. [CrossRef]
9. Deák, B.; Kovács, B.; Rádai, Z.; Apostolova, I.; Kelemen, A.; Kiss, R.; Lukács, K.; Palpurina, S.; Sopotlieva, D.; Báthori, F.; et al. Linking environmental heterogeneity and plant diversity: The ecological role of small natural features in homogeneous landscapes. *Sci. Total Environ.* **2021**, *763*, 144199. [CrossRef]
10. European Commission. Green Infrastructure (GI)—Enhancing Europe's Natural Capital, Communication from the Commission to the European Parliament, the Council, the European Economic and Social Committee and the Committee of Regions. 2013. Available online: https://eur-lex.europa.eu/legal-content/EN/TXT/?uri=CELEX:52013DC0249 (accessed on 19 July 2021).
11. Moysiyenko, I.; Zachwatowicz, M.; Sudnik-Wójcikowska, B.; Jabłońska, E. Kurgans help to protect endangered steppe species in the Pontic grass steppe zone, Ukraine. *Wulfenia* **2014**, *21*, 83–94.

12. Sudnik-Wójcikowska, B.; Moysiyenko, I.I.; Zachwatowicz, M.; Jabłońska, E. The value and need for protection of kurgan flora in the anthropogenic landscape of steppe zone in Ukraine. *Plant Biosyst.* **2011**, *145*, 638–653. [CrossRef]
13. Deák, B.; Tölgyesi, C.; Kelemen, A.; Bátori, Z.; Gallé, R.; Bragina, T.M.; Yerkin, A.I.; Valkó, O. The effects of micro-habitats and grazing intensity on the vegetation of burial mounds in the Kazakh steppes. *Plant Ecol. Divers.* **2017**, *10*, 509–520. [CrossRef]
14. Deák, B.; Valkó, O.; Nagy, D.D.; Török, P.; Torma, A.; Lőrinczi, G.; Kelemen, A.; Nagy, A.; Bede, Á.; Mizser, S.; et al. Habitat islands outside nature reserves—Threatened biodiversity hotspots of grassland specialist plant and arthropod species. *Biol. Conserv.* **2020**, *241*, 108254. [CrossRef]
15. Dembicz, I.; Moysiyenko, I.I.; Kozub, Ł.; Dengler, J.; Zakharova, M.; Sudnik-Wójcikowska, B. Steppe islands in a sea of fields: Where island biogeography meets the reality of a severely transformed landscape. *J. Veg. Sci.* **2021**, *32*, e12930. [CrossRef]
16. Duelli, P.; Obrist, M.K. In search of the best correlates for local organismal biodiversity in cultivated areas. *Biodivers. Conserv.* **1998**, *7*, 297–309. [CrossRef]
17. Sudnik-Wójcikowska, B.; Moysiyenko, I.I. Ukrainian kurgans as refugia of steppe flora and their role in steppe restoration. In *Steppenlebensräume Europas—Gefährdung, Erhaltungsmaßnahmen und Schutz*; Baumbach, H., Pfützenreuter, S., Eds.; Thüringer Ministerium für Landwirtschaft, Forsten, Umwelt und Naturschutz: Erfurt, Germany, 2013; pp. 201–210.
18. Deák, B.; Valkó, O.; Török, P.; Kelemen, A.; Bede, Á.; Csathó, A.I.; Tóthmérész, B. Landscape and habitat filters jointly drive richness and abundance of specialist plants in terrestrial habitat islands. *Landsc. Ecol.* **2018**, *33*, 1117–1132. [CrossRef]
19. Czerepko, J.; Gawryś, R.; Szymczyk, R.; Pisarek, W.; Janek, M.; Haidt, A.; Kowalewska, A.; Piegdoń, A.; Stebel, A.; Kukwa, M.; et al. How sensitive are epiphytic and epixylic cryptogams as indicators of forest naturalness? Testing bryophyte and lichen predictive power in stands under different management regimes in the Białowieża forest. *Ecol. Indic.* **2021**, *125*, 107532. [CrossRef]
20. Velev, S. Climatic regioning. In *Geography of Bulgaria*; Kopralev, I., Ed.; Publishing House ForCom: Sofia, Bulgaria, 2002; pp. 155–156.
21. Bohn, U.; Golub, G.; Hettwer, C. (Eds.) *Karte der Naturlichen Vegetation Europas. Legende/Map of the Natural Vegetation of Europe, Part 2: Legend (German/English)*; Bundesamt für Naturschutz: Bonn, Germany, 2003.
22. Meshinev, T. Vegetation and phytogeography: A brief characteristic. In *Biogeography and Ecology of Bulgaria*; Monographie Biologicae; Fet, V., Popov, A., Eds.; Springer: Dordrecht, The Netherlands, 2007; Volume 82, pp. 581–588.
23. Tzonev, R.; Roussakova, V.; Dimitrov, M. The Western-Pontic steppe vegetation in Bulgaria. *Hacquetia* **2006**, *5*, 5–23.
24. Euro+Med. Euro+Med—The Information Resource for Euro-Mediterranean Plant Diversity. 2006–2021. Available online: http://ww2.bgbm.org/EuroPlusMed/ (accessed on 20 September 2021).
25. The Plant List. Version 1.1. 2013. Available online: http://www.theplantlist.org/ (accessed on 1 September 2021).
26. Hill, M.O.; Bell, N.; Bruggeman-Nannenga, M.A.; Brugués, M.; Cano, M.J.; Enroth, J.; Flatberg, K.I.; Frahm, J.-P.; Gallego, M.T.; Garilleti, R.; et al. An annotated checklist of the mosses of Europe and Macaronesia. *J. Bryol.* **2006**, *28*, 198–267. [CrossRef]
27. Nimis, P.L.; Hafellner, J.; Roux, C.; Clerc, P.; Mayrhofer, H.; Martellos, S.; Bilovitz, P.O. The lichens of the Alps – an annotated checklist. *MycoKeys* **2018**, *31*, 1–634. [CrossRef] [PubMed]
28. *Flora of Republic of Bulgaria*; Academic Publishing "Prof. Marin Drinov": Sofia, Bulgaria, 2012.
29. Delipavlov, D.; Cheshmedzhiev, I.; Popova, M.; Terziyski, D.; Kovachev, I. *Guide to Vascular Plants in Bulgaria*; Academic Publishing of the Agricultural University: Plovdiv, Bulgaria, 2003.
30. Mucina, L.; Bültmann, H.; Dierßen, K.; Theurillat, J.P.; Raus, T.; Čarni, A.; Šumberová, K.; Willner, W.; Dengler, J.; García, R.G.; et al. Vegetation of Europe: Hierarchical floristic classification system of vascular plant, bryophyte, lichen and algal communities. *Appl. Veg. Sci.* **2016**, *19*, 3–264. [CrossRef]
31. Assyov, B.; Petrova, A. (Eds.) *Conspectus of the Bulgarian Vascular Flora. Distribution Maps and Floristic Elements*; Bulgarian Biodiversity Foundantion: Sofia, Bulgaria, 2012.
32. Ganeva, A.; Düll, R. A contribution to the Bulgarian bryoflora. Checklist of Bulgarian bryophytes. In *Contributions to the Bryoflora of Former Yugoslavia and Bulgaria*; Düll, R., Ganeva, A., Martincic, A., Pavletic, Z., Eds.; IDH: Bad Münstereifel, Germany, 1999; pp. 111–199.
33. Wirth, V. *Die Flechten Baden-Württemmbergs*; E. Ulmer GMBH Co: Stuttgart, Germany, 1995.
34. Nimis, P.L. *ITALIC—The Information System on Italian Lichens*; Version 6.0; University of Trieste, Department of Biology: Berlin, Germany, 2016; Available online: http://dryades.units.it/italic (accessed on 1 August 2021).
35. Petrova, A.; Vladimirov, V.; Georgiev, V. *Invasive Alien Plant Species in Bulgaria*; Institute of Biodiversity and Ecosystem Research, BAS: Sofia, Bulgaria, 2012.
36. Petrova, A.; Vladimirov, V. Balkan endemics in the Bulgarian flora. *Phytol. Balc.* **2010**, *16*, 293–311.
37. Petrova, A.; Vladimirov, V. Red list of Bulgarian vascular plants. *Phytol. Balc.* **2009**, *15*, 63–94.
38. Biological Diversity Act. State Gazette, Sofia, Issue 77, August 2002, Last Change and Addition Issue 58, July 2016. Available online: http://eea.government.bg/bg/legislation/biodiversity/ZBR_en_26_07_2016.pdf (accessed on 27 October 2021).
39. Council Directive 92/43/EEC of 21 May 1992 on the Conservation of Natural Habitats and of Wild Fauna and Flora. OJ L 206. 22 July 1992, pp. 7–50. Available online: https://eur-lex.europa.eu/legal-content/EN/TXT/?uri=CELEX:31992L0043 (accessed on 22 February 2021).
40. Convention on International Trade in Endangered Species of Wild Fauna and Flora (CITES), 1973. Signed at Washington, D.C.; on 3 March 1973, Amended at Bonn, on 22 June 1979, Amended at Gaborone, on 30 April 1983. Available online: https://checklist.cites.org/#/en (accessed on 1 September 2021).

41. Clarke, K.R. Non-parametric multivariate analyses of changes in community structure. *Austral. Ecol.* **1993**, *18*, 117–143. [CrossRef]
42. PRIMER-E Ltd. PRIMER (Plymouth Routines In Multivariate Ecological Research), Version 7. 2017. Available online: http://www.primer-e.com (accessed on 19 December 2021).
43. Hammer, Ø.; Harper, D.; Ryan, P. PAST: Paleontological Statistics Software Package for Education and Data Analysis. *Palaeontol. Electron.* **2001**, *4*. Available online: https://www.nhm.uio.no/english/research/infrastructure/past/ (accessed on 18 January 2022).
44. Dell Inc. *Dell Statistica (Data Analysis Software System), Version 13*. 2016. Available online: http://www.dell.com (accessed on 18 January 2022).
45. Peev, D.; Delcheva, M. Florogeographical elements and evolutionary trends in the Bulgarian Flora. In *Biogeography and Ecology of Bulgaria*; Monographie Biologicae; Fet, V., Popov, A., Eds.; Springer: Dordrecht, The Netherlands, 2007; Volume 82, pp. 571–579.
46. Arnaudov, N. *The Moss Flora of Vitosha Mt*; Ann. de l'Universite de Sofia: Sofia, Bulgaria, 1908; Volume 3–4, pp. 1–37.
47. Petrova, A.; Vladimirov, V. Recent progress in floristic and taxonomic studies in Bulgaria. *Bot. Serb.* **2018**, *42*, 35–69. [CrossRef]
48. Bartha, D.; Király, G.; Schmidt, D.; Tiborcz, V.; Barinza, Z.; Csiky, J.; Jakab, G.; Lesku, B.; Schmotzer, A.; Vidéki, R.; et al. (Eds.) *Magyarország Edényes Növényfajainak Elterjedési Atlasza [Distribution Atlas of Vascular Plants of Hungary*; Nyugat-magyarországi Egyetem Kiadó/University of West Hungary: Sopron, Hungary, 2015.
49. Mosyakin, S.L.; Fedoronchuk, N.M. *Vascular Plants of Ukraine: A Nomenclatural Checklist*; National Academy of Sciences of Ukraine-MG Kholodny Institute of Botany: Kyiv, Ukraine, 1999.
50. Sudnik-Wójcikowska, B.; Moysiyenko, I. Zonal character of the flora of kurgans in central and southern Ukraine. *Biodiv. Res. Conserv.* **2010**, *17*, 47–52. [CrossRef]
51. Barczi, A. Data for the botanical and pedological surveys of the Hungarian kurgans (Great Hungarian Plain, Hortobágy). *Thaiszia J. Bot.* **2003**, *13*, 113–126.
52. Cwener, A. Ros´liny naczyniowe kurhano´w w dorzeczu dolnej Szreniawy i Nidzicy (Wy_zyna Małopolska, południowa Polska) [The vascular plants of the mounds in the lower Szreniawa and Nidzica river-basin (Małopolska Upland, South Poland)]. *Fragm. Florist. Geobot. Pol.* **2004**, *11*, 27–40.
53. Deák, B.; Valkó, O.; Török, P.; Tóthmérész, B. Factors threatening grassland specialist plants—A multi-proxy study on the vegetation of isolated grasslands. *Biol. Conserv.* **2016**, *204*, 255–262. [CrossRef]
54. Erdős, L.; Méri, A.; Bátori, Z.; Gallé, R.; Kőrmőczi, L. North–south facing vegetation gradients in the Villány Mountains: A case study on the population and the community level. *Pak. J. Bot.* **2012**, *44*, 927–932.
55. Batič, F.; Primožič, K.; Surina, B.; Trošt, T.; Mayrhofer, H. Contributions to the lichen flora of Slovenia X. Lichens from the Slovenian Julian Alps. *Herzogia* **2003**, *16*, 143–154.
56. Coppins, B.J.; Orange, A. Arthopyrenia A. Massal. In *The lichens of Great Britain and Ireland*, 2nd ed.; Smith, C.W., Aptroot, A., Coppins, B.J., Fletcher, A., Gilbert, O.L., James, P.W., Wolseley, P.A., Eds.; The British Lichen Society: London, UK, 2009; pp. 136–138.
57. Richardson, D.H.S. *Pollution Monitoring with Lichens. Naturalists' Handbooks 19*; The Richmond Publishing: Slough, UK, 1992.
58. van Haluwyn, C.; van Herk, C.M. Bioindication: The community approach. In *Monitoring with Lichens—Monitoring Lichens*; Nimis, P.L., Scheidegger, C., Wolseley, P., Eds.; Kluwer Academic: Dordrecht, The Netherlands, 2002; pp. 39–64.
59. Valkó, O.; Zmihorski, M.; Biurrun, I.; Loos, J.; Labadessa, R.; Venn, S. Ecology and conservation of steppes and semi-natural grasslands. *Hacquetia* **2016**, *15*, 5–14. [CrossRef]
60. Teleki, B.; Sonkoly, J.; Erdős, L.; Tóthmérész, B.; Prommer, M.; Török, P. High resistance of plant biodiversity to moderate native woody encroachment in loess steppe grassland fragments. *Appl. Veg. Sci.* **2020**, *23*, 175–184. [CrossRef]
61. Boch, S.; Müller, J.; Prati, D.; Fischer, M. Low-intensity management promotes bryophyte diversity in grasslands. *Tuexenia* **2018**, *38*, 311–328. [CrossRef]
62. Chytrý, M.; Pyšek, P.; Tichý, L.; Knollová, I.; Danihelka, J. Invasions by alien plants in the Czech Republic: A quantitative assessment across habitats. *Preslia* **2005**, *77*, 339–354.
63. Celesti-Grapow, L.; Blasi, C. The Role of Alien and Native Weeds in the Deterioration of Archaeological Remains in Italy. *Weed Technol.* **2004**, *18*, 1508–1513.
64. Sudnik-Wójcikowska, B.; Moysiyenko, I. The synanthropic flora of kurgans within three steppe zones in southern Ukraine. *Biodiv. Res. Conserv.* **2008**, *11–12*, 41–48.
65. CICES—The Common International Classification of Ecosystem Services. Available online: https://cices.eu (accessed on 18 January 2022).
66. Matsler, M.; Meerow, S.; Mell, I.; Pavao-Zuckerman, M. A 'green' chameleon: Exploring the many disciplinary definitions, goals, and forms of "green infrastructure". *Landsc. Urban Plan.* **2021**, *214*, 104145. [CrossRef]

plants

MDPI

Article

Biodiversity in Urban Areas: The Extraordinary Case of Appia Antica Regional Park (Rome, Italy)

Duilio Iamonico

Ce.R.S.I.Te.S., University of Rome Sapienza, Viale XXIV Maggio 7, 04100 Latina, Italy; d.iamonico@yahoo.it

Abstract: The first inventory of the flora of Appia Antica Regional Park (Italy), one of the largest protected urban areas in Europe (4580 ha), its biological, ecological and biogeographical composition, and notes of the vegetation physiognomies and landscape are presented; physical characteristics of the territory (geomorphology, lithotypes, and phytoclimate) are also given. The landscape is defined by an agricultural matrix with natural and seminatural areas as patches, and riparian vegetation communities as corridors. The vegetation physiognomies are represented by types linked to the Mediterranean climate (mixed, Mediterranean, and riparian forests; scrubby, rocky, aquatic, and helophytic vegetation; anthropogenic communities). The floristic list includes 714 taxa (104 families and 403 genera). Therophytes prevail over hemicryptophytes; woody flora comprises about 30% of alien species. As regards chorotypes, together with a considerable number of Mediterranean species, there are many exotic species with wide distribution areas testifying to a long-lasting anthropic impact. Floristic novelties (european, national, and regional levels) for 21 taxa are reported. The extraordinary species diversity discovered (43% of flora of Rome and 20% of regional flora) is linked to the landscape heterogeneity, the characteristics of which are: (1) persistence of residual natural patches, (2) occurrence of quite well-preserved aquatic habitats and humid meadows, (3) a rich anthropogenic flora, (4) an interesting flora of archeological sites, (5) occurrence of species not common in Latium, (6) occurrence of populations of aliens in crops (which cause economic impact), (7) presence of aliens on archeological ruins (which cause economic-social impacts). The extensive set of data provided represents a general base framework for guiding future research efforts and landscape action plans consistent with environmental sustainability.

Keywords: alien status; biological records; Europe; Latium; Mediterranean flora; Rome; urban biodiversity

Citation: Iamonico, D. Biodiversity in Urban Areas: The Extraordinary Case of Appia Antica Regional Park (Rome, Italy). *Plants* **2022**, *11*, 2122. https://doi.org/10.3390/plants11162122

Academic Editors: Robert Philipp Wagensommer and Daniel Sánchez-Mata

Received: 11 July 2022
Accepted: 11 August 2022
Published: 15 August 2022

1. Introduction

Biodiversity loss is a phenomenon mainly related to the intensification of land use and management, as well as the conversion of natural lands to agricultural, forestry, and building areas [1]. This phenomenon is particularly intense in urban areas where several processes, e.g., loss of soil, biological invasions or pollution, are highly exacerbated due to human activities and have a significant impact on the native flora and residual fragments of natural vegetation [2–4]. In addition, as widely known, urbanization has been accelerating for several decades at an alarming rate around the world, and most of the world's population is now concentrated in urban areas [5]. Urbanization is one of the major causes of plant diversity loss at the local and regional scales [6].

Urban areas are "hotspots" of biological invasions, being not only key points of entry for many non-native species, but also playing an important role in the secondary spread of aliens towards surrounding territories, i.e., rural or natural landscapes [3,7]. However, at the same time, urban parks and natural areas in cities are important as detailed-scale biodiversity hotspots [8]. So, maintenance of the biodiversity in urban areas represents a very important conservation issue [9,10]. With this aim in mind, the first fundamental step for sustainable actions is to improve the floristic knowledge of these areas by preparing inventories of taxa [9,11]. Floristic catalogues provide useful data for subsequent studies, e.g., distributional patterns [12], or for future actions aimed, for instance, at managing non-native taxa [13].

The flora of Mediterranean cities still remains poorly known [9]. Concerning Italy, most urban floras refer to cities of northern Italy, e.g., Milan in Lombardy region [14] or Trieste in Friuli-Venezia Giulia region [15]. Concerning central and southern Italy, few contributions have been published, e.g., for Naples in Campania region [16] or Rome in Lazio region [9].

Rome, and its administrative territory, is one of the largest and most populated urban areas in southern Europe, and it has been investigated since the 19th century from a botanical point of view. The first published floristic works of the Roman area were by Sebastiani in 1813 and 1815 [17,18] and Sebastiani and Mauri in 1818 [19], whereas the more recent list of spontaneous plants growing in Rome was published about 10 years ago [9]. The territory of Rome Municipality covers 1287 km^2, and it is characterized by very high landscape heterogeneity dependent primarily on its location in the center of the Mediterranean basin. This location favors, in turn, the influence by various environmental factors, e.g., biogeographic effects (from western and eastern Europe), proximity of the Tyrrhenian Sea (on the west) and Apennine Mountains (on the east), different types of phytoclimates, 3000 years of human impact, etc. [9,20]. These environmental features contribute to a high floristic richness (1649 taxa according to [9]) as well as the occurrence of many types of natural vegetation communities and potential types, the dominant ones being the sub-Mediterranean deciduous oak woods with *Quercus cerris* L., *Q. frainetto* Ten., *Q. robur* L., and *Q. pubescens* Willd. Despite the quite extensive knowledge of flora of the Roman area, detailed studies on specific areas (e.g., those protected by regional or national laws) are still partially lacking. Some of these areas have been studied in recent years, e.g., the Monte Mario Natural Reserve in the northwestern part of the Rome Municipality [21], the Laurentino-Acqua Acetosa Natural Reserve in the southwestern part [22], or the Augazzano Urban Park [23] and Nomentum Natural Reserve [24] in the eastern part.

Appia Antica Regional Park, which is the topic of the present work, was not investigated in detail from a floristic point of view, despite being one of the larger protected areas of the Rome Municipality and, as an urban park, in the whole of Europe (with about 4580 ha). There is only a floristic catalogue on the Caffarella Valley that occupies only 190 ha (about 4% of the total area), but it was published 22 years ago [25].

As a part of ongoing studies on the flora of urban areas, with special attention to the Rome territory [21–23,26], I here present the flora of Appia Antica Regional Park, also giving notes on abiotic factors (climate, geology, geomorphology, etc.) as well as landscape structure and main vegetation physiognomies.

2. Materials and Methods

This work was based mainly on field surveys that were carried out during the period 2010–2022. Collected material was deposited mainly at the Herbarium RO (secondly at FI and HFLA). Further specimens, kept at A, AC, BM, BR, CAS, COL, F, FI, G, GH, GOET, HFLA, JE, K, MICH, MO, NA, P, RO, RSA, SI, UC, US, URT, and YU, were examined (acronyms of the herbaria follow *Index Herbariorum* [27]). Relevant botanical literature was also analyzed (citations occur throughout the text).

The taxa (species, subspecies, and variety ranks) were identified using the new edition of the *Flora of Italy* [28]; recent monographs were also considered for critical genera (e.g., *Amaranthus* [29] or *Vitis* [30]). The nomenclature follows the Italian Checklists of both native and non-native flora and the subsequent updates [31–33]. Concerning the genera *Amaranthus* and *Cyanus*, I considered, respectively, my recent monograph [29] and the new edition of the *Flora of Italy* [28]. Biological forms and chorotypes of each taxon follow the new edition of the *Flora of Italy* [28]; grouped chorotypes follow [26], except for the category "Aliens", which includes, in the present paper, all the non-native taxa. The alien status and definitions of categories follow the recent Italian Checklist of non-native flora [31] and Pyšek and collaborators [34].

In the floristic list (Appendix A), the systematic order of the families follows Italian Checklists of both native and non-native flora [31,32]. Within each family, the taxa are ordered alphabetically. For each taxon, after the accepted scientific name, the following information is reported: endemic, cryptogenic, alien status.

Geological and geomorphological information and the pertinent nomenclature refer mainly to [35,36] and [37,38], respectively. Further, field observations were conducted.

Climatic and phytoclimatic data derive from the climate classification by Zepner and collaborators [39] and the map of the Roman area by Blasi and Michetti [40].

Landscape remarks are based on [41,42], but also on direct observations in the field.

Vegetation physiognomies were observed directly in the field, lacking a detailed phytosociological study of the park area (see also [41]). Nomenclature of the main syntaxonomic ranks follows the recent classification system for European vegetation [43].

Photographs are original and produced by the author of the present manuscript, except for the bottom picture of Figure 8 ("Cava di Fioranello"), which was produced using a drone and kindly provided by Lucio Virzì (Rome).

3. Study Area

3.1. Geographical Context

The Appia Antica Regional Park is located mostly in the southeastern part of the urban area of Rome Municipality (less than 5% of the territory is included in the Municipalities of Ciampino and Marino, toward the south) and was established in 1988 by the Regional Law No. 66. Later, in 2002, by the Regional Law 31 May 2002, an extension of the original area was proposed and approved. Currently, the park covers about 4580 ha; altitude ranges from about 15 m in the north sector of the park to 189 m a.s.l. in the south (Frattocchie locality, Marino Municipality).

The general shape of the park is a long wedge defined by one main axis, the Appia Antica street, which runs for about 16 km from northeast (Numa Pompilio square, in the vicinity of the archeological area of *Caracalla* thermal baths, in the historic center of Rome) to southwest (Ciampino and Marino Municipalities). Coordinates of the park are: 41°50′00″ latitude N, 12°33′00″ longitude E. The main boundaries of the park are (see Figure 1):

- on the west: Terme di Caracalla street up to the Scott Park in *Ardeatino* district (toward the south); Ardeatina street up to the hamlet *Falcognana* (part of the Municipality of Rome located about 4 km from the Great Ring Junction toward the south) with the exclusion of an area around Fioranello street. In addition, the area named *Tor Marancia* (about 220 ha) occurs to the west of Ardeatina street (*Ardeatino* district, north of the park);
- on the south: Falcognana street between the hamlets *Falcognana* (to the west) and *Poggio Le Mole* (Municipality of Marino, to the east), with the exclusion of the hamlet *Santa Maria delle Mole*;
- on the east: Appia Nuova street from the hamlet *Frattocchie* (Municipality of Marino) to the Roman urban area named *Arco di Travertino*; Latina street (with the exclusion of part of the district *Appio Latino* between Antonio Coppi street and the railway) up to Numa Pompilio square. In addition, two archeological areas occur to the east of Appia Nuova street, named *Acquedotti* (about 240 ha) and *Latin tombs* (about 2 ha);
- on the north: Numa Pompilio square.

Figure 1. Map of Appia Antica Regional Park (boundary in red line) with main localities and streets; yellow line: Appia Antica street. Scale bar = 2 km.

3.2. *Geological Features*

The area of Appia Antica Regional Park had a geological history strictly linked with the activity of the Colli Albani volcano (currently known as *Castelli Romani*), which started about 600,000 years ago [26]. During the first phase (0.60–0.36 Ma), violent eruptions occurred, causing the formation of a large volcanic structure. This later collapsed, giving rise to the Tuscolano-Artemisian caldera. A second volcanic phase (0.27–0.10 Ma) was marked by the rise of a second volcanic structure, internal to the previous one. Finally, there was a third phase (0.10–0.01 Ma), which was characterized by intense hydromagmatic activity, creating the lakes Albano and Nemi plus several minor basins, drained in the past by man for agricultural purposes (Vallericcia, Laghetto, Valle Marciana, Prata Porci,

Pantano Secco). The territory of Appia Antica Regional Park, which is located from the north-western slope of *Castelli Romani* to the historic center of Rome (toward the north), is covered by lithotypes related to volcanic rocks with alkaline-potassic chemism [26,35]. Three main types occur: lavas, tuffs, and unconsolidated pyroclastic deposits (named also as *Pozzolane*). The former is related to the volcanic activity of the above-mentioned second phase, when just one erupted material flowed out of the Tuscolano-Artemisian caldera toward the current urban area of Rome. This lava was named *Colata di Capo di Bove* (literally "Lava flow of Bove Capo") and it appears in various parts of the park (Figure 2). The remaining area is covered by various types of tuffs and unconsolidated pyroclastic deposits (Figure 3). In addition to lavas, tuffs, and *Pozzolane*, lahar deposits occur in the *Acquedotti* area (east of the park) being part of the Ciampino Plain, the origin of which represents the most recent activity of Colli Albani volcano (25 Ka; [36]).

Figure 2. Lava *Colata di Capo di Bove* (south of Caffarella valley, north of the park).

Figure 3. Unconsolidated pyroclastic deposits (*Pozzolane*) in the northwestern sector of Caffarella valley (north of the park).

Holocene deposits are represented by alluvials and colluvials of sands, silty sand, or silty clayey sand related to the fluvial processes [35].

3.3. Geomorphological Features

As a whole, the natural landscape in which the city of Rome has developed was mainly molded by fluvial processes. In addition, polygenetic, structural, and gravitational forms also occurred and are widespread; finally, there are many landforms deriving from the millennian human activities [38].

The study area includes all four main types of landforms that can be found in the whole Roman area [38], i.e.,:

➢ Fluvial landforms: the study area is characterized by a more or less structured hydrographic network composed mainly of channels, mostly with steady water flows, natural ponds, and freshwater springs. The most complex networks can be found in the *Caffarella* (north of the park) and *Acquedotti* (east) areas.

The *Caffarella* area is a typical V-shaped valley molded by the river Almone and its tributaries (channels, locally named *Marrane*). Almone is a left-side tributary of the river Tevere originated in the Colli Albani volcano and joining the river Tevere in the *Ostiense* district (southwestern Rome); the total length of Almone is about 21 km, and its drainage basin is about 51 km^2 [44]. In particular, the river Almone, which has a mean river flow of 1.7 m^2/s, forms on some parts of its alluvial plain a series of meanders with deposition of material on the inside of each bend and erosion of the outside bank of the bend (Figure 4).

The *Acquedotti* area is flat with a main channel (named *Acqua Mariana*, that originates in Molara Valley in the Castelli Romani Regional Park [45] about 20 km toward the south) and secondary channels (Figure 4). *Acqua Mariana* is an artificial channel created during the 12th century by redirecting a tributary of the natural channel *Acqua Crabra* [36,44]. The hydrographic network was affected by erosion of lahar deposits distributed in the Ciampino plain (<23 Ma) [46].

Figure 4. Meander of river Almone (yellow line: flow direction): deposition of material on the inside bank of the bend (white arrow) and erosion of the outside bank (red arrow) (left side photo); secondary channel in Caffarella valley (right side photo).

- Structural landforms: these forms are the results of the fluvial erosion cut of the flat ignimbritic plateau generated after the eruption of the Colli Albani volcano (middle Pleistocene); surfaces are often bordered by cliffs affected by various weathering and denudation processes [37]. Some structural landforms can be observed in the Caffarella valley, where the volcanic plateau was cut by the Almone River (Figure 5).

Figure 5. Flat surface on east side of the Caffarella valley. Flood plain is at about 20 m a.s.l., top of the structural landform (red line) is at about 43 m a.s.l. Yellow line: base of the slope (45–80°).

- Gravitational landforms: two main types were observed, i.e., falls, which occur especially in the Caffarella valley and Tor Marancia locality where there are some slopes with a high gradient (more than 80%, even vertical) (Figure 6) (no data occur in the map of the Italian landslide for the study area [47]) and sinkholes originating from the collapse of underground cavities (see below under "Man-made landforms") and the subsequent subsidence of the shallower layers of the soil [48]. The latter gravitational landforms have dimensions ranging from 1–6 m of depth and 1–12 m of diameter (pers. obs.) (Figure 6).

Figure 6. Gravitational landforms: fall on the east side of Caffarella valley (top photo); sinkhole, diameter about 12 m, depth about 5 m (bottom photo).

- Man-made landforms: the study area has been affected by several human activities mostly beginning thousand years ago. The marks of these activities are often juxtaposed with those related to the natural processes [37].

The most widespread human-made landforms are represented by a dense network of underground cavities, which are especially concentrated in the Caffarella area (north of the park). These cavities (Figure 7) were firstly used to extract material for construction of buildings and catacombs, i.e., lithic tuffs and unconsolidated pyroclastic deposits (*Pozzolane*), and to distribute and collect water [48]. The more recent use of these cavities (up to the 1990s) was as mushroom patches (genera *Pleurotus* (Fr.) P. Kumm. and *Agaricus* L. (*A. bisporus* (J.E.Lange) Imbach, named "champignon")).

Figure 7. Entrance of an underground cavity (Caffarella valley).

Further landforms occurring in Appia Antica Park are the mines (currently inactive). They started during the 6th–5th centuries BC [49] and are characterized by both straight scarps and step-like slopes. In some cases, the mines are no longer readily visible, being covered by the vegetation and/or subjected to naturalistic engineering operations ([50]; Figure 8). A famous mine (named *Cava di Fioranello*; Figure 8) occurs near the Ciampino airport (south of the park) and was used to extract basalt to prepare a type of cobblestone resembling truncated and square-based pyramids (locally named *sanpietrini*) that represents the traditional pavement of many streets and squares in the city of Rome [51]; Fioranello's mine is currently used for climbing sport activity.

Figure 8. Mines: abandoned mine of Pozzolana located near the Latina street (northeast of the park); its slope (yellow lines) is covered by natural vegetation (top photo); mine *Cava di Fioranello* of basalt, near Ciampino airport, south of the park (bottom photo).

In addition to underground cavities and mines, there are further man-made elements that deserve to be mentioned since they locally changed the landscape from both environmental and social points of view. These man-made landforms are the artificial lakes occurring in Caffarella valley and Acquedotti locality (Figure 9):

- the first lake (locally named *Laghetto della Caffarella* = Caffarella's small lake) was created in 2004 for the natural regeneration of aquatic habitats. Caffarella's lake has one tributary and one emissary (linked with the nearby channel), covers an area of about 2000 m^2 (including the surrounding humid meadows), and has a maximum depth of 1.5 m;
- the second lake (locally named *Laghetto degli Acquedotti* = Acquedotti's small lake) was redeveloped about 10 years ago under planning for both the body of water and the adjacent channel *Acqua Mariana*, which serves as both tributary and emissary. The Aquedotti's lake covers an area of about 800 m^2 and has a maximum depth of 0.5 m.

Figure 9. Artificial lakes occurring in Caffarella valley (top photo) and Acquedotti locality (bottom photo).

3.4. Climate and Phytoclimate

According to the online ClimateCharts.net (accessed on 11 July 2022) database, which is based on the Köppen-Geiger's climate classification [39], the study area would be included in the Temperate type ("C"), which is defined by a temperature of the hottest month of $\geq$10 °C and temperature of the coldest month ranging between 0 and 18 °C. In particular, the subtype is "Temperate without dry season" (code "Cfa"), defined by precipitation in the driest month in summer of >40 mm rain and temperature in the hottest month of $\geq$22 °C.

From the phytoclimatic point of view, the whole Roman area belongs to the Mediterranean region and the Meso Mediterranean type [40]. Appia Antica Regional Park is included in the following two subtypes:

- Mesomediterranean Subhumid-Thermomediterranean Subhumid: this subtype (area outside of the Great Ring Junction) is characterized by a mean annual precipitation of 680–820 mm, mean summer precipitation of 82.23–96.34 mm, mean annual temperature of 14.60–15.21 °C, maximum annual temperature of 19.95–21.39 °C, minimum annual temperature of 9.27–10.41 °C. The potential natural vegetation concerns four vegetation series: Turkey oak Series [*Teucrio siculi-Quercion cerris* Ubaldi 1988 (*Crataego levigatae-Quercion cerridis* Arrigoni 1997 *sensu* [43])], Downy and Turkey oaks Series [*Ostryo-Carpinion orientalis* Horvat 1959 (*Fraxino orni-Ostryon* Tomažič 1940 *sensu* [43]) and *Lonicero etruscae-Quercion pubescentis* Arrigoni et Foggi ex Foggi et al. 1990 (*Crataego laevigatae-Quercio cerridis* Arrigoni 1997 *sensu* [43])], Holm oak Series (*Quercion ilicis* Br.-Bl. ex Molinier 1934), and Hornbeam Series [*Doronico-Fagion* Ubaldi et al. 1990 (*Geranio striati-Fagion* Gentile 1970 *sensu* [43])]. As a meteorological reference point, the thermo-pluviometric station is that named *Ciampino* (129 m a.s.l.); the thermo-pluviometric diagram (Figure 10) shows mean temperatures of the coldest (January) and the hottest (August) months of, respectively, 7.3 and 24.9 °C; mean annual rainfall is 792.8 mm [maximum monthly value in November (112.6 mm), minimum in July (22.1 mm)]; aridity is about 3 months;
- Mesomediterranean Subhumid-Thermomediterranean Dry: this subtype (interior area to the Great Ring Junction) is characterized by a mean annual precipitation of 650–820 mm, mean summer precipitation of 56.9–76.6 mm, mean annual temperature of 14.60–15.21 °C, maximum annual temperature of 18.88–21.16 °C, minimum annual temperature of 9.27–10.41 °C. Potential natural vegetation regards three vegetation series: Turkey oak Series [*Teucrio siculi-Quercion cerris* (*Crataego laevigatae-Quercio cerridis sensu* [43])], Downy and Turkey oak Series [*Ostryio-Carpinion orientalis* (*Fraxino orni-Ostryon sensu* [43]) and *Lonicero-Quercion pubescentis* (*Crataego laevigatae-Quercio cerridis sensu* [43])], and Holm oak Series (*Quercion ilicis*). The meteorological reference point is the thermo-pluviometric station of *Monte Mario* (143 m a.s.l.); the thermo-pluviometric diagram (Figure 10) shows mean temperatures of the coldest (January) and the hottest (July) months of, respectively, 7.4 and 24.0 °C; mean annual rainfall is 766.0 mm [maximum monthly value in November (113.0 mm), minimum value in July (17.0 mm)]; aridity is about 2 months.

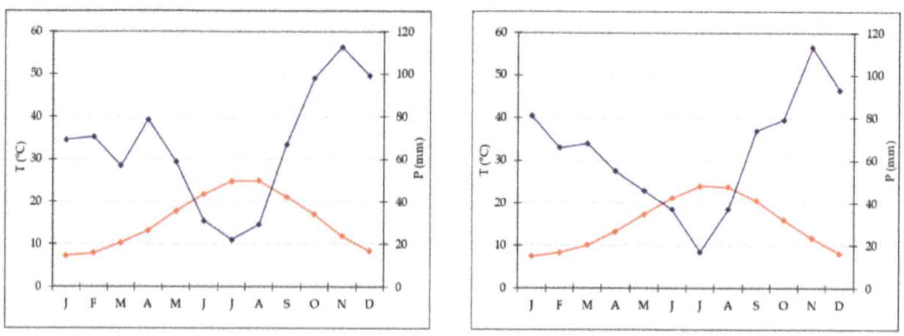

Figure 10. Thermo-Pluviometric Bagnolous-Gaussen diagrams of the stations Ciampino (left side diagram) and Monte Mario (right side diagram). Blue lines refer to rainfalls; red lines refer to temperatures. Axis x refers to months (e.g., "J" = January).

3.5. *Landscape Remarks and Actual Vegetation Physiognomies*

The landscape of Appia Antica Regional Park is defined by an agricultural matrix (primarily wheat fields; Figure 11) that covers more than 50% of the total area. Natural patches are mostly *remnant*-type according to Forman and Godron [52], being caused by widespread disturbance from the matrix. These patches are represented in the study area by shrubs or forest residual areas (Figure 12) that resulted after both agricultural activities and grazing (sheep and cows) (the so-called *Anthropic determinism sensu* [53]). Other patches related to human activities, but not residual, are the mesophilous meadows used as pastures (Figure 12) and/or for recreational activities. In further cases, patches exist since the natural environmental conditions did not allow easy and favorable land uses (the so-called *Natural determinism sensu* [53]). Examples are the humid meadow areas occurring in the Almone valley (north of the park) which remain due to the near-surface aquifer (Figure 12). Concerning the landscape corridors, they are represented mainly by channels (locally named *Marrane*) with banks covered by both herbaceous or shrubby vegetation only and riparian forests (Figure 13).

In addition to the natural and seminatural patches, the landscape of the park is characterized by many historical and archeological elements (Figures 14 and 15). The former are represented mainly by farmhouses (e.g., *Vaccareccia* (16th century), *Ex Mulino* and *Vigna Cardinali* (19th century)). The archeological elements consist of many types of monuments, e.g., aqueducts (e.g., *Aqua Claudia* (38–52 a.C.)), *Felix Aqueduct* (around half of the 1st century a.C.), *Antoniano Aqueduct* (around half of 2nd century a.C.)), catacombs (e.g., *San Callisto, San Domitilla, San Sebastiano* (half of 2nd century a.C.)], churchs [*Domine Quo Vadis* (medieval), *Sant'Urbano* (6th century a.C.)), nymphaeum (*Egeria nymphaeum*, 2nd century a.C.), tombs and sepulchres (e.g., *Latin tombs* (1st–2nd century a.C.), *Cecilia Metella sepulchre* (30–10 a.C.)), palaces (e.g., *Circus* and *Massenzio's imperial palace*, 2nd–3rd century a.C.), temples (e.g., *temple of God Redicolo*, second half of the 2nd century a.C.), walls (e.g., *Aureliane's wall* with *San Sebastiano door* (270–275 a.C.)), ways (e.g., *Appia Antica* (end of 4th century a.C.), *Latina* (end of 4th century a.C. to beginning of 3rd century a.C.)); medieval towers or fortifications (e.g., Tor Fiscale and Valca towers) also occur.

As a whole, this landscape composition reveals a high landscape fragmentation that occurred during the past, which caused a decrease in the environmental quality of the territory. In addition, the ecologic connectivity appears to be low, especially in the southern part of the park, where few corridors occur and the matrix occupies a higher percentage of the total area than in the northern part.

Figure 11. Landscape elements of Appia Antica Regional Park: matrix. Wheat field in locality Acquedotti.

Figure 12. Landscape elements of Appia Antica Regional Park: patches. Shrubs adjacent to crops (top photo); pasture (central photo); humid meadow (bottom photo).

Figure 13. Landscape elements of Appia Antica Regional Park: corridors. Riparian forest in Caffarella valley.

Figure 14. Historical farmhouses of Appia Antica Regional Park landscape: *Vaccareccia* (top photo); *Vigna Cardinali* (bottom photo).

Figure 15. Archeological elements of Appia Antica Regional Park landscape. First row (from left to right): *Aqua Claudia* aqueduct; *Druso*'s Arch (*Antoniano* aqueduct); catacomb *San Sebastiano* (entrance). Second row: church *Sant'Urbano*; *Egeria* nymphaeum; *Cecilia Metella* sepulcher. Third row: *Massenzio* ruins; temple of *God Redicolo*; *Aureliane*'s wall. Fourth row: *San Sebastiano*'s door; *Latina* way; medieval tower *Valca*.

The natural vegetation of Appia Antica Regional Park is represented by several types, more or less linked to the Mediterranean climate that characterizes the study area. Although a detailed vegetation study of the park is lacking, ongoing surveys (Iamonico in prep.) allow presenting a general view of the main vegetation physiognomies occurring in the park.

- Zonal vegetation:
  - Vegetation of the nemoral forest zone:
    - Zonal temperate broad-leaved forests:
      1. Quercetea pubescentis Doing-Kraft ex Scamoni et Passarge 1959 (Figure 16): mixed forest communities of deciduous species mainly occurring in northern areas of the park; common species are: *Quercus ilex* L. subsp. *ilex*, *Q. pubescens*, *Q. petraea* (Matt.) Liebl., *Fraxinus ornus* L., *Acer campestre* L. among trees, and *Crataegus monogyna* Jacq., *Euonymus europaeus* L., *Cornus sanguinea* L. subsp. *sanguinea*, and *Viburnum tinus* L. among shrubs;

Figure 16. Broad-leaved forests (*Quercetalia pubescenti-petraeae*) in Caffarella valley.

2. Quercetea robori-petraeae Br.-Bl. et Tx. ex Oberd. 1957: a small forest dominated by *Quercus robur* L. subsp. *robur* occurs in the northern area of the park (*Cartiera Latina* locality);

> Intrazonal scrub and woodlands of the nemoral zone:

3. Robinietea Jurko ex Hadac et Sofron 1980: it comprises anthropogenic woody vegetation (Figure 17) characterized by an high presence of *Robinia pseudoacacia* L. and or *Ailanthus altissima* (Mill.) Swingle. Other common species are: *Sambucus nigra* L., *Ulmus minor* Mill. subsp. *minor* (among trees), *Rubus ulmifolius* Schott (among shrubs); herb layer is dominated by nitrophilous taxa, e.g., *Galium aparine* L. and *Urtica dioica* L. These types of vegetation occur sparsely throughout the study area;
4. Crataego-Prunetea Tx. 1962: scrub vegetation occurring as patches among the cultivated fields or along the margins of the forests. The main communities occurring in the study area are those with *Cornus sanguinea* subsp. *sanguinea, Crataegus monogyna, Euonymus europaeus, Prunus spinosa* L. subsp. *spinosa, Rhamnus alaternus* L. subsp. *alaternus, Rosa canina* L., *Rubus ulmifolius*, etc. (Figure 18), whereas in the areas more affected by human pressures, monophytic communities with *Rubus ulmifolius* occur (Figure 18). Further shrub communities are those dominated by *Paliurus spina-christi* Mill. (which can be found in Caffarella valley in areas that are not or only marginally affected by pasture; Figure 19), *Spartium junceum* L. (often on the top of cliffs; Figure 19), and groups of species (*Cornus sanguinea* subsp. *sanguinea, Ligustrum vulgare* L., *Euonymus europaeus, Crataegus monogyna, Ulmus minor* subsp. *minor*) that occupy the anthropic sinkholes originating from the collapse of underground cavities;

Figure 17. Anthropogenic woody vegetation (Robinieta): *Robinia pseudoacacia* dominated community (Caffarella valley).

Figure 18. Scrub vegetation of the nemoral zone (Crataego-Prunetea): mixed scrub, locality Caffarella (top photo); monophytic communities with *Rubus ulmifolius*, locality Acquedotti (bottom photo).

Figure 19. Scrub vegetation in the nemoral zone (Crataego-Prunetea): *Paliurus spina-christi* community, locality Caffarella (top photo); *Spartium junceum* community, locality Tor Marancia (bottom photo).

> Intrazonal boreo-temperate grasslands and heath:

5. Molinio-Arrhenatheretea Tx. 1937: anthropogenic managed pastures, meadows and tall-herb meadow fringes on fertile deep soils at low and mid-altitudes.

    - Cool Temperate Group of Alliances:

        - Cynosurion cristati Tx. 1947 (Arrhenatheretalia elatioris Tx. 1931): mesophilous grasslands, grazed and mown once, growing on well-drained mineral/nutrient-rich soils. These communities are common throughout the park, particularly in the areas affected by human recreational use (Figure 20).

Figure 20. *Lolium perenne* dominated community, Cynosurion cristati (Divino Amore locality).

- Vegetation of the mediterranean zone:

    > Zonal mediterranean forests and scrub:

    6. Quercetea ilicis Br.-Bl. ex A. Bolós et O. de Bolós in A. Bolós y Vayreda 1950: thermo-mesomediterranean oak forests and associated Mediterranean macchia. Small forest patches with *Quercus ilex* subsp. *ilex* (Figure 21) or *Q. suber* L. and Mediterranean macchia occur. Holm oak forests are present in Caffarella and Tor Marancia localities. Only one patch of corn oak forest remains (the so-called *Boschetto Farnese* = Farnese's wood, included in the private farmstead "Farnesiana"); this forest is represented by a mixed evergreen/deciduous species where *Q. suber* grows together with *Q. pubescens* and *Q. ilex* subsp. *ilex*, whereas the shrub layer is composed by *Ulmus minor* subsp. *minor*, *Rubus ulmifolius*, *Crataegus monogyna*, *Prunus spinosa* subsp. *spinosa*, *Euonymus europaeus*, *Rhamnus alaternus* subsp. *alaternus*, *Cornus sanguinea* subsp. *sanguinea*, and *C. mas* L. Finally, the residual patches of Mediterranean macchia, which are dominated by *Arbutus unedo* L., *Phillyrea latifolia* L., and *Quercus ilex* subsp. *ilex* (Figure 21).

Figure 21. Quercetea ilicis: Olm Hoak forest in Caffarella valley (top photo); Mediterranean macchia in Acquedotti locality (bottom photo).

- Azonal vegetation:
    > Alluvial forests and scrub:
    7. Alno glutinosae-Populetea albae P. Fukarek et Fabijanic 1968: Mediterranean riparian communities on soils with high water table; this type is well represented in the park along rivers and channels, especially in the central and north zones of Caffarella valley and Tor Marancia locality, where forests are dominated by *Populus nigra* L., *Salix alba* L. subsp. *alba*, and *Fraxinus angustifolia* Vahl. subsp. *oxycarpa* (M.Bieb ex Willd.) Franco & Rocha Afonso (Figure 22).

Figure 22. Riparian forest along river Almone (Caffarella locality).

- Vegetation of rock crevices and screes:
  > Adiantetea Br.-Bl. et al. 1952: communities dominated by *Adiantum capillus-veneris* L. and bryophytes that grow on siliceous dripping cliffs; they are common along deeper channels and springs (Figure 23).

    8. Cymbalario-Parietarietea diffusae Oberd. 1969: thermo-nitrophilous Mediterranean chasmophytic vegetation of walls and cliffs; quite distributed on both natural (volcanic cliffs) and anthropic (aqueducts, walls) surfaces. Frequent species are: *Antirrhinum majus* L., *Capparis orientalis* Veill., *Cymbalaria muralis* G.Gaertn., B.Mey. & Schreb. subsp. *muralis, Ficus carica* L., *Fumaria capreolata* L., *Parietaria judaica* L., *Reichardia picroides* Roth, *Sonchus tenerrimus* L., *Umbilicus rupestris* (Salisb.) Dandy (Figure 24).

Figure 23. Adiantetea on the volcanic cliffs of channel Acqua Mariana (Acquedotti locality).

Figure 24. Cymbalario-Parietarietea diffusae on the ancient Roman Claudio aqueduct (Acquedotti locality).

- Freshwater aquatic vegetation:
    9. Lemnetea O. de Bolos et Masclans 1955: pleustophytic vegetation that colonizes fresh waters; in the territory of the park, these communities mainly occur on weakly flowing waters (channels or lakes) where the alien *Lemna minuta* Kunth dominates the autochthonous *L. minor* L. (Figure 25).

Figure 25. *Lemna minuta* community, Lemnetea (Acquedotty locality).

- Vegetation of freshwater springs, shorelines and swamps.
    10. Phragmito-Magnocaricetea Klika in Klika et Novak 1941: perennial helophytic communities colonizing lacustrine and fluvial areas on eu- to mesotrophic soils of freshwater bodies; the communities mostly occur in the northern sector of the park along channels or around ponds and artificial lakes (Figure 26). Common species are: *Alisma plantago-aquatica* L., *Apium nodiflorum* (L.) Lag., *Arundo donax* L. (which often forms monophytic communities; Figure 27), *Equisetum telmateja* Ehrh., *Lymniris pseudacorus* (L.) Fuss., *Nasturtium officinalis* R.Br., *Phragmites australis* (Cav.) Tin. ex Steud., *Symphytum officinale* L., *Typha latifolia* L. (which sometimes forms monophytic communities; Figure 28), *Veronica anagallis-aquatica* L.

Figure 26. Phragmito-Magnocaricetea on channel Acqua Mariana (Acquedotti locality) with *Alisma planta-go-aquatica* (top photo) and *Veronica anagallis-aquatica* dominated (bottom photo).

Figure 27. *Arundo donax* community (Divino Amore locality).

Figure 28. *Typha latifolia* community (Caffarella valley).

- Anthropogenic vegetation: secondary vegetation communities that derive from direct or indirect results of human action; they are represented by several types occurring throughout the study area; most of these types are meadow for-mations. The main types observed are listed as follows:

11. Papaveretea rhoeadis S. Brullo et al. 2001 (= Secalinetea Br.-Bl. In Br.-Bl. et al. 1952 = Stellarietea mediae Tx. et al. in Tx. 1950): annual weed segetal vegetation of arable crops on base-rich soils; it is widely distributed in Appia Antica Regional Park since crops represent the landscape matrix (Figure 29).

Figure 29. Segetal vegetation dominated by *Glebionis segetum* (L.) Fourr. and *Papaver rhoeas* L., Papaveretea rhoeadis (Acquedotty locality).

12. Sisymbrietea Gutte et Hilbig 1975: anthropogenic vegetation of animal shelters and disturbed ruderal sites (Figure 30).

Figure 30. *Trisetaria panicea* dominated community, Sisymbrietea (Caffarella locality).

13. Polygono-Poetea annuae Rivas-Mart. 1975: nitrophilous pioneer vegetation of trampled habitats. This type of vegetation is common, especially on roadsides and crevices of paved roads (Figure 31).

Figure 31. Polygono-Poetea annuae: therophyte communities on crevices of paved Appia Antica street; occurred taxa: *Amaranthus retroflexus* L., *Cynodon dactylon* (L.) Pers., *Euphorbia prostrata* Aiton, *Herniaria glabra* L. subsp. *glabra*, *Polycarpon tetraphyllum* subsp. *tetraphyllum*, *Portulaca oleracea* L. subsp. *oleracea*, *Solanum nigrum* L.

14. Artemisietea vulgaris Lohmeyer et al. in Tx. ex von Rochow 1951: perennial meso-xerophilous ruderal vegetation. The most common communities found in the park are dominated by *Silybum marianum* (L.) Gaertn. (margins of fields and uncultivated lands; Figure 32), *Conium maculatum* L. (banks of water courses; Figure 32), and *Sambucus ebulus* L. (humid and disturbed soils near rivers and channels; Figure 32).

Figure 32. Artemisietea vulgaris: *Silybum marianum* community, Caffarella valley (top photo); *Conium maculatum* community, Acquedotti locality (central photo); *Sambucus ebulus* community, Acquedotti locality (bottom photo).

15. Chenopodietea Br.-Bl. in Br.-Bl. et al., 1952: weed segetal and ruderal vegetation of man-made habitats. This group includes many types in the park, the most common are the Mediterranean annual grasslands of Hordeion murini Br.-Bl. in Br.-Bl. et al. 1936 and Securigero securidacae-Dasypyrion villosi Cano-Ortiz, Biondi et Cano in Cano-Ortiz et al. ex Di Pietro in Di Pietro et al., 2015. The former is represented by ruderal grasslands occurring in nutrient-rich soils characterized by a high presence of humans (Figure 33). The second group (Securigero securidacae-Dasypyrion villosi) includes the therophytic anthropogenic grasslands in fallow-land habitats of the central regions of the Apennine Peninsula; this type of vegetation is dominated by *Dasypyrum villosum* (L.) P.Candargy (Figure 33), representing one of the most widespread anthropogenic steppe formations in the Roman countryside [54].

Figure 33. Chenopodietea: Hordeion-murini (Acquedotty locality; top photo); Securigero securidacae-Dasypyrion villosi (Acquedotti locality; bottom photo).

16. Epilobietea angustifolii Tx. et Preising ex von Rochow 1951 (=Galio-Urticetea Passarge 1967): tall-herb semi-natural perennial vegetation on disturbed forest edges and nutrient-rich riparian fringes (Figure 34).

Figure 34. *Galega officinalis* L. dominated community, Epilobietea angustifolii (Acquedotti locality).

4. Results

4.1. General Statistics of the Flora

The flora of Appia Antica Regional Park is composed by 714 taxa (including species and subspecies; see Appendix A), belonging to 104 families and 403 genera. These include 11 Lycopodiophyta and ferns and allies (six families and seven genera), three gymnosperms (*Pinus pinea* L., *P. halepensis* Mill. and *Cupressus sempervirens* L., as aliens), one Magnoliidae (*Laurus nobilis* L.), 557 angiosperm eudicots (81 families, 308 genera), and 142 angiosperm monocots (13 families, 85 genera).

The richest families are (Figure 35): Asteraceae Bercht. & J.Presl (81 taxa; 54 genera), Fabaceae Lindl. (74 taxa; 26 genera), and Poaceae Barnhart (69 taxa; 43 genera), followed by Caryophyllaceae Juss. (29 taxa; 12 genera), Lamiaceae Martinov (28 taxa; 14 genera), Brassicaceae Burnett (24 taxa; 19 genera), Rosaceae Juss. (23 taxa; 14 genera), and Apiaceae Lindl. (21 taxa; 16 genera). Eleven families include 10 (Chenopodiaceae Vent. and Euphorbiaceae Juss.) to 17 (Plantaginaceae Juss.) taxa. Fifty families comprise two to nine taxa. Finally, 34 families are monospecific for the flora.

The richest genera are (Figure 36): *Trifolium* (20 taxa); *Silene* and *Vicia* (10); *Amaranthus* and *Medicago* (8); *Carex*, *Euphorbia*, and *Quercus* (7); *Allium*, *Geranium*, *Ranunculus*, and *Veronica* (6); *Crepis*, *Erodium*, *Lathyrus*, and *Rumex* (5).

According to definition given by Fanfarillo and collaborators [54], the segetal flora of Appia Antica Regional Park includes 348 taxa (corresponding to 49.01% of the total flora), of which 14 (1.97% of the total flora) are strictly segetal (species that only occur in segetal habitats), whereas 25 (3.52% of the total flora) are characteristic segetal (species characteristic of segetal habitats that also commonly colonize other habitats) (Table 1). Three species (*Chenopodium vulvaria* L., *Tribulus terrestris* L., and *Xanthium spinosum* L.) are new additions to the Italian segetal flora, being not listed by [55]. The richness (percentage) of the segetal flora of Appia Antica Park is much higher than that of the entire segetal flora of Italy, i.e., 10.10% (percentage of Italian segetal flora over the Italian vascular flora), 1.61% (percentage of characteristic segetal), and 0.59% (percentage of strictly segetal). These data reveal a high rate of agricultural areas in the park, which, in fact, occupy more than 50% of the territory (see above under the Section "3.5. Landscape Remarks and Actual Vegetation Physiognomies").

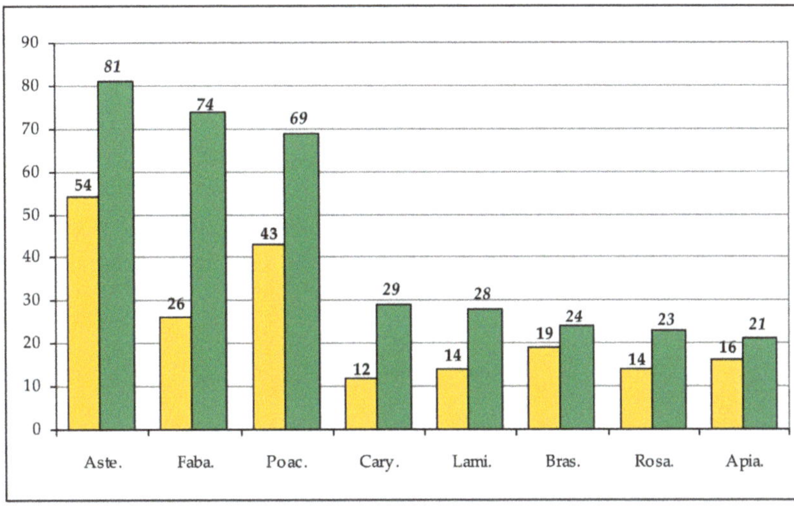

Figure 35. Number of genera (orange columns) and taxa (dark-green columns) per family; only the richest families (more than 20 taxa (species and subspecies)) are displayed. Abbreviations: Aste. = Asteraceae; Faba. = Fabaceae; Poac. = Poaceaeae; Cary. = Caryophyllaceae; Lami. = Lamiaceae; Bras. = Brassicaceae; Rosa. = Rosaceae; Apia. = Apiaceae.

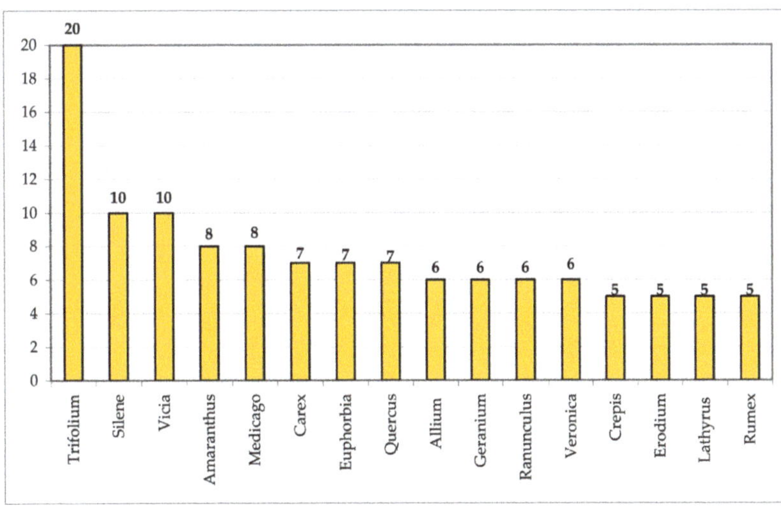

Figure 36. Number of taxa per richest genera [more than four taxa (species and subspecies) per genus).

Table 1. Strictly (marked with an asterisk) and characteristic segetal species occurring in the territory of Appia Antica Regional Park (names in alphabetical order).

* *Alopecurus myosuroides* Huds, subsp. *myosuroides*
Anisantha diandra (Roth) Tutin ex Tzvelev
Anthemis arvensis L. subsp. *arvensis*
Anthemis cotula L.
* *Aphanes arvensis* L.
Avena sterilis L. subsp. *sterilis*
* *Ballota nigra* L. subsp. *meridionalis* (Bég.) Bég.
Buglossoides arvensis (L.) I.M.Johnst.
* *Cyanus segetum* L.
* *Delphinium consolida* L. subsp. *consolida*
* *Ervilia hirsuta* (L.) Opiz
* *Ervum tetraspermum* L.
* *Euphorbia exigua* L. subsp. *exigua*
* *Euphorbia falcata* L. subsp. *falcata*
Filago germanica (L.) Huds.
Gladiolus italicus Mill.
Herniaria glabra L.
Herniaria hirsuta L. subsp. *hirsuta*
Lamium purpureum L.
Lathyrus annuus L.
Lathyrus aphaca L. subsp. *aphaca*
* *Legousia speculum-veneris* (L.) Chaix subsp. *speculum-veneris*
Lysimachia arvensis (L.) U.Manns & Anderb. subsp. *arvensis*
Matricaria chamomilla L.
Muscari comosum (L.) Mill.
Myosotis arvensis (L.) Hill subsp. *arvensis*
Neslia paniculata (L.) Desv. subsp. *thracica* (Velen.) Bornm.
Orobanche crenata Forssk.
* *Papaver dubium* L.
* *Papaver hybridum* L.
* *Papaver rhoeas* L.
Phalaris paradoxa L.
Rapistrum rugosum (L.) All.
Sinapis alba L. subsp. *alba*
Sinapis arvensis L. subsp. *arvensis*
* *Spergula arvensis* L.
Valerianella eriocarpa Desv.
Veronica arvensis L.
Veronica polita Fr.
Vicia bithynica (L.) L.

4.2. Life Form Analysis

The vascular flora of Appia Antica Regional Park is characterized by a prevalence of therophytes (287 taxa, corresponding to 40.20% of the total flora), followed by the hemicryptophytes (201 taxa, 28.15%) (Figure 37). The ratio T/H is 1.41, being quite high in comparison with other areas of Rome municipality (Figure 38). These data highlight the xeric conditions of Appia Antica Regional Park in pastures, uncultivated lands, and synanthropic environments that represent the most common habitats. A similar environmental situation occurs in Laurentino and Aguzzano Parks [high therophyte/hemicryptophyte (T/H) ratios], whereas in Monte Mario, Pineto, and Veio Parks (which occur on the north of the Rome area), mesophilous habitats are well represented and the T/H ratio is, therefore, lower due to the higher percentage of hemicryptophytes in comparison with that of the floras of Appia Antica, Laurentino, and Aguzzano Parks. To note, the T/H ratio for Appia Antica Park is also higher than that of the Caffarella valley (1.32), which is included in the studied area, representing about 4% (ca. 190 ha) of the total area.

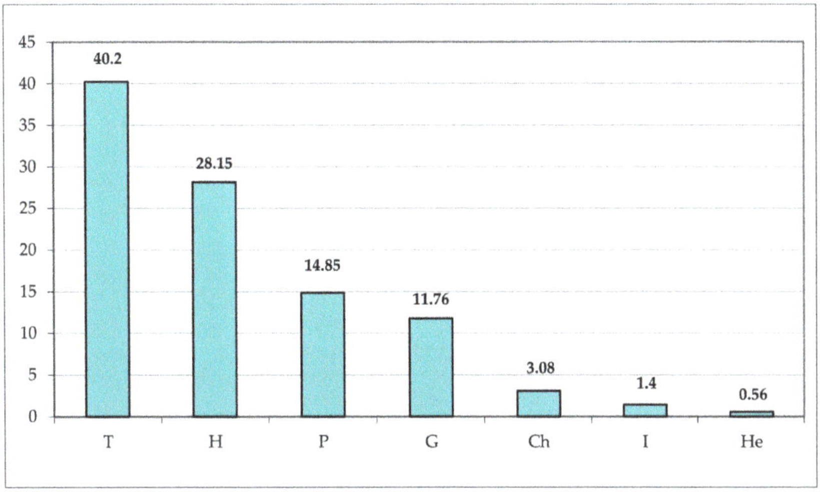

Figure 37. Percentage (axis x) of plant life form spectra of the vascular flora of Appia Antica Regional Park. T: therophytes; H: hemicryptophytes; P: phanerophytes; G: geophytes; Ch: chamaephytes; I: idrophytes; He: helophytes.

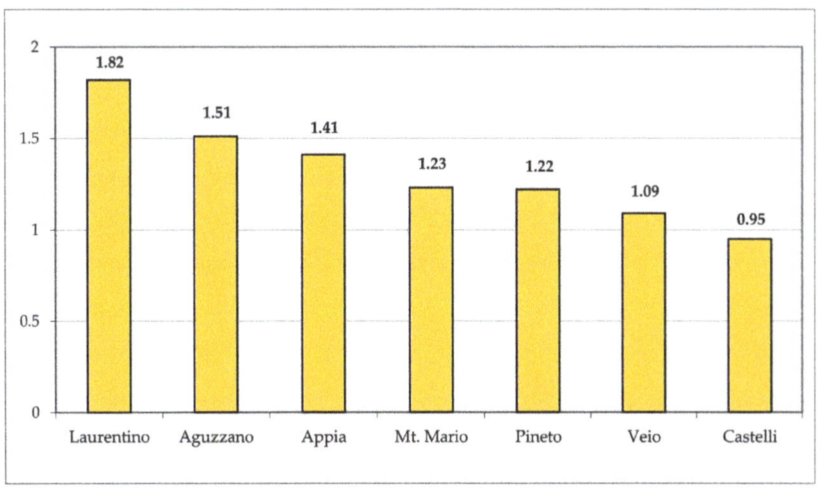

Figure 38. T/H ratio (y axis) of some parks (x axis) in Rome Province.

The phanerophytes in Appia Antica Regional Park, reaching 14.85% (106 taxa) of the total flora, together with a fair percentage of rhizomatous geophytes (42 taxa, 5.88%), typical of forest habitats, highlight the rather rich flora of woody habitats. However, about 1/3 of the phanerophytes (5.04%) are aliens (Table 2), and most of them do not occur in wooded areas [e.g., *Agave americana* L., *Campsis radicans* (L.) Bureau, *Eucalyptus camaldulensis* Dehnh. subsp. *camaldulensis*, *Lantana camara* L, *Maclura pomifera* (Raf.) C.K. Schneid., *Malus domestica* L., *Passiflora caerulea* L., *Ziziphus jujuba* Mill.]. In fact, as a whole, forests represent only residual areas in the territory of the park, and several patches are actually shrubs (see Section "3.5. Landscape Remarks and Actual Vegetation Physiognomies").

Table 2. Phanerophytes (P) occurring in the territory of Appia Antica Regional Park.

	Native Taxa		Alien Taxa	
Biological Form	**N° of Taxa**	**Percentage**	**N° of Taxa**	**Percentage**
P scapose	30	4.20	11	1.54
P caespitose	22	3.08	10	1.40
P lianose	7	0.98	6	0.84
P succulent	0	0	1	0.14
Nano-phanerophytes	11	1.54	6	0.84
Total	70	9.80	36	5.04

Helophytes and hydrophytes represent 1.96% of the total flora (4 and 10 taxa, respectively) highlighting the presence of aquatic habitats (especially in the northern sector of the park). Of note, among rhizomatous geophytes, several grow in these habitat [e.g., *Adiantum capillus-veneris* L., *Eleocharis palustris* (L.) Roem. et Schult., *Limniris pseudacorus* (L.) Fuss, *Symphytum tuberosum* L. subsp. *angustifolium* (A.Kerner) Nyman, etc.], thus revealing the relevant occurrence of aquatic and strictly related habitats.

Finally, the low percentage of chamaephytes (3.08%, 22 taxa) can be explained by the scarcity of rocky habitats (more widespread in the mountain belt and on windy peaks in Latium region) and the absence of garrigues (occurring especially along coasts).

4.3. Geographical Analysis

The chorological spectrum (Figure 39) shows a prevalence of Eurimediterranean species (162 taxa, corresponding to 22.69% of the total flora). By considering the Mediterranean component *sensu lato* (Euri- plus Stenomediterranean), the percentage reaches 38.94% [288 taxa (162 euri-, 22.69%; 116 steno-, 16.25%)], which is congruent with the phytoclimatic background of the studied area (Meso-Thermomediterranean climate type and oak series as potential vegetation). This fact is also confirmed by both the high value of the Eurimediterranean/Eurasian species ratio (1.91, which is high in comparison with the ratios of other Parks included in Rome Province; Figure 40), and the low percentage of Nordic species (27 taxa, 3.78%); of note, the Eurimediterranean/Eurasian ratio for Appia Antica Park is also higher than that of the Caffarella valley (1.35), which is included in the studied area, representing about 4% (ca. 190 ha) of the total area. The percentage of the wide distribution species (Cosmopolitan, Subcosmopolitan, and Tropical; 110 taxa) plus the aliens (102 taxa) is high, namely 29.70% (15.41% and 14.29%, respectively); this value is mainly due to human impact, which causes the occurrence and spreading of r-selected species. Finally, the eastern component (SE-European, Pontic, and Turanian s.lat.), typical of the flora of Central Italy and particularly of Lazio, totals 5.32% (38 taxa); similar floristic backgrounds were observed in the whole region (see [26]). Finally, the endemic taxa are very few (only 3 taxa, for 0.42%), probably due to the lack of habitats characterized by biogeographical insularity.

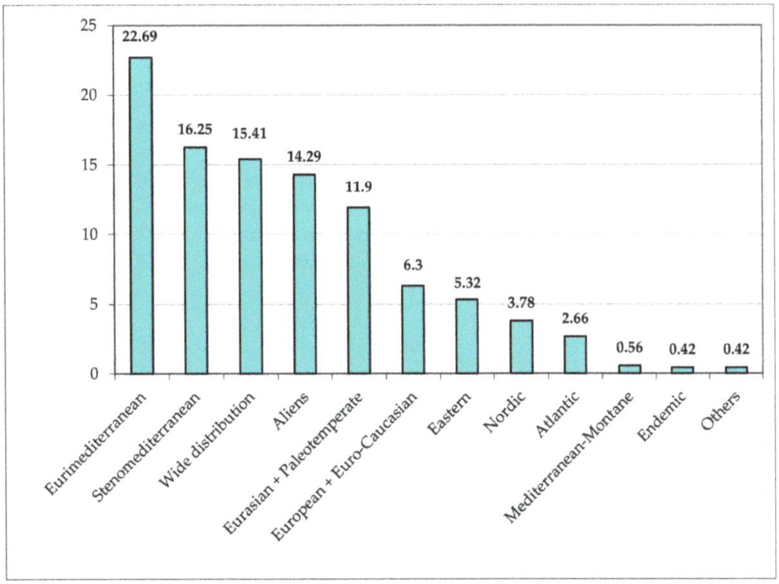

Figure 39. Chorological spectrum of the vascular flora of Appia Antica Regional Park.

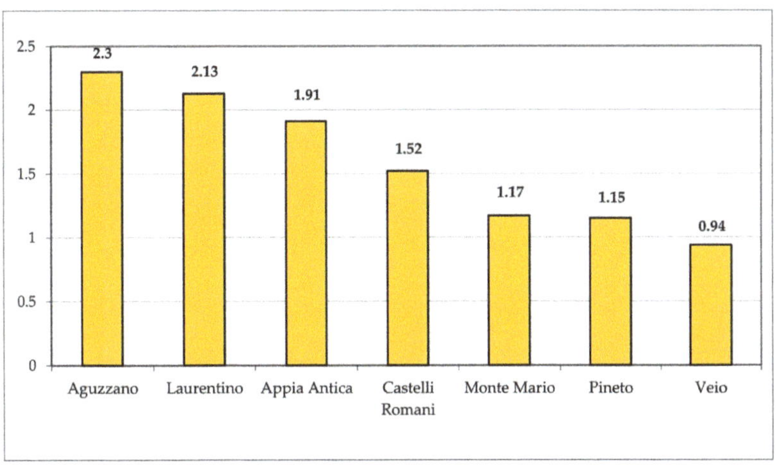

Figure 40. Eurimediterranean/Eurasian ratio (y axis) of some parks (x axis) in Rome Province.

Concerning the alien taxa, they total 102, corresponding to 14.29% of the total flora (native taxa account for 85.71%). Of 52 families, each one contains at least one non-native taxon; 13 of these 46 families are aliens, including one species only (Basellaceae Raf., Bignoniaceae Juss., Cactaceae Juss., Cannaceae Juss., Cupressaceae Gray, Cleomaceae Bercht. & J. Presl, Ebenaceae Gürke, Meliaceae Juss., Nyctaginaceae Juss., Passifloraceae Juss. ex Roussel, Phytolaccaceae R.Br., Pittosporaceae R.Br., Platanaceae T.Lestib.). The families rich in alien species include Asteraceae (10 taxa), followed by Poaceae and Amaranthaceae Juss. (7), Fabaceae (5), Rosaceae and Solanaceae Juss. (4), Araceae Juss., Asparagaceae Juss., Convolvulaceae Juss., Moraceae Gausich., and Oxalidaceae (3), and Apocynaceae Juss., Arecaceae Bercht. & J.Presl, Brassicaceae, Chenopodiaceae Vent., Euphorbiaceae Juss., Pinaceae Spreng. ex F.Rudolphi, and Vitaceae (2); the other families include one alien taxon.

A total of 84 genera include at least one alien taxon. *Amaranthus* is the richest one, with 6 alien species, followed by *Erigeron* L., *Oxalis* L., and *Solanum* L. (3), and *Cuscuta* L., *Dysphania* R.Br., *Euphorbia* L., *Galinsoga* Riuz. & Pav., and *Pinus* L. (2); the other 75 genera each include one alien taxon.

Most of the exotic taxa are native to America (51 taxa, representing 51.00% of the total aliens), followed by the Asia (19 taxa, 19.38%); Europe and Africa, including, respectively, 11 (11.22%) and 8 (8.16%) taxa, and two (2.04%) taxa native to Australia (Figure 41). Concerning the alien status, most of the non-native taxa are casual (50, corresponding to 49.02% of the total aliens) and naturalized (40, 39.22%), whereas only 12 species (11.76%) can be considered as invasive (Figure 42).

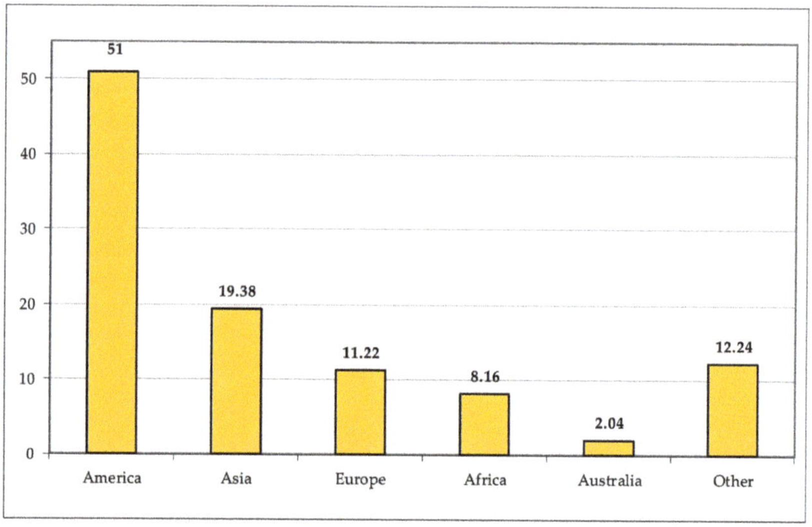

Figure 41. Share of aliens plant species and subspecies (percentages in axis y) by thier per origin.

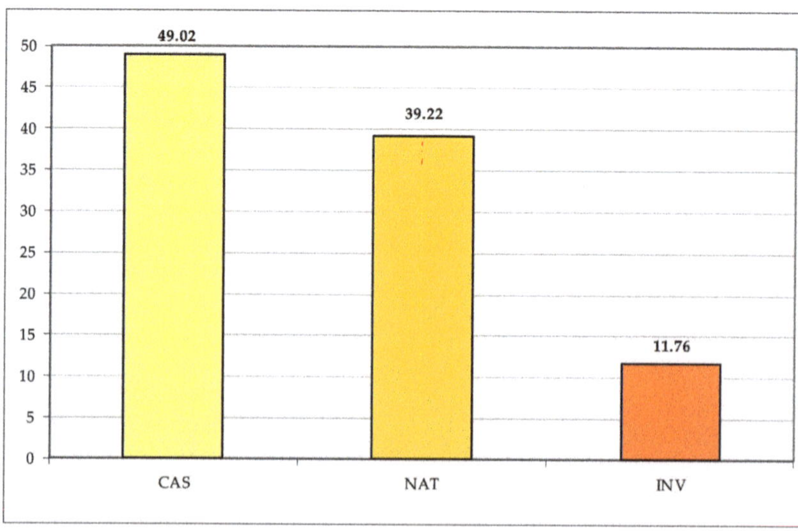

Figure 42. Share of casual (CAS), naturalized (NAT), and invasive (INV) alien plant species and subspecies (percentages in axis y) occurring in Appia Antica Regional Park.

4.4. Floristic Notes

More than 10 years of field surveys allowed the discovery of several floristic novelties or confirmations at the European (1 species), national (3 species), regional (10 species), and local (7 species) levels; all of these novelties refer to alien species (Table 3; see Appendix B).

Table 3. Floristic novelties discovered during field surveys. Names are ordered alphabetically per type of novelty.

Type of Novelty	Scientific Name
First record for **Europe** and first one out of the native range	*Denisophytum bessac* (Choiv.) E.Gagnon & G.P.Lewis
First confirmed records for **Italy**	*Euphorbia pulcherima* Willd. ex Klotzsch *Rosa chinensis* Jacq. var. *semperflorens* (Curtis) Koehne
First naturalized occurrence for **Italy**	*Aloe maculata* All. subsp. *maculata*
First records for **Latium region**	*Heliotropium amplexicaule* Vahl. *Hydrangea macrophylla* (Thunb.) Ser. *Ruellia simplex* C.Wright *Trachelospermum jasminoides* (Lindl.) Lem.
Change in alien status for **Latium region**, from casual to naturalized	*Campsis radicans* (L.) Bureau *Canna indica* L. *Cyperus alternifolius* L. subsp. *flabelliforme* Kük. *Kalanchoe daigremontiana* Raym.
Confirmation as alien in **Latium region**	*Melia azedarach* L. *Punica granatum* L.
New records for the flora of **Rome**	*Bidens subalternans* DC. *Chlorophytum comosum* (Thumbs.) Jacques *Zantedeschia aetiopica* (L.) Spreng.
Confirmations for the flora of **Rome**	*Anreedera cordifolia* (Ten.) Steenis *Diospyrus kaki* L. *Papaver somniferum* L. *Passiflora caerulea* L.

4.4.1. New Record for Europe and First One Out of the Native Range

***Denisophytum bessac* (Choiv.) E.Gagnon & G.P.Lewis** (≡ *Caesalpinia bessac* Chiov.): a population of this species (monophytic shrub community; Figure 43) was first observed during the spring of 2016 on the west side of Caffarella valley. I initially identified this population as belonging to the genus *Caesalpinia* L., mainly based on the flower and leaf morphologies [56]. However, on the basis of a recent taxonomic work [57], a new classification of the *Caesalpinia* group was proposed, recognizing 26 genera (some newly described, other ones resurrected or redelimitated). By using the diagnostic key (genus rank) provided by Gagnon and collaborators [57], the identification of the Roman population (erect shrubs armed, with leaves bipinnate terminating with a pair of pinnae, flowers yellow, and fruits dehiscent and unarmed) was restricted to *Caesalpinia* or *Denisophytum* R. Vig. Note that these two genera are clearly separated from the molecular point of view (see Figure 3A in [57]) and the resurrection of *Denisophytum* is well supported. On the other hand, these two genera are morphologically similar, as highlighted by the authors ("no reliable diagnostic characters have been found to differentiate these two genera" [57] pag. 45). The only character considered to distinguish *Denisophytum* and *Caesalpinia* is the color of the flowers, which are yellow (sometimes with red markings on the standard, i.e., the median petal) in *Denisophytum* and orange, red, green, or white (rarely yellow or pink) in *Caesalpinia* ([57] pag. 27). Furthermore, on the basis of the detailed emended descriptions given, the two genera differ by the length of the fruits (pods), which are 18–49 mm long in *Denisophytum* ([57] pag. 45) and 34–120 mm long in *Cesalpinia* ([57] pag. 43). Flowers in the Roman population are all yellow, with minute red spots on the standard (Figure 43); pods (pers. obs.) are up to 50 mm long (never less than 40 mm). All things considered, I here identify the Roman population as a *Denisophytum* species.

The identification at species rank was quite difficult. Currently, *Denisophytum* comprises eight species, but unfortunately, no diagnostic key at species rank was provided by Gagnon and collaborators [57]. Furthermore, no further comprehensive key of *Denisophytum* taxa was found in the literature, and the descriptions of the taxa, when existing, are not very detailed. Hence, I decided to check the protologues of all eight of these species and examine all of the original material that I was able to trace. On the basis of this research (the complete work is still ongoing), I assembled the following data:

➢ *D. bessac* (Choiv.) E. Gagnon & G.P. Lewis (≡ *Caesalpinia bessac* Chiov.): Chiovenda described this species (sub *Caesalpinia bessac*) in his *Flora Somala* [58], providing a detailed description that matches the plants found in Caffarella valley. The same author described also *C. eriantherum* Chiov. (see below) as morphologically similar to *C. bessac*. These two species differ from each other by the size of their leaflets, which are longer in *C. bessac* (basal leaflets 9–12 mm long vs. 4–5 mm long in *C. eriantherum*; distal leaflets 10–20 mm long vs. 7–8 mm long in *C. eriantherum*).

Syntype found: FI001388.

➢ *D. buchii* (Urb.) E. Gagnon & G.P. Lewis (≡*Denisophytum buchii* Urb.): this species displays the calyx as 7–8 mm long according to the protologue [59], whereas plants found in Caffarella have a calyx 3–5 mm long; furthermore, the leaflets are emarginate at the apex in *D. buchii* ("foliolis... apice emarginatis"), while leaflets in my specimens are obtuse to rounded.

➢ *D. eriantherum* (Chiov.) E. Gagnon & G.P. Lewis (≡ *Caesalpinia eriantherum* Chiov.): on the basis of the original description by Chiovenda ([58] sub *Caesalpinia eriantherum*), plants found in Caffarella valley differ by both the size of leaflets [those basal 4–5 mm long (protologue) vs. 8–12 mm long (Caffarella plants); those distal 7–8 mm long (protologue) vs. 15–20 mm long (Caffarella plants) and the number of flowers per inflorescence, i.e., 10 (protologue) vs. up to 30 (Caffarella plants)]. Further, a var. *pubescens* (Brenan) E. Gagnon & G.P. Lewis (≡ *Caesalpinia erianthera* var. *pubescens* Brenan) is currently accepted, but it is characterized in having leaflets that are densely pubescent [60], whereas Roman plants have leaves that are glabrous.

Syntypes found: FI001390 (var. *eriantherum*), FI001392, and K000232357 (var. *pubescens*).

➢ *D. madagascariense* R. Vig.: it is the only unarmed species in *Denisophytum* [57] pag. 45, while Caffarella plants have curved thorns along the shoot.

Syntypes found: P00131739.

➢ *D. pauciflorum* (Griseb.) E. Gagnon & G.P. Lewis (≡ *Libidibia pauciflora* Griseb.): this species is different from all other species in having few flowers per raceme ("recemis simplicibus laxis paucifloris (v. pedunculis 1floris)" in the protologue [61]) and corolla slightly longer than the calyx ("calycis... corolla paullo superantibus"), whereas my specimens have many flowers per raceme (up to 30, never less than 20), and corollas are 5–7 times longer than the calyx.

Syntypes found: GH00065814, P02142660, P02142661, UC936921, US00382832, and YU001398;

➢ *D. rosei* (Urb.) E. Gagnon & G.P. Lewis (≡ *Caesalpinia rosei* Urb.): it has leaflets glaucous in the abaxial surface ("foliole... in siccu supra obscure viridia, subtus valde pallida, glaucescentia" in the protologue [62]). Plants found in Rome show leaves green on both surfaces.

Syntypes found: NY00022764 and US00479309;

➢ *D. sessilifolium* (S.Watson) E. Gagnon & G.P. Lewis (≡ *Caesalpinia sessilifolia* S.Watson): it is characterized in having sessile leaves, each one with two or three pairs of pinnae according to the protologue [63]. Caffarella plants have leaves that are petiolate with mostly five pairs of pinnae (rarely four).

Syntypes found: A00061947, AC00319854, BR0000005110933, CAS0001542, COL000092321, F0057403F, GH00059873, GOET004917, JE00004880, MICH1107159, MO125071, NA0026234, P02940720, PH00010119, RSA0003187, SI001822, US00344744, and US00345006.

- *D. stuckertii* (Hassl.) E. Gagnon & G.P. Lewis (≡ *Caesalpinia stuckertii* Hassl.): the diagnosis and description given in the protologue [64] are congruent with the morphology of the plants found in Caffarella valley, except for the fruit, which was described as "oblongum basi et apice acutum", whereas fruits in Caffarella's plants have apexes that are obtuse-rounded (Figure 41); also, the illustration given by Gagnon and collaborators [57] pag. 44, Figure 12 displays a typical fruit of *D. stuckertii*, confirming that Roman plants cannot be assigned to this *Denisophytum* species. Further, Gagnon and collaborators' illustration shows stipules that are clearly different from those of the Roman plants (foliaceous with two lobes vs. narrow, not lobed).

Syntype found: G00364837.

All things considered, the population found in Appia Antica Regional Park is identifiable as *Denisophytum bessac*, a species native to central Somalia and currently unknown elsewhere [65]. This record represents, therefore, not only the first one for Europe, but also the first discovery outside the native range of this Somalian species.

Figure 43. *Denisophytum bessac* on the western side of Caffarella valley; flower and fruit (bottom-left insets), pinna (bottom-right inset).

4.4.2. New Records or Changes in Alien Status for Italy

Aloe maculata* All. subsp. *maculata: this species, currently considered as casual in Italy [33], was recorded as casual in Latium for the first time in 2012 on the basis of a population found in Caffarella valley [66]. It is currently still considered as casual alien for the region [31,67]. After continuous monitoring over the years, it was observed that this population blooms regularly, is able to maintain itself by both vegetative and sexual reproduction (Figure 44), and has spread. I here consider *Aloe maculata* as naturalized in the Latium region, and this status is reported here for the first time at the national level.

Euphorbia pulcherrima **Willd. ex Klotzsch**: this species was recorded for Italy only in the Campania region based on a single population found in Naples city [16,68]. Recently, *Euphoribia pulcherrima* was excluded from the region (and Italy) since the site in Naples was destroyed (a new garage was built) [69]. My discovery on a riverbed of channel Acqua Mariana (Acquedotti locality) represents the only Italian site in which the species (casual) certainty occurs (Figure 45).

Rosa chinensis **Jacq. var.** ***semperflorens*** **(Curtis) Koehne**: I first observed, in spring 2015 in Caffarella valley, a small population of a rose cultivar growing in a shrub community dominated by *Rubus ulmifolius*. During the subsequent years, I again observed this population and found another one not far from the former (Figure 46). As a whole, these two populations seem to flourish occasionally in the park, and, therefore, their presence can be considered as casual.

Concerning the identification of this rose, I note first that it is a so-called modern rose having flowers with 17–20 petals [70]. On the basis of the treatment of the genus in *Flora of China* [71], as well as the *European Garden Flora* [72], plants found in Caffarella are identifiable as *Rosa chinensis* s.l. showing the following morphology: shrubs, leaves evergreen, each one with 3–5 leaflets, stipules adnate to petiole, hypanthium globose, flowers double, red-scarlet, usually solitary, up to 10 cm in diameter, not fragrant. Three varieties were recognized by Ku and Robertson [71], i.e., var. *chinensis*, var. *spontanea* (Rehder & E. H. Wilson) T. T. Yu & T. C. Ku, and var. *semprerflorens* (Curtis) Koehne. Var. *spontanea* differs from the other two by the flowers, which are single, whereas var. *chinensis* and var. *semperflorens* have flowers double or semi-double. Morphological differences between these two latter varieties refer to branches (robust in var. *chinensis* vs. slender in var. *semperflorens*), flowers (several and rarely solitary vs. solitary, rarely in fascicles of two or three), and petals (red, pink, or white vs. deep red or deep purple). The plants I found display slender branches and flowers solitary with petals deep red. Therefore, they are identifiable as var. *semperflorens* according to *Flora of China*.

According to Ku and Robertson [71], *Rosa chinensis* var. *semperflorens* has a cultivated origin and it is widely used in China. POWO [73] does not recognize infraspecific taxa of *Rosa chinensis*, recording it as native in South-Central China (it corresponds to var. *spontanea* in *Flora of China*, which is the only native variety from Guizhou, Hubei, and Sichuan) and as alien in other parts of Asia (India, Korea, Kazakhstan, Laos, Nepal, Pakistan, Uzbekistan, and Vietnam), North America (Alabama), Central America (Guatemala), Australia (Queensland and Western Australia), Europe (Belgium, Bulgaria, and Greece) and Oceania (Cook Islands and Guinea Islands); further (not reported in POWO), it was recorded as casual in Slovakia [74]. The occurrences in Europe are based on [75] for Belgium, [76] for Greece, and [77] for Bulgaria. Concerning Italy [33], *R. chinensis* is doubtfully recorded in Elba island (Tuscany, Central Italy) where the species "was possibly cultivated" [78]. So, my discovery in Rome represents the first certain record of *R. chinensis* s.l. in Italy.

Figure 44. *Aloe maculata* All. subsp. *maculata* (Caffarella valley).

Figure 45. *Euphorbia pulcherrima* on riverbed of channel Acqua Mariana (Acquedotti locality).

Figure 46. *Rosa chinensis* var. *semperflorens* in Caffarella valley.

4.4.3. New Records or Confirmations for Latium

Heliotropium amplexicaule **Vahl.**: it is recorded in northern and peninsular Italy and Sicily [33]. Four sites are on the Italian peninsula, one in Tuscany (Pisa Province [79]) and three in Campania (Naples and Salerno Provinces [80]). For Latium, it was indicated as no longer recorded in Rome based on an old collection (April 1928, herbarium RO; [67,81]). I found in 2017 a population in the central reservation of Appia Nuova street (Figure 47), and observed it again in 2018 and 2022. The species is casual for Latium, representing the fifth record for the Italian peninsula.

Figure 47. *Heliotropium amplexicaule* in the central reservation of Appia Nuova street.

***Hydrangea macrophylla* (Thunb.) Ser.**: this species was recorded in Italy [33] in the north (Lombardy, Trentino-Alto Adige, and Veneto regions, as casual), center (Tuscany region, as naturalized), and South (Campania region, as casual). The population found (first observation in 2015) grows along the channel Acqua Mariana (Acquedotti locality), blooms regularly, and sustains itself especially by vegetative reproduction; of note, the plants are regularly pruned (one or two times per year), but they rapidly re-grow and flowers appear after about 2 months (Figure 48). It can be considered a naturalized species and represents the first record for the Latium region.

Hydrangea macrophylla is a species native to Japan, and it is characterized in having high morphological variability [82,83], which led to the publication of several infraspecific names [84]; in fact, various authors accepted the recognition of infraspecific taxa (subspecies, varieties, and forms; see e.g., [82,85,86]), but the infraspecific variability is still incompletely known. Furthermore, a species related to *H. macrophylla*, i.e., *H. serrata* (Thunb.) Ser. ex DC., shows, in turn, an high phenotypic variability [87] and the relationship between these two species and their infraspecific taxa would need further study [82–87]. Lacking final conclusions about this group, I here prefer to avoid the use of infraspecific taxa and consider *H. macrophylla* as separate from *H. serrata*, according to POWO [84].

Figure 48. *Hydrangea macrophylla* along Acqua Mariana channel (Acquedotti locality).

***Ruellia simplex* C. Wright**: this species is currently known in Italy as casual and recorded only in two regions, i.e., Apulia (southern italian peninsula) and on the island of Sardinia [33]. These findings refer to single localities, one per region, i.e., Otranto (Lecce Province) for Apulia [88] and Serramanna (Medio Campidano Province) for Sardinia [89]. The population found in Appia Antica Regional Park (Acquedotti locality, along the channel Acqua Mariana; Figure 49) represents the first record for the Latium region and the third one at the national level. The few individuals found were first observed in 2020, and they do not seem to be able to spread. So, *Ruellia simplex* is here considered as casual for Latium.

Figure 49. *Ruellia simplex* along channel Acqua Mariana (Acquedotti locality).

***Trachelospermum jasminoides* (Lindl.) Lem.** (≡ *Rhynchospermum jasminoides* Lindl.): this species is recorded in Italy as casual and only in two regions [33], i.e., Lombardy (just one site in Mantova Province; [90]) and Sardinia [just one site in Cagliari Province [91]). A small population was found in the Acquedotty locality on the cliff of channel *Acqua Mariana* (Figure 50), and it represents the first record for the Latium region and peninsular Italy (casual species) and the third one for Italy.

Figure 50. *Trachelospermum jasminoides* on cliff along Acqua Mariana channel (Acquedotti locality).

4.4.4. Changes or Confirmation of Alien Status for Latium

***Campsis radicans* (L.) Bureau**: this species was reported as casual alien for Latium in *Flora of Italy* [33], whereas in volume no. 1 of *Atlante della flora vascolare del Lazio* [67], it was considered as naturalized. Actually, the first indication of naturalization of *Campsis radicans* in Latium was in 2014 [66] on the basis of a population found in Caffarella valley. Of note, the *Flora of Rome* [9] does not list this species. I here confirm the occurrence and naturalization of *Campsis radicans* in Appia Antica Regional Park in Caffarella locality. Furthermore, a new population was recently found in the Acquedotti locality of the park (Figure 51).

Figure 51. *Campsis radicans* in Acquedotti locality.

***Canna indica* L.**: this species is currently considered as casual alien in the Latium region [33,67]. I found many populations of *Canna indica*, especially in the northern sector of the park (Caffarella valley and Acquedotti locality; Figure 52) where, during the years, the number of individuals increased, thus showing that the populations are able to spread. All of the plants found bloom regularly. *C. indica* is here considered as naturalized in Latium.

***Cyperus alternifolius* L. subsp. *flabelliforme* Kük.**: this taxon was considered as a casual alien in Latium [33,67], and also indicated for Rome (sub *Cyperus involucratus* Rottb.). In Appia Antica Regional Park there are many individuals occurring especially in the northern sector (Tor Marancia, Caffarella, and Acquedotti localities), which have highly increased in number over the years. All of the plants found bloom regularly and the populations have spread (Figure 53). *C. alternifolius* subsp. *flabelliforme* is a naturalized species throughout the whole study area.

Figure 52. *Canna indica* along channel Acqua Mariana, near Claudio's aqueduct (Acquedotti locality).

Figure 53. *Cyperus alternifolius* subsp. *flabelliforme* along channel Acqua Mariana as part of holophyte community of Phragmito-Magnocaricetea (Acquedotti locality).

***Kalanchoe daigremontiana* Raym.**: species reported as casual for Latium [33,67], but not cited in *Flora of Rome* [9]. The populations found in Appia Antica Regional Park are not only able to sustain themselves, but they have spread (Figure 54). *Kalanchoe daigremontiana* is a naturalized non-native species in Latium.

Figure 54. *Kalanchoe daigremontiana* in *Arundo donax* community (Caffarella valley).

Melia azedarach **L.**: this species was indicated as doubtfully spontaneous in *Atlante della flora vascolare del Lazio* [67], whereas it is casual for the region in Italian and Roman floras [9,33]. I here confirm the occurrence of *Melia azedarach* as casual in Latium, having found individuals in shrubs and forest communities and along paths (Figure 55).

Figure 55. *Melia azedarach* (red arrow) in Caffarella valley (left and top-right photos); young individual in Acquedotti locality (bottom-right photo).

***Punica granatum* L.**: it is currently considered as casual in Latium [33,67] and Rome [9], whereas in *Flora vascolare del Lazio* by Anzalone and collaborators [92], *Punica granatum* is reported as naturalized, with various localities listed (including Rome). In Caffarella valley, I traced a population along a channel consisting of well-developed individuals (2–5 m tall, with truck of 15–20 cm in diameter) that have bloomed and fruited regularly for more than 10 years (Figure 56). Further scattered individuals were observed in Caffarella and Acquedotti localities. The species is naturalized in the park.

Figure 56. *Punica granatum* in Caffarella valley.

4.4.5. New Records or Confirmations for Rome

***Anreedera cordifolia* (Ten.) Steenis**: this species was not listed in the *Flora of Rome* [9], whereas in *Atlante della flora vascolare del Lazio* [67] it was indicated generically for "Roma città" (= Rome city). Two populations growing in a *Rubus ulmifolium* dominated community (Caffarella valley and Acquedotti locality; Figure 57) were found some years ago, and they appear to be well-established; the species is naturalized. This discovery confirms the occurrence of *Anreedera cordifolia* in Rome.

***Bidens subalternans* DC.**: this species was not listed in either the *Flora of Lazio* by Anzalone and collaborators [92] or the *Flora of Rome* [9], whereas the more recent *Atlante della flora vascolare del Lazio* [67] indicates *Bidens subalternans* DC. as casual in Tivoli and Genzano cities and the Aurunci Mountains. Recently, the species was considered as naturalized in Latium in Frosinone and Rome Provinces [93]. In any case, the species was not reported for Rome city. My discovery (a dense population growing on a little bridge of channel *Acqua Mariana*) represents, therefore, the first one for the city, and it can be considered as casual for the moment (Figure 58).

Figure 57. *Anreedera cordifolia* on *Rubus ulmifolium* dominated community (Acquedotti locality).

Figure 58. *Bidens subalternans* in Acquedotti locality.

***Chlorophytum comosum* (Thumbs.) Jacques**: this species was recently recorded for the first time in Latium (Viterbo Province, Orte Scalo locality) as casual [91]. My discovery on the banks of channel *Acqua Mariana* (Acquedotti locality; Figure 59) represents the first one for Rome Province and the second at the regional level. I first observed the population in 2017, and it persisted during the years (last observation in June 2022), but the population does not spread and is, therefore, considered as casual.

Figure 59. *Chlorophytum comosum* along channel Acqua Mariana (Acquedotti locality).

***Diospyrus kaki* L.**: this species was reported for Latium in both *Flora of Italy* [33] and, along the river Tevere, in *Flora of Lazio* [92]; contrarily, it was not listed either in *Flora of Rome* [9] or in *Atlante della flora vascolare del Lazio* [67]. Some individuals, which bloom and fruit regularly, were traced in Caffarella valley (Figure 60), thus confirming the occurrence of *Diospyrus kaki* in the city.

***Papaver somniferum* L.**: this species was listed in the *Flora of Rome* as "doubtfully alien" [9], but not later reported for the Italian capital in *Atlante della flora vascolare del Lazio* [67]. Scattered individulas were found in Caffarella valley, thus confirming the occurrence of *P. somnifermum* in Rome (Figure 61).

***Passiflora caerulea* L.**: this species was reported in the *Flora of Rome* as casual [9], but not later indicated for the city in *Atlante della flora vascolare del Lazio* [67]. I found various individuals growing on the Claudio's aqueduct (Figure 62), thus confirming the occurrence (as casual) of *Passiflora caerulea* in Rome.

***Zantedeschia aetiopica* (L.) Spreng.**: this African species is not currently reported for Rome [9,67]. In Appia Antica Regional Park, individuals were found along channels in Caffarella valley and, especially, in the Acquedotty locality, where the species is able to spread along the channel Acqua Mariana (Figure 63). I hereby consider this alien as naturalized.

Figure 60. *Diospyrus kaki* along a channel in Caffarella valley.

Figure 61. *Papaver somniferum* in Caffarella valley.

Figure 62. *Passiflora caerulea* (blue arrow) associated with *Capparis orientalis* (red arrow) on the ancient Roman Claudio's aqueduct.

Figure 63. *Zantedeschia aetiopica* along channel Acqua Mariana (Acquedotti locality).

4.4.6. Species No Longer Recorded

Literature analysis and herbaria investigations allow the verification of nine species collected in the past in Appia Antica Regional Park, but no longer recorded (see Appendix B). I never found any population of these species during the field surveys. Seven species are native, and two (*Ehrhata erecta* Lam. and *Tarenaya spinosa* (Jacq.) Raf.) are allochthonous (native, respectively, to South America and Africa).

***Astragalus glycyphyllos* L.**: this species was recorded in Caffarella valley in the *Centuriae XII* of *Florae Romanae Prodromus* [19] pag. 240 and not listed among the flora of Caffarella valley [25]. No specimen collected in the territory of Appia Antica Regional Park was traced. *A. glycyphyllos* is not a common species in the territory of Rome Municipality, although it is very common in the Latium region as a whole [92]. Although the study area was investigated in depth for more than 10 years through field surveys, further field investigations are needed to verify its occurrence in the park.

***Catabrosa aquatica* (L.) P.Beauv.**: this species was recorded in Caffarella valley in *Egeria nymphaeum* ("nella Grotta della Ninfa Egeria" = in the cave of Nymph Egeria) in the *Centuriae XII* of *Florae Romanae Prodromus* sub *Aira aquatica* L., [19] pag. 38. *C. aquatica* was not listed in the flora of either Caffarella valley [25] or Rome city, in the *Flora of Lazio* [92]. No specimen collected in Caffarella valley was traced. However, I traced one specimen (included in Montelucci's Herbarium at RO; collection number 5621) bearing a plant collected by G. Montelucci in April, 20 (year 1944) at "Via Appia Nuova, lungo fossetto tra Ciampino e S. Maria delle Mole. Luoghi aquitrinosi" (= Appia Nuova street, along channel between Ciampino and S. Maria delle Mole. swampy sites). The area around these two localities (Ciampino and S. Maria delle Mole) would represent the southernmost part of Appia Antica Regional Park, at least partially (the part on the west side of Appia Nuova street). However, the landscape configuration has deeply changed in this sector of the park over time, especially with new buildings. Therefore, the "swampy sites" cited by G. Montelucci in the label, which could be at that time, were probably destroyed; I will verify the absence of this type of habitat during field trips. The species is potentially extinct in the territory of the park.

***Ehrharta erecta* Lam.**: this species is listed in the flora of Caffarella valley [25] as no longer recorded. This indication was based on a specimen (deposited at RO; Figure 64) collected by G. G. B. Cuboni on 17 March 1876 at "Valle della Ninfa Egeria" (= Valley of Ninfa Egeria). *E. erecta* is an alien species for Italy [31], and its ability to reproduce is probably very low in Latium; this could be the reason why this species was no longer recorded.

***Linaria pelisseriana* (L.) Mill.**: this species was recorded in Caffarella valley in the *Centuriae XII* of *Florae Romanae Prodromus* [19] pag. 203; it was not cited in the flora of Caffarella valley [25]. Only one specimen (deposited at RO) collected in the territory of Appia Antica Regional Park (Ardeatina street on 15 May 1892) was traced (Figure 65). *L. pelisseriana* is a therophyte with a very short life cycle (flowering time March–April [28]), being also very rare in the territory of Rome Municipality. However, it is common in Latium [92]. Although the study area was investigated in depth for more than 10 years through field surveys, I cannot consider this species as extinct (or potentially so), and I think that further field trips are necessary to verify its presence in the park.

***Parapholis cylindrica* (Willd.) Romero Zarco**: this species was recorded in Caffarella valley (sub *Rottboellia subulata* Savi) in the *Centuriae XII* of *Florae Romanae Prodromus* ([19], pag. 62]), but not reported in the flora of Caffarella valley [25] and Rome [9], or Rome city in the *Flora of Lazio* [92]. Only one specimen (deposited at RO) collected in the territory of Appia Antica Regional Park (Caffarella on XIX century) was traced (Figure 66). *P. cylindrica* is a species growing mainly along Mediterranean coasts (rarely on inlands) on subsaline and clay soils [28]. Subsaline soils are not present in the territory of the park to date, and it is possible that the habitat was lost or destroyed. The species, which is considered rare in Italy as a whole [28], is potentially extinct in Appia Antica Regional Park.

***Polycnemum heuffelii* Láng**: the occurrence in Italy of this European species (distributed from Poland to Greece) is based on just one specimen (deposited at RO; Figure 67), collected by A. Cacciato along Appia Pignatelli street on 6 August 1966 [94]. The Flora of Italy [33] report this species as cryptogenic. The habitat in which this species was discovered (roadsides) was exposed to various types of human activities during the time; this factor could have caused the extinction of the Roman population.

***Silene gallinyi* Rchb.**: this species was recorded in Caffarella valley (sub *Silene trinervia* Sebast. & Mauri) in *Florae Romanae Prodromus* [19]. *S. trinervia* was there firstly described with *Locus classicus* including the locality Caffarella ("Alla Caffarella presso il Fonte di Egeria..." = At Caffarella near Egeria's spring). This *Silene* species was not listed in the flora of Caffarella valley [25]. After Herbaria checking (including RO, where original collections of A. Sebastiani and E. Mauri are currently deposited [95,96]), I verified that the last collections of this species were dated June–July 1980 (three specimens in Anzolone's Herbarium at RO; Figure 68). *S. gallinyi* is a species that can be easily distinguished from the other *Silene* taxa occurring in the territory of the park [*S. bellidifolia* Jacq., *S. conica* L., *S. gallica* L., *S. italica* (L.) Pers. subsp. *italica*, *S. latifolia* Poir., *S. nocturna* L., *S. pendula* L., *S. vulgaris* (Moench) Garcke subsp. *tenoreana* (Colla) Soldano & F.Conti, *S. vulgaris* (Moench.) Garcke subsp. *vulgaris*]. Despite this, anf the more than 10 years of field surveys, some carried out specifically to search for *S. gallinyi*, I did not yield positive results. My opinion is that this species is extinct throughout the park.

Stachys germanica* L. subsp. *germanica: this species was recorded in Caffarella valley in *Egeria nymphaeum* ("presso la Grotta di Egeria" = near the cave of Egeria) in the *Centuriae XII* of *Florae Romanae Prodromus* ([19] pag. 194). *S. germanica* s.str. was not listed in the flora of Caffarella valley [25]. Five specimens (at RO) collected in the territory of Appia Antica Regional Park [Caffarella valley (May 1829) and in uncultivated lands near Cecilia Metella sepulcher (3 June 1922)] were traced (Figure 69). *S. germanica* subsp. *germanica* is not a common species in the territory of Rome Municipality, although it is very common in the Latium region as a whole [92]. Plants of this species are easy to see in field due both to their size (30 to 60 cm on average) and their hairiness (densely white-woolly). Despite these facts, I never found it during >10 years of field surveys. The species is potentially extinct in the territory of the park.

***Tarenaya spinosa* (Jacq.) Raf.**: this species is recorded in the Latium region based on a specimen (deposited at RO; Figure 70) collected by A. Cacciato along Appia Pignatelli street in 1966 [78,92]. It was not reported in the *Flora fo Rome* [9]. I never found any individual of *Tarenaya spinosa* during field surveys. *T. spinosa* is no longer found for two possible reasons: a low ability to reproduce and modification of its habitat (roadsides) by humans.

Figure 64. Specimen of *Ehrharta erecta* collected in Caffarella valley on 17 March 1876 (RO-Herbarium Romano no. 51234).

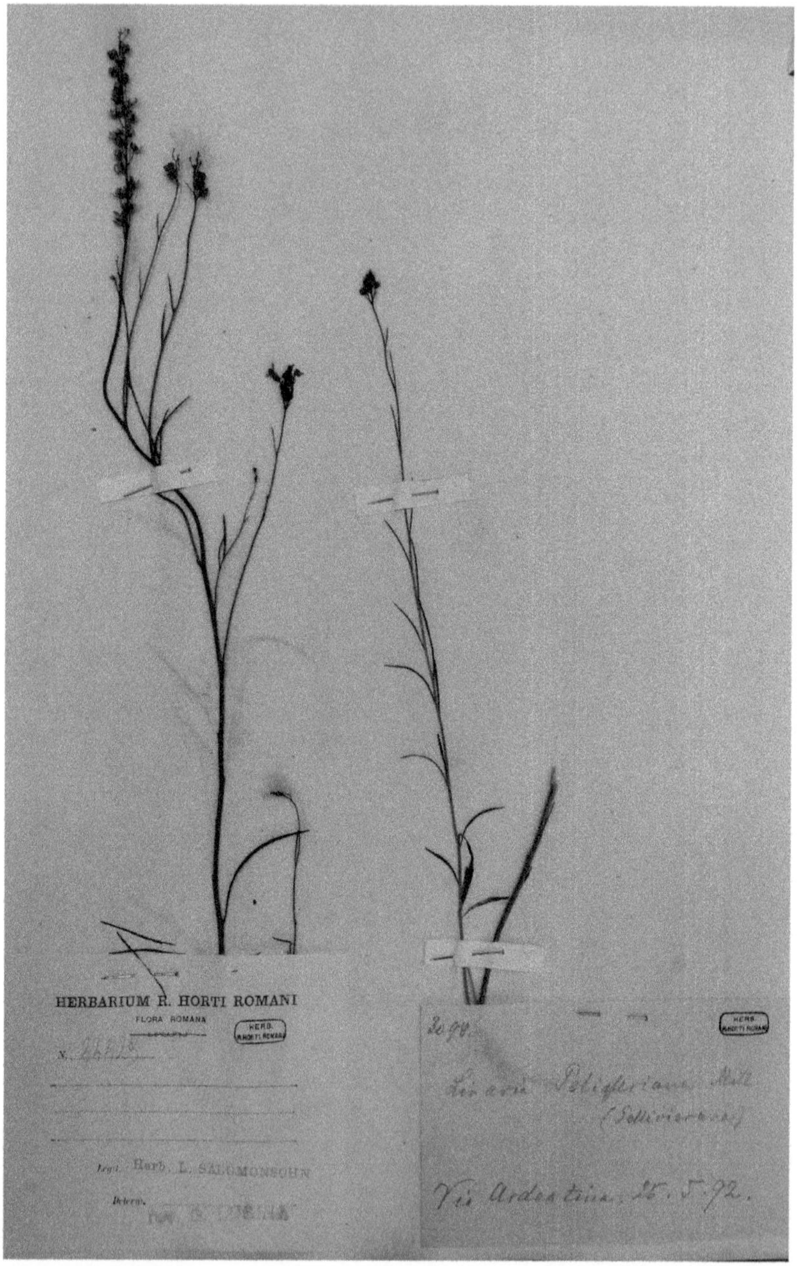

Figure 65. Specimen of *Linaria pelisseriana* collected along Ardeatina street on 15 May 1892 (RO-Herbarium Romano no. 34429).

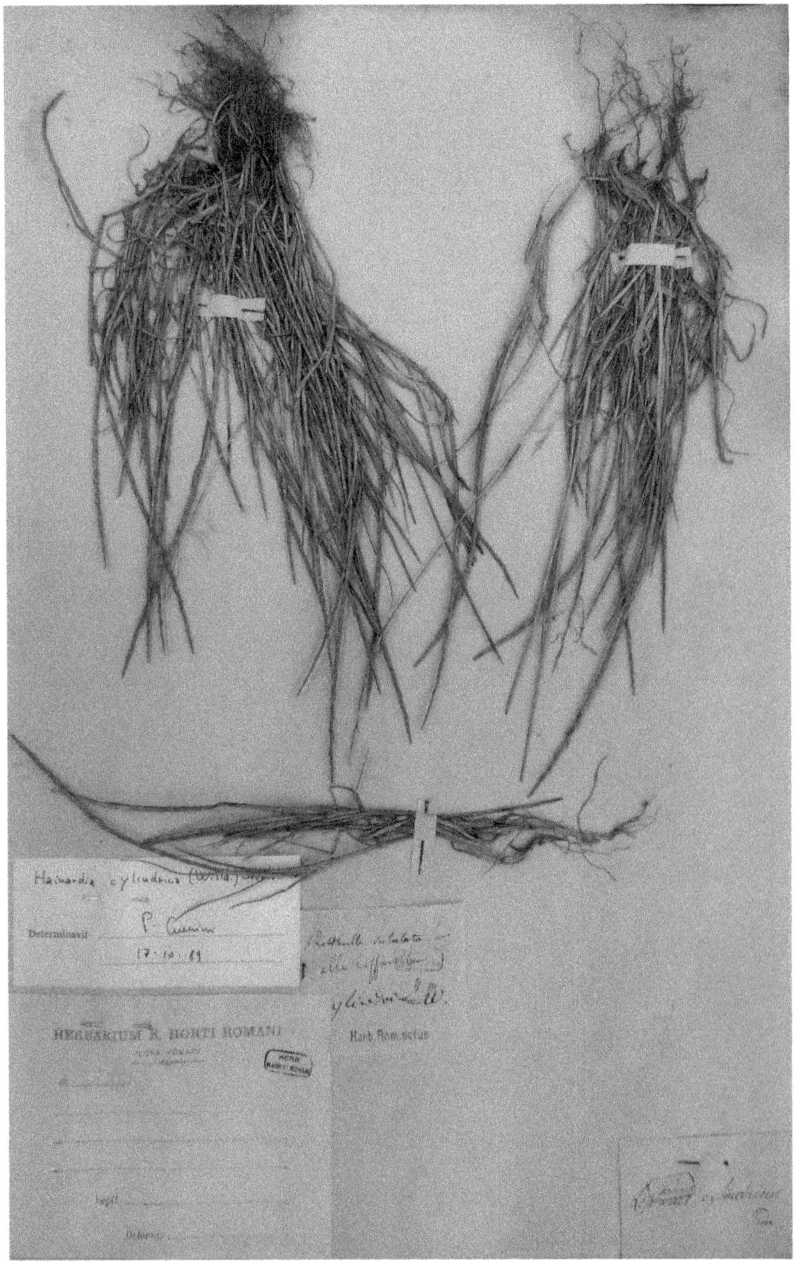

Figure 66. Specimen of *Parapholis cylindrica* collected in Caffarella valley in the 19th century (RO-Herbarium Romano no. 55564).

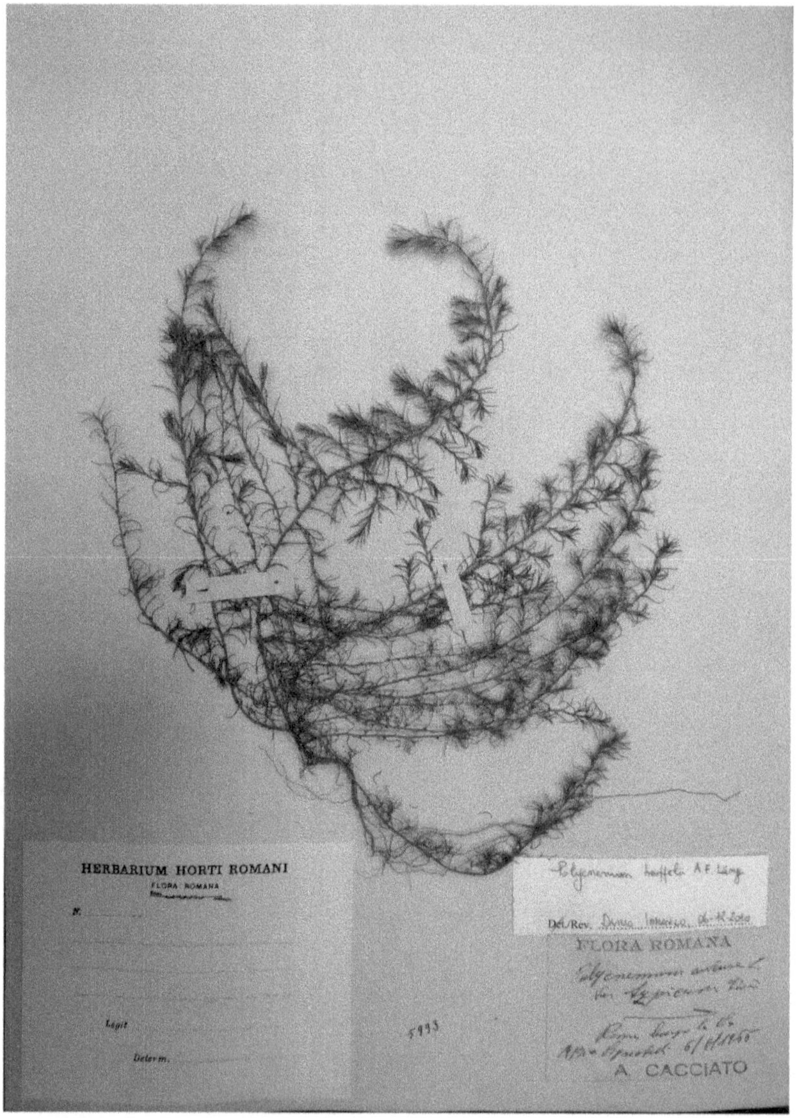

Figure 67. Specimen of *Polycnemum heuffelii* collected along Appia Pignatelli street in august 1966 (RO-Herbarium Romano no. 5993).

Figure 68. Specimen of *Silene gallinyi* (sub *S. trinervia*) collected in Caffarella valley in July 1980 (RO-Herbarium Anzalone no. 13276).

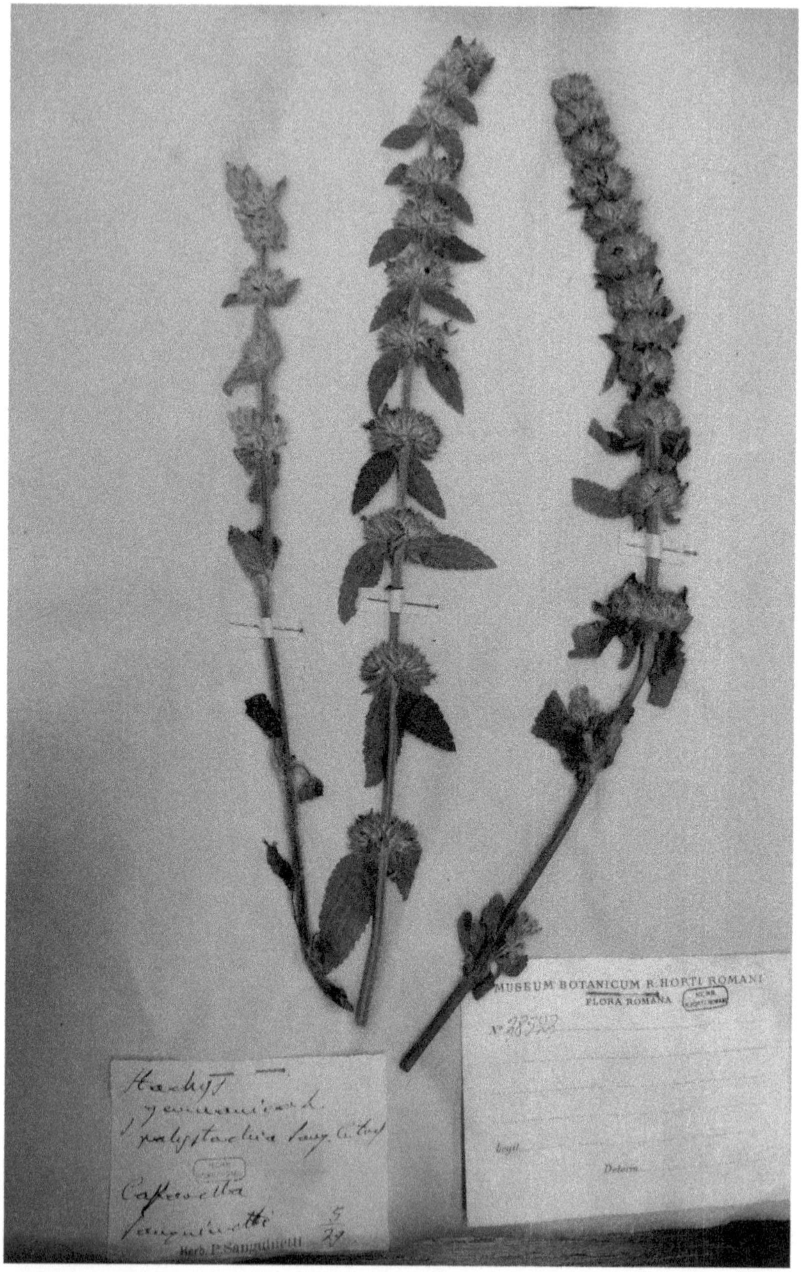

Figure 69. Specimen of *Stachys germanica* L. subsp. *germanica* collected in Caffarella valley in May 1829 (RO-Herbarium Romano no. 38523).

Figure 70. Specimen of *Tarenaya spinosa* collected along Appia Pignatelli street in august 1965 (RO-Herbarium Romano s.n.).

4.4.7. Species Having Loci Classici and/or Nomenclatural Types Collected in Appia Antica Regional Park

Biarum tenuifolium* (L.) Schott subsp. *tenuifolium: the basionym *Arum tenuifolium* L. was described by Linnaeus in the first edition of his *Species Plantarum* [97], where the provenance "*Habitat in* Romae" was reported. The Linnaean name was recently lectotypified by Iamonico [98] on a specimen preserved at BM (barcode BM000647349); further, an epitype collected in the Caffarella valley (20 August 2015) was designated (the specimen kept at HFLA; Figure 71) to serve as an interpretative type according to the current concept in *Arum* (see e.g., [99]).

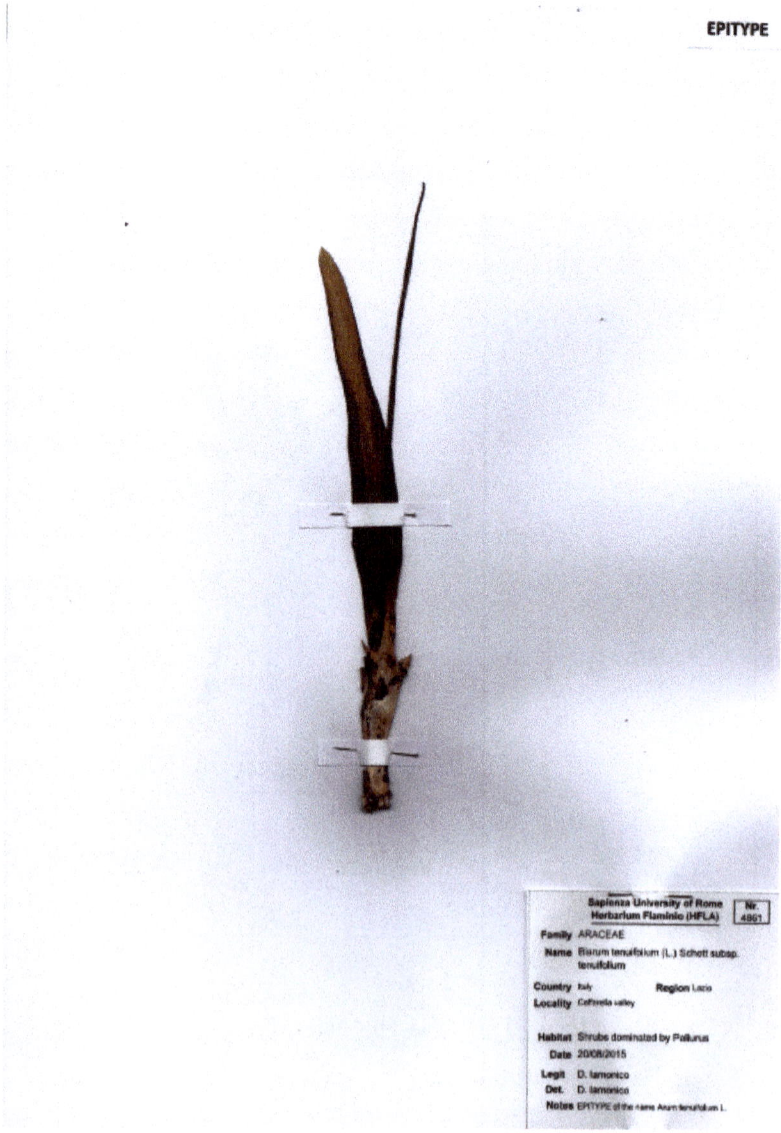

Figure 71. Epitype of *Biarum tenuifolium* subsp. *tenuifolium* (HFLA no. 4861).

Epilobium lanceolatum **Sebast. & Mauri**: this species was described in 1818 in *Florae Romanae Prodromus* [19] pag. 138. *Locus classicus* includes the locality Caffarella ("In umbrosis, ad oras nemorum, sepes circa Romam frequens. Copiosamente intorno ai boschetti della Caffarella presso la Grotta di Egeria..." = In the shades, at the edge of the woods, hedges about Rome. Copiously around small forest of Caffarella near Egeria's cave). Iamonico and collaborators [100] designated as lectotype a specimen (deposited at RO; Figure 72) collected by E. Mauri and A. Sebastiani on 3 June 1812.

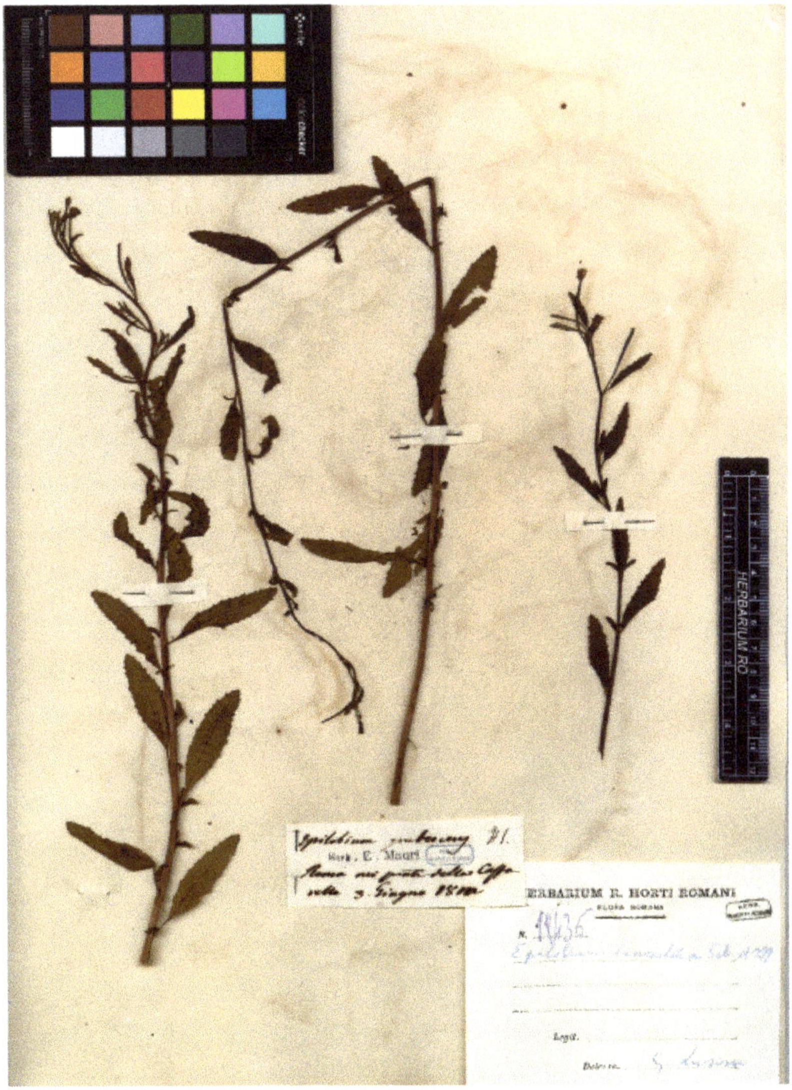

Figure 72. Lectotype of *Epilobium lanceolatum* (RO no. 19436).

Silene trinervia **Sebast. & Mauri**: this species was described in 1818 in *Florae Romanae Prodromus* ([19] pag. 152), and *locus classicus* includes the locality Caffarella ("Alla Caffarella presso il Fonte di Egeria..." = At Caffarella near Egeria's spring). Lacking specimens of original material, Iamonico [101] designated an iconography (Table II in [19]) as lectotype; further, a specimen (deposited at RO; Figure 73) collected by E. Mauri in Caffarella locality in July 1832 was designated as epitype. *Silene trinervia* is currently a heterotypic synonym of *S. gallinyi* Rchb.

Figure 73. Epitype of *Silene trinervia* (RO-Herbarium Romano no. 6762).

***Typha latifolia* L.**: on the basis of a recent study on some Linnaean names of aquatic plants by Iamonico and Iberite [102], an epitype was designated for this Linnaean name. The exsiccatum, deposited at RO (Figure 74), was collected in Caffarella valley by D. Iamonico on 14 October 2020. The population from which the epitype was collected is that shown in Figure 28 of the present paper.

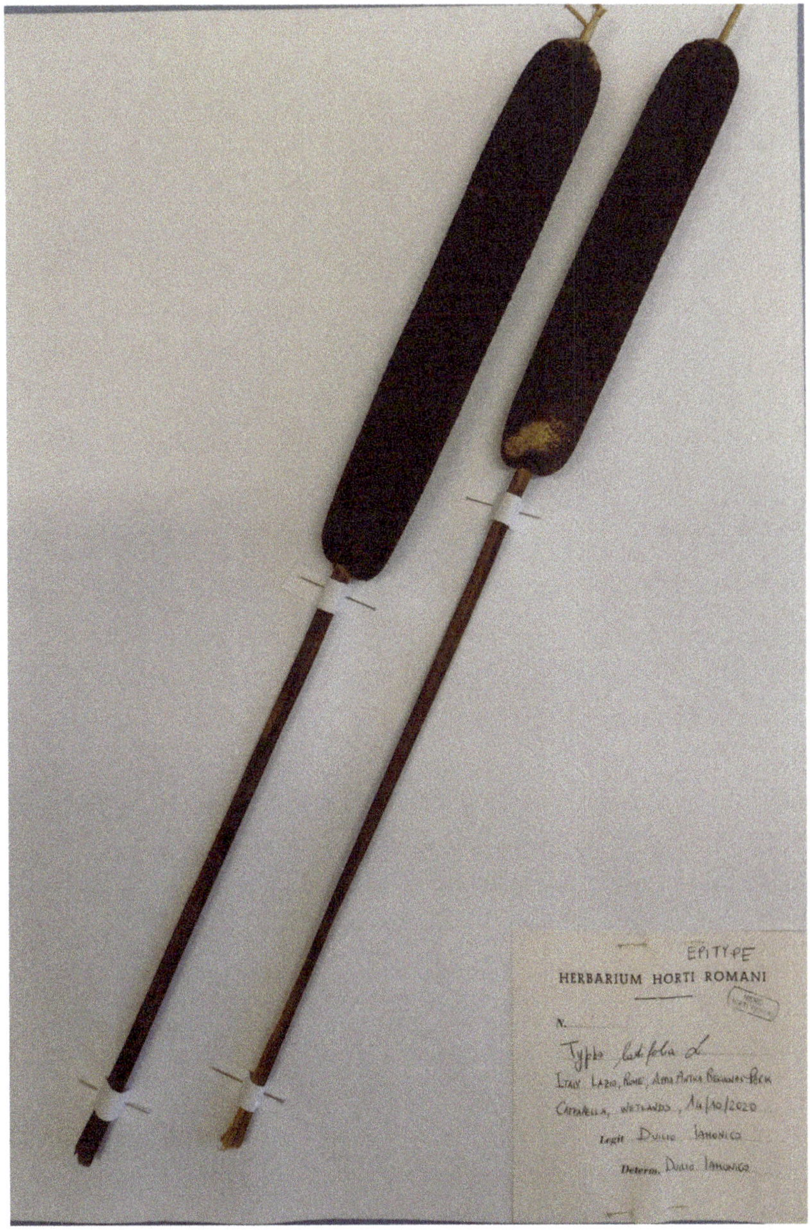

Figure 74. Epitype of *Typha latifolia* (RO s.n.).

4.4.8. Other Notable Species

Amaranthus hypochondriacus **L.**: this species, indicated as casual alien for Italy [33], was recently reported as naturalized in the Latium region [103] based on two populations occurring in Roma and Frosinone Provinces. The population in Rome refers to 10–15 individuals growing in the territory of Appia Antica Regional Park, i.e., in the Acquedotti locality along the channel Acqua Mariana (Figure 75). The presence of *A. hypochondriacus* has been documented for 6 years.

Figure 75. *Amaranthus hypochondricus* in Acquedotti locality.

Colocasia esculenta **(L.) Schott**: this tropical/subtropical Asian species was recently recorded in Latium for the first time as naturalized (first alien status for Italy). The population was found in the Acquedotti locality along channel *Acqua Mariana* (first observation in 2015, last one in 2019 according to Iamonico [104]). The author, by comparing the climate features of Rome and the native distribution range of *C. esculenta*, showed that the occurrence of this species in Rome is probably linked to micro-climatic factors, i.e., (1) soil (sandy sediments submerged during autumn and winter seasons, well-drained, and partially soaked during spring and summer); (2) brightness (low light intensity, which characterizes the site over almost the entire day); (3) air (high humidity related to both the morphology of the site (a gorge) and the close occurrence of a small waterfall (height: 2.5–3.0 m).

I continue to control the population and found it also during the next years (2021 and 2022; Figure 76).

Figure 76. *Colocasia esculenta* along channel Acqua Mariana (Acquedotti locality).

***Lemna minuta* Kunth**: this South American species was first discovered in the Latium region based on collections made in 2007, of which one refers to the Caffarella valley [105]. I am continuing to control the various populations, founding many others both in Caffarella valley and in other parts (mostly in Acquedotti and Tor Marancia localities) where the abundance of the autochthonous *L. minor* decreased over time. Note that the populations of *L. minuta* even survived well through snowfall (Figure 77).

Figure 77. *Lemna minuta* in aquatic habitat of *Egeria nymphaeum* during snowfall in February 2018 (Caffarella valley).

Lupinus albus **L. subsp.** ***graecus*** **(Boiss. et Spruner) Franco & Pinto da Silva**: 11 localities of this taxon currently occur in the Latium region, one referring to Caffarella valley [25,106]. I monitored this population over the years, but in 2018 the volcanic slope on which it occurred was destroyed to build paths (Figure 78). During the next 2 years, a part of the area previously occupied by *L. albus* subsp. *graecus* was recolonized by the South American alien *Nassella neesiana* (Trin. et Rupr.) Barkworth (Figure 78); the population of *L. albus* subsp. *graecus* recorded by [25,106] is extinct. Fortunately, I found another population not far from the lost one (never seen before 2018) composed of tens of individuals blooming and fruiting regularly (Figure 79).

Figure 78. *Lupinus albus* subsp. *graecus* in Caffarella valley: population lost (**top** photo, dated 2017); *Nassella neesiana* dominated community (**bottom** photo, dated 2022).

Figure 79. *Lupinus albus* subsp. *graecus* in Caffarella valley: new population found.

***Plumbago auriculata* Lam.**: this species was recently recorded in Latium (year 2016) as casual [107]. My discovery in the northeastern sector of Caffarella valley represents the second record for Latium. The population found occurs in shrubby vegetation dominated by *Rubus ulmifolium* and *Sambucus nigra* (Figure 80).

Figure 80. *Plumbago auriculata* in shrubby vegetation (Caffarella locality).

5. Discussion and Conclusions

The present study provides the first comprehensive inventory of the flora of Appia Antica Regional Park, one of the largest protected urban areas in Europe.

The data presented revealed an extraordinary species diversity of the flora, which comprises 714 taxa representing about 43% of the flora of Rome (1649 taxa according to [9]) and about 20% of the flora of the Latium region (3593 taxa according to [108]). This notable datum is mainly linked to the high landscape heterogeneity of the park, which comprises several types of habitats, from those strictly natural (e.g., broad-leaved forests, Mediterranean macchia, Mediterranean riparian woods, humid meadows, helophytic vegetation, etc.) to those strictly anthropogenic (i.e., the segetal and ruderal communities and floras of, respectively, arable crops and disturbed/human-made sites). The main landscape environmental characteristics of the studied area can be, therefore, summarized as:

(1) the persistence of residual patches of natural vegetation (especially in the northern sector of the park), notably concerning those forests (Figure 15) that reveal a rich woody flora (106 taxa of Phanerophytes, corresponding to 14.85% of the total flora). Some patches also have high cultural-historical value, e.g., the so-called "Bosco Sacro" (= Sacread Wood) with various centenarian individuals of *Quercus ilex* subsp. *ilex* (see also [25]); other trees, found throughout the park, are very large and have, therefore, high nature conservation value (Figure 81);

Figure 81. Large individual of *Quercus ilex* subsp. *ilex* in Caffarella valley.

(2) the occurrence of quite well preserved aquatic habitats and humid meadows (mostly in Caffarella valley and Tor Marancia locality; Figure 82) that include taxa and vegetation communities not so common in urban areas where their general decline is a widely acknowledged trend across not only Rome [9], but also worldwide [109,110]. Of note, some of these habitats are included in the Annex I of the Habitat Directive of 21 May 1992 (formally known as Council Directive 92/43/EEC on the Conservation of natural habitats and of wild fauna and flora), e.g., that named "*Salix alba* and *Populus alba* galleries" (code 92A0; see Figure 22).

Figure 82. Humid meadow dominated by *Ranunculus repens* L. (Caffarella valley).

(3) a rich anthropogenic flora occuring in cultivated lands (segetal taxa; Figure 29) and human-made habitats (ruderal taxa), e.g., those trampled (Figure 31) and/or grasslands and pastures occurring on nutrient-rich soils (e.g., Figures 33 and 34). Notably, this type of flora plays an important role as a source of floristic richness in metropolitan areas [9,111–113]. It is also noteworthy that some species of conservation interest grow in anthropogenic environments, as crop wild relatives [114,115].

The anthropogenic flora includes, in addition to native species well-adapted to human disturbance, exotic taxa that decrease the quality of the biodiversity and can negatively impact the natural vegetation. Some naturalized and invasive species are particularly dangerous in the territory of the park and threaten native ones; examples are:

- *Robinia psudoacacia* and *Ailathuts altissima*: these two trees (both invasive in the park) are common along edges of forests and into shrubby mantels, as well as in human-made habitats and on ruins. In various cases, *R. psudoacacia* and *A. altissima* even form more or less dense woody communities (Figures 17 and 85);
- *Lemna minuta*: a North American natant hydrophyte, invasive in the park (Figure 83), which caused the decrease (or even the local extinction) of several populations of the autochthonous *L. minor* (see also [116]);
- *Lonicera japonica* Thumb.: this species is considered to be a naturalized alien in the park, occurring in several sites. In some cases (e.g., along banks of channel *Acqua Mariana*, in the Acquedotti locality), it forms dense populations that cover shrubby native vegetation communities of *Rubus ulmifolius* Schott, *Cornus sanguinea* L., *Crataegus monogyna* Jacq., *Phillyrea latifolia* L. and *Viburnum tinus* L. (Figure 86), or climbs on young trees of *Populus nigra* L.;
- *Fallopia baldschuanica* (Regel) Holub: this climbing species is naturalized in the park, where it forms often dense populations growing on shrubs (Figure 84) and competing with autochthonous liana taxa, e.g., *Clematis vitalba* L., *Humulus lupulus* L., or *Lonicera etrusca* Santi.

Figure 83. *Lemna minuta* monophytic community on a pond in Caffarella valley.

Figure 84. *Fallopia baldschuanica* on shrubby vegetation along a path running near Appia Nuova street.

Figure 85. Anthropogenic woody vegetation dominated by *Ailanthus altissima* (Caffarella valley).

Figure 86. *Lonicera japonica* on *Crataegus monogyna* (**left** photo) and *Viburnum tinus* (**right** photo) (Acquedotti locality).

Concerning the segetal flora, some allochthonous species impact the wheat crops from an economic point of view, since populations densely grow on fields, imposing significantly increased costs for their management and reducing wheat yields [29,117]. *Amaranthus retroflexus* L. and *A. hybridus* L. are the main dangerous species observed in the park (Figure 87).

Figure 87. Dense population of young individuals of *Amaranthus sp. pl.* on a crop (Acquedotti locality).

(4) the flora of archeological sites: these areas have a significant role in enhancing plant diversity in cities, being refuges of natural flora and vegetation in the urban ecosystems [9,118–120].

Appia Antica Regional Park is rich in archeological elements (e.g., aqueducts, catacombs, churches, tombs, sepulchers, temples, etc.) that also significatively contribute to the landscape structure (see Section "3.5. Landscape remarks and actual vegetation physiognomies"). Several taxa occur on both the top and vertical walls of the ruins (see also [120]), and they consist of herbs (therophytes, hemicryptophytes, and geophytes), subshrubs (chamaephytes), shrubs (caespitose phanerophytes), or even trees (scapose phanerophytes). Various species have cliffs or rocky places as their natural habitat and found the same dry environmental conditions by growing on archeological remains, which, therefore, can be considered as secondary habitats for them; examples of species with different biological forms that occur in the park are (Figure 88): *Fumaria officinalis* L. subsp. *officinalis* (therophyte), *Umbilicus rupestris* (Salisb.) Dandy (bulbose geophyte), *Micromeria graeca* (L.) Benth. ex Rchb. (suffruticose chamaephyte), *Capparis orientalis* (nano-phanerophyte), *Hedera helix* L. subsp. *helix* (liana), and *Olea europea* L (scapose phanerophyte).

In some cases, the archeological sites allow the presence of species not common in the whole region of Latium. An example is *Parietaria lusitanica* L. subsp. *lusitanica*, which occurs in Rome territory mostly in the northern sector, whereas only three scattered populations are currently recorded toward the south, and one of them is located in Appia Antica Regional Park on walls along Appia Antica street near Cecilia Metella sepulcher [121].

Finally, alien species also occur on these sites and they can inflict damage on their structures, mainly due to their roots, which induce both chemical and mechanical forms of deterioration [122]. Examples for Appia Antica Park are *Ailanthus altissima*, *Phoenix canariensis* H.Wildpret, and *Robinia pseudoacacia*, which can be observed at the base or on the vertical walls of ancient Roman aqueducts (Figure 89).

Figure 88. *Fumaria officinalis* subsp. *officinalis* on Felix's Aqueduct (top-left photo); *Umbilicus rupestris* on the vertical wall of the aqueduct located in front of the *Egeria* nymphaeum, in Caffarella valley (top-right photo); *Micromeria graeca* on Claudio's Aqueduct (central-left photo); *Capparis orientalis* on Felix's Aqueduct (central-right photo); *Hedera helix* subsp. *helix* on Felix's Aqueduct (bottom-left photo); *Olea europaea* on Claudio's Aqueduct (bottom-right photo).

Figure 89. *Ailanthus altissima* (**top** photo) and *Phoenix canariensis* (**bottom** photo) at the base of Felix's Aqueduct (Tor Fiscale and Acquedotti localities, respectively).

In conclusion this study, by providing (1) an extensive set of floristic data on the species diversity of Appia Antica Regional Park, including its structure in biological, ecological and biogeographical terms and floristic novelties at the regional, national and European levels, (2) an overview of the landscape structure and vegetation physiognomies, and (3) an emphasis on the non-native flora and its ecological, social and economic impacts on autochthonous flora, archeological ruins and crops, gives a general base framework for guiding future scientific and applied researches and landscape action plans. Concerning basic scientific studies, the first one should investigate the vegetation communities in detail (based on the physiognomies listed in the present paper and the Land Use map published by Iamonico [40]) carried out on phytosociological plots. With regard to applied research and landscape action plans, there are many possibilities, e.g., eradication of dangerous alien species (see e.g., [123]), planning of education paths for natural flora and vegetation (see e.g., [124]), urban forestry (see e.g., [125]), etc. All of these research efforts and actions have, as their final aim, the conservation of biodiversity in terms of environmental sustainability [126], providing, in turn and by green infrastructure and ecosystem services implementation, benefits for urban residents in the form of improved human health and well-being (see e.g., [127,128]).

Funding: No external funding was obtained for this research.

Institutional Review Board Statement: Not applicable.

Informed Consent Statement: Not applicable.

Data Availability Statement: Not applicable.

Acknowledgments: I acknowledge the Directors and Curators of the herbaria mentioned for their support during my visits or for loans of specimens/photographs, with special thanks to D. Magri, M. Iberite, A. Tilia, and M. Pierfranceschi (Herbarium RO). Thanks are due M. Del Monte (Univerisity of Rome Sapienza, Italy) for the information on geomorphological literature; to E. Fanfarillo (University of Siena, Italy) for the discussion about the segetal flora; to M. Arianoutsou (University of Athens, Greece) and F. Verloove (Meise Botanic Garden, Belgium) for the information about the occurrence of *Rosa chinensis* in Greece and Belgium; to F. Luebert (University of Bonn) for the help in identification of *Heliotropium amplexicaule*. I also extend my thanks to L. Virzì (Rome) for permission to reproduce his picture of the mine "Cava di Fioranello" (Figure 8).

Conflicts of Interest: The author declares no conflict of interest.

References

1. Knapp, S.; Aronson, M.F.J.; Carpenter, E.; Herrera-Montes, A.; Jung, K.; Kotze, D.J.; La Sorte, F.A.; Lepczyk, C.A.; Macgregor-Fors, I.; Macivor, J.S.; et al. A Research Agenda for Urban Biodiversity in the Global Extinction Crisis. *Bioscience* **2021**, *71*, 268–279. [CrossRef]
2. Fineschi, S.; Loreto, F. A Survey of Multiple Interactions Between Plants and the Urban Environment. *Front. For. Glob. Change* **2020**, *3*, 30. [CrossRef]
3. Gaertner, M.; Wilson, J.R.U.; Cadotte, M.W.; Maclvor, J.S.; Zeni, R.D.; Richardson, D.M. Non-native species in urban environments: Patterns, processes, impacts and challenges. *Biol. Invasions* **2017**, *19*, 3461–3469. [CrossRef]
4. Corlett, R.T. Plant diversity in a changing world: Status, trends, and conservation needs. *Plant Divers.* **2016**, *38*, 10–16. [CrossRef]
5. United Nations. Department of Economic and Social Affairs, Population Division. *World Population Prospects 2019*; United Nations, Department of Economic and Social Affairs, Population Division: New York, NY, USA, 2019.
6. Wang, M.; Li, J.; Kuang, S.; He, Y.; Chen, G.; Huang, Y.; Song, C.; Anderson, P.; Łowicki, D. Plant Diversity Along the Urban–Rural Gradient and Its Relationship with Urbanization Degree in Shanghai, China. *Forests* **2020**, *11*, 171. [CrossRef]
7. Jovanović, S.; Glišić, M. Review of research on urban flora and vegetation in Southeast Europe. *Acta Bot. Croat.* **2021**, *80*, 74–81. [CrossRef]
8. Talal, M.L.; Santelmann, M.V. Plant Community Composition and Biodiversity Patterns in Urban Parks of Portland, Oregon. *Front. Ecol. Evol.* **2019**, *7*, 201. [CrossRef]
9. Celesti-Grapow, L.; Capotorti, G.; Del Vico, E.; Lattanzi, E.; Tilia, A.; Blasi, C. The vascular flora of Rome. *Plant Biosyst.* **2013**, *147*, 1059–1087. [CrossRef]
10. Kowarik, I. Novel urban ecosystems, biodiversity, and conservation. *Environ. Pollut.* **2011**, *159*, 1974–1983. [CrossRef]
11. Sharrock, S. *GSPC, Global Strategy for Plant Conservation. A Guide to the GSPC. All the Targets, Objectives and Facts*; Botanic Gardens Conservation International: Richmond, VA, USA, 2011.

12. Ricotta, C.; Di Nepi, M.; Guglietta, D.; Celesti-Grapow, L. Exploring taxonomic filtering in urban environments. *J. Veg. Sci.* **2008**, *19*, 229–238. [CrossRef]
13. Ziller, S.R.; de Sá Dechoum, M.; Silveira, R.A.D.; da Rosa, H.M.; Motta, M.S.; da Silva, L.F.; Oliveira, B.C.M.; Zenni, R.D. A priority-setting scheme for the management of invasive non-native species in protected areas. *NeoBiota* **2020**, *62*, 591–606. [CrossRef]
14. Banfi, E.; Galasso, G. La flora spontanea della città di Milano alle soglie del terzo millennio e i suoi cambiamenti a partire dal 1700. *Mem. Soc. Ital. Sci. Nat. Mus. Civ. Stor. Nat. Milano* **1998**, *27*, 267–387.
15. Martini, F. La flora vascolare spontanea della città di Trieste (Italia nordorientale). *Webbia* **2006**, *61*, 57–94. [CrossRef]
16. De Natale, A.; La Valva, V. The Urban Flora of Naples. *Webbia* **2000**, *54*, 271–375. [CrossRef]
17. Sebastiani, A. *Romanarum Plantarum Fasciculus Primus*; Typis De Romanis: Roma, Italy, 1813.
18. Sebastiani, A. *Romanarum Plantarum Fasciculus Alter. Accedit Enumeratio Plantarum Sponte Nascentium in Ruderibus Amphitheatri Flavii*; Typis Pauli Salviucci et Filii: Roma, Italy, 1815.
19. Sebastiani, A.; Mauri, E. *Florae Romanae Prodromus Exhibens Centuriae xii Plantarum Circa Romam et in Cisapenninis Pontificiae Ditionis Provinciis Sponte Nascentium Sexuali Systemate Digestas*; Vincenzo Poggioli, Reverendae Camerae ApostolicaeTypographum: Roma, Italy, 1818.
20. Blasi, C.; Frondoni, R. Modern perspectives for plant sociology: The case of eco-logical land classification and the ecoregions of Italy. *Plant Biosyst.* **2011**, *145*, 30–37. [CrossRef]
21. Iamonico, D.; Lorenzetti, R. La Flora vascolare della Riserva Naturale di Monte Mario (Roma): Studio di base ed analisi comparativa con la flora di altre aree verdi di Roma. *Riv. Piemont. Stor. Nat.* **2008**, *29*, 141–168.
22. Iamonico, D.; Lorenzetti, R. La Flora vascolare della Riserva Naturale del Laurentino–Acqua Acetosa (Roma). *Ecol. Urbana* **2008**, *20*, 3–20.
23. Iamonico, D.; Lorenzetti, R. Il Parco Urbano di Aguzzano (Roma): Studio floristico e considerazioni preliminari sul ruolo nella rete ecologica romana. *Ecol. Urbana* **2010**, *22*, 19–32.
24. Buccomino, G.; Lerzio, M.; Leporatti, M.G. Contributo alla conoscenza della Flora vascolare della Riserva naturale di Nomentum (Lazio). *Acta Plant. Notes* **2017**, *5*, 86–102.
25. Buccomino, G.; Stanisci, A. Contributo alla conoscenza floristica della valle della Caffarella (Roma). *Inform. Bot. Ital.* **2000**, *32*, 3–15.
26. Abbate, G.; Bonacquisti, S.; Giovi, E.; Iamonico, D.; Iberite, M.; Lorenzetti, R. The vascular flora of the Castelli Romani Regional Park (Rome, Central Italy): Recent observations and early herbarium surveys. *Webbia* **2009**, *64*, 47–74. [CrossRef]
27. Thiers, B. Index Herbariorum: A Global Directory of Public Herbaria and Associated Staff. New York Botanical Garden's Virtual Herbarium. Available online: http://sweetgum.nybg.org/ih/ (accessed on 8 April 2022).
28. Pignatti, S.; Guarino, R.; La Rosa, M. *Flora d'Italia*, 2nd ed.; Edagricole: Milano, Italy, 2017–2019; Volume 1–4.
29. Iamonico, D. Taxonomic revision of the genus *Amaranthus* (*Amaranthaceae*) in Italy. *Phytotaxa* **2015**, *199*, 1–84. [CrossRef]
30. Ardenghi, N.M.G.; Galasso, G.; Banfi, E.; Zoccola, A.; Foggi, B.; Lastrucci, L. A taxonomic survey of the genus *Vitis* L. (Vitaceae) in Italy, with special referenceto Elba Island (Tuscan Archipelago). *Phytotaxa* **2014**, *166*, 163–198. [CrossRef]
31. Bartolucci, F.; Peruzzi, L.; Galasso, G.; Albano, A.; Alessandrini, A.; Ardenghi, N.M.G.; Astuti, G.; Bacchetta, G.; Ballelli, S.; Banfi, E.; et al. An updated checklist of the vascular flora native to Italy. *Plant Biosyst. Int. J. Deal. All Asp. Plant Biol.* **2018**, *152*, 556–592. [CrossRef]
32. Galasso, G.; Conti, F.; Peruzzi, L.; Ardenghi, N.M.G.; Celesti-Frapow, L.; Albano, A.; Alessandrini, A.; Bacchetta, G.; Ballelli, S.; Bandini Mazzanti, M.; et al. An updated checklist of the vascular flora alien to Italy. *Plant Biosyst.* **2018**, *152*, 556–592. [CrossRef]
33. Portal to the Flora of Italy. Available online: http:/dryades.units.it/floritaly (accessed on 8 April 2022).
34. Pyšek, P.; Richardson, D.M.; Rejemánek, M.; Webster, G.L.; Williamson, M.; Kerschner, J. Alien plants in checklists and floras: Towards better communication between taxonomists and ecologists. *Taxon* **2004**, *53*, 131–143. [CrossRef]
35. Faccenna, C.; Funiciello, R.; Marra, F. Inquadramento geologico strutturale dell'area romana. In *La Geologia di Roma*; Funiciello, R., Ed.; Istituto Poligrafico e Zecca di Stato: Roma, Italy, 1995; pp. 31–47.
36. Funiciello, R.; Giordano, G.; De Rita, D. The Albano maar lake (Colli Albani Volcano Italy): Recent volcanic activity and evidence of pre-Roman Age catastrophic lahar events. *J. Volcanol. Geotherm. Res.* **2003**, *123*, 43–46. [CrossRef]
37. Del Monte, M. *Geomorphology of Rome*; Sapienza Università Editrice: Roma, Italy, 2020.
38. Del Monte, M.; D'Orefice, M.; Luberti, G.M.; Marini, R.; Pica, A.; Vergari, F. Geomorphological classification of urban landscapes: The case study of Rome (Italy). *J. Maps* **2016**, *12*, 178–189. [CrossRef]
39. Zepner, L.; Karrasch, P.; Wiemann, F.; Bernard, L. ClimateCharts.net–an interactive climate analysis web platform. *Int. J. Digit. Earth* **2021**, *14*, 338–356. [CrossRef]
40. Blasi, C.; Michetti, L. Carta del Fitoclima dell'area romana. *Inform. Bot. Ital.* **2001**, *33* (Suppl. 1), 14–18.
41. Iamonico, D. Multitemporal analysis of landscape of the Appia Antica Regional Park (Rome). *Eur. J. Remote Sens.* **2008**, *40*, 27–37. [CrossRef]
42. Iamonico, D.; Piccari, F.; Bonamico, A. Carta della vegetazione del Parco Regionale dell'Appia Antica (Lazio, Roma). In Proceedings of the 11ª Conferenza Italiana Utenti Esri, Roma, Auditorium del Massimo, Via Massimiliano Massimo 1, Roma, Italy, 21–22 Maggio 2008; 2008; Volume 30.
43. Mucina, L.; Bultmann, H.; Dierßen, K.; Theurillat, J.-P.; Raus, T.; Carni, A.; Sumberov, K.; Willner, W.; Dengler, J.; Gavil, R.; et al. Vegetation of Europe: Hierarchical floristic classification system of vascular plant, bryophyte, lichen, and algal communities. *Appl. Veg. Sci.* **2016**, *19* (Suppl. 1), 3–264. [CrossRef]

44. Chiocchini, U. The Roman cistern of road Cristoforo Colombo, Rome, Italy: Operating mode and connection with the old Appian Way. *Archaeol. Anthropol. Sci.* **2018**, *11*, 321–336. [CrossRef]
45. Capelli, G. La Marrana dell'acqua Mariana. Un corso d'acqua al servizio dei Papi. *Ital. J. Groundw.* **2015**, *2015*, 79–82. [CrossRef]
46. Funiciello, R.; Giordano, G.; De Rita, D.; Carapezza, M.L.; Barberi, F. L'attività recente del cratere del Lago Albano di Castelgandolfo. *Rend. Lincei. Sci. Fis. e Nat.* **2002**, *13*, 113–143. [CrossRef]
47. IFFI. Inventario dei Fenomeni Franosi in Italia. Cartografia on Line. Available online: https://www.progettoiffi.isprambiente.it/cartografia-on-line/ (accessed on 14 April 2022).
48. Ciotoli, G.; Nisio, S.; Serafini, R. Analisi della suscettibilità ai sinkholes antropogenici nel centro urbano di Roma. *Mem. Descr. Carta Geol. It.* **2015**, *XCIX*, 167–188.
49. Cifani, G. *Architettura Romana Arcaica: Edilizia e Società Tra Monarchia e Repubblica*; L'Erma di Bretschneider: Roma, Italy, 2008.
50. Chiocchini, U. Le voragini connesse all'attività estrattiva nella valle del Fosso Almone, Roma. *Mem. Descr. Carta Geol. It.* **2015**, *XCIX*, 111–138.
51. Fredi, P.; Ciccacci, S. A Route of Fire in Central Italy: The Latium Ancient Volcanoes. In *Landscapes and Landforms of Italy*; Soldati, M., Marchetti, M., Eds.; World Geomorphological Landscapes; Springer: Cham, Switzerland, 2017; pp. 303–315. [CrossRef]
52. Forman, R.T.T.; Godron, M. Patches and Structural Components for a Landscape Ecology. *Bioscience* **1981**, *31*, 733–740.
53. Blasi, C.; Milone, M.; Guida, D.; De Filippo, G.; Di Gennaro, A.; La Valva, V.; Nicoletti, D. Ecologia del paesaggio e qualità ambientale del Parco Nazionale del Cilento e Vallo di Diano. *Doc. Del Territ.* **2000**, *45*, 20–30.
54. Fanelli, G. *Dasypyrum villosum* vegetation in the territory of Rome. *Rend. Lincei. Sci. Fis. Nat.* **1998**, *9*, 149–170. [CrossRef]
55. Fanfarillo, E.; Latini, M.; Iberite, M.; Bonari, G.; Nicolella, G.; Rosati, L.; Salerno, G.; Abbate, G. The segetal flora of winter cereals and allied crops in Italy: Species inventory with chorological, structural and ecological features. *Plant Biosyst.* **2020**, *154*, 935–946. [CrossRef]
56. Gagnon, E.; Lewis, G.P.; Solnge Sotuyo, J.; Hughes, C.E.; Bruneau, A. A molecular phylogeny of Caesalpinia sensu lato: Increased sampling reveals new insights and more genera than expected. *S. Afr. J. Bot.* **2013**, *89*, 111–127. [CrossRef]
57. Gagnon, E.; Bruneau, A.; Hughes, C.E.; De Queiroz, L.P.; Lewis, G.P. A new generic system for the pantropical *Caesalpinia* group (*Leguminosae*). *PhytoKeys* **2016**, *71*, 1–160. [CrossRef] [PubMed]
58. Chiovenda, E. *Flora Somala*; Arti Grafiche: Roma, Italy, 1929; Volume 1.
59. Urban, I. *Symbolae Antillanae, Seu, Fundamenta Florae Indiae Occidentalis*; Fratres Borntraeger: Lipsiae, Germany, 1911–1913; Volume 7.
60. Brenan, J.P.M. Notes on African Caesalpinioïdeae. *Kew Bull.* **1963**, *17*, 197–218. [CrossRef]
61. Grisebach, G. *Catalogus Plantarum Cubensium Exhibens Collectionem Wrightianam Aliasque Minores ex Insula Cuuba Missas*; Gulielmum Engelmann: Lipsiae, Germany, 1866.
62. Fedde, F. *Repertorium Specierum Novarum Regni Vegetabilis*; Fasc, X.V., Ed.; Selbstverlag des Herausgebers: Berlin, Germany, 1919.
63. Watson, S. Contributions to american botany. *Proc. Amer. Acad. Arts* **1886**, *21*, 414–468. [CrossRef]
64. Hassler, E. Novitates Argentinae. I. *Repert. Spec. Nov. Regni Veg.* **1913**, *12*, 201–202.
65. POWO. *Denisophytum Bessac* (Choiv.) E.Gagnon & G.P.Lewis. Plant of the World Online. Available online: https://powo.science.kew.org/taxon/urn:lsid:ipni.org:names:77158013-1 (accessed on 21 June 2022).
66. Iamonico, D.; Iberite, M.; Nicolella, G. Aggiornamento alla flora esotica del Lazio (Italia centrale). II. *Inorfm. Bot. Ital.* **2014**, *46*, 331–368.
67. Lucchese, F. *Atlante Della Flora Vascolare Del Lazio. Cartografia, Ecologia e Biogeografia. Parte Generale e Flora Alloctona*; DigitaliaLab: Roma, Italy, 2017; Volume 1.
68. Del Guacchio, E. Integrazioni, aggiornamenti e note alla flora esotica della Campania. *Inform. Bot. Ital.* **2015**, *47*, 147–154.
69. Motti, R.; Esposito, A.; Stinca, A. New additions to the exotic vascular flora of Campania (southern Italy). *Ann. Bot.* **2018**, *8*, 75–85.
70. Bendahmane, M.; Dubois, A.; Raymond, O.; Le Bris, M. Genetics and genomics of flower initiation and development in roses. *J. Exp. Bot.* **2013**, *64*, 847–857. [CrossRef]
71. Ku, T.C.; Robertson, C.R. *Rosa, L. Flora of China*; Wu, Z.Y., Raven, P.H., Eds.; Science Press: Beijing, China, 2003; Volume 9, pp. 339–381.
72. Cullen, J.; Alexander, J.C.M.; Brady, A.; Brickell, C.D.; Green, P.S.; Heywood, V.H.; Jorgensen, P.-M.; Jury, S.L.; Knees, S.G.; Leslie, A.C.; et al. (Eds.) *The European Grande Flora. A Manual for the Identification of Plants Ciultivated in Europe, Boyh Out-Doors and Under Glass*; Cambridge University Press: Cambridge, UK, 1995.
73. POWO. *Rosa Chinensis* Jacq. Plant of the World Online. Available online: https://powo.science.kew.org/taxon/urn:lsid:ipni.org:names:732029-1 (accessed on 16 June 2022).
74. Medvecká, J.; Kliment, J.; Májeková, J.; Halada, L'.; Zaliberová, M.; Gojdičová, E.; Feráková, V.; Jarolímek, I. Inventory of the alien flora of Slovakia. *Preslia* **2012**, *84*, 257–309.
75. Waarnemingen.be. Rosa Chinese Jacq. Available online: https://waarnemingen.be/observation/98646133/ (accessed on 21 June 2022).
76. Zieliński, J.; Biel, B.; Tan, K. The Roses of Amorgos (Kiklades, Greece): Garden escapes. *Phytol. Balc.* **2016**, *22*, 351–353.
77. Delipavlov, D.; Cheshmedzhiev, I. (Eds.) *Opredelitel na Rasteniiata v Bulgariia*; Akademy of the Agrarian University: Plovdiv, Bulgaria, 2011.
78. Carta, A.; Forbicioni, L.; Frangini, G.; Pierini, B.; Peruzzi, L. An updated inventory of the vascular flora of Elba island (Tuscan Archipelago, Italy). *Ital. Bot.* **2018**, *6*, 1–22. [CrossRef]

79. Criscuoli, M.; Pierini, B.; La Rosa, M.; Peruzzi, L. *Heliotropium amplexicaule* Vahl. In *Notulae Alla Flora Esotica D'italia 5 (90–114)*; Barberis, G., Nepi, C., Peccenini, S., Peruzzi, L., Eds.; Società Botanica Italiana: Firenze, Italy, 2011; Volume 43, pp. 374–375.
80. Del Guacchio, E.; La Valva, V. The non-native vascular flora of Campania (southern Italy). *Plant Biosyst.* **2017**, *152*, 767–779. [CrossRef]
81. Cecchi, L.; Selvi, F. A synopsis of Boraginaceae subfam. Hydrophylloideae and Heliotropioideae in Italy. *Plant Biosyst.* **2014**, *148*, 2–12. [CrossRef]
82. Ohba, H.; Akiyama, S. A Revision of the Species of *Hydrangea* (Hydrangeaceae) Described by Siebold and Zuccarini, Part 2. *Bull. Natl. Mus. Nat. Sci. Ser. B Bot.* **2014**, *40*, 29–45.
83. Uemachi, T.; Mizhuara, Y.; Degchi, K.; Shinjo, Y.; Kajino, E.; Ohba, H. Phylogenetic Relationship of Hydrangea macrophylla (Thunb.) Ser. and H. serrata (Thunb.) Ser. Evaluated Using RAPD Markers and Plastid DNA Sequences. *J. Jpn. Soc. Hort. Sci.* **2014**, *83*, 163–171. [CrossRef]
84. POWO. *Hydrangea macrophylla* (Thunb.) Ser. Plant of the World Online. Available online: https://powo.science.kew.org/taxon/urn:lsid:ipni.org:names:791637-1 (accessed on 23 June 2022).
85. McClintock, E. A monograph of the genus *Hydrangea*. *Proc. Calif. Acad. Sci.* **1957**, *29*, 14–256.
86. Ohba, H.; Akiyama, S. A Revision of the Species of *Hydrangea* (*Hydrangeaceae*) Described by Siebold and Zuccarini, Part 1. *Bull. Natl. Mus. Nat. Sci. Ser. B Bot.* **2013**, *39*, 17–194.
87. De Smet, Y.; Granados Mendoza, C.; Wanke, S.; Gowtghebeur, P.; Samain, M.-S. Molecular phylogenetics and new (infra)generic classification to alleviate polyphyly in tribe *Hydrangeeae* (Cornales: *Hydrangeaceae*). *Taxon* **2015**, *64*, 741–753. [CrossRef]
88. Galasso, G.; Domina, G.; Adorni, M.; Ardenghi, N.M.G.; Banfi, E.; Bedini, G.; Bertolli, A.; Brundu, G.; Calbi, M.; Cecchi, L.; et al. Notulae to the Italian alien vascular flora: 1. *Ital. Bot.* **2016**, *1*, 17–37. [CrossRef]
89. Lazzeri, V.; Sammartino, F.; Campus, G.; Caredda, A.; Mascia, F.; Mazzoncini, V.; Testa, N.; Gestri, G. Note floristiche tosco-sarde ii: Novità regionali e locali e considerazioni tassonomiche per le regioni Sardegna e Toscana. *Ann. Del Mus. Civ. Di Rovereto. Sez. Archeol. Stor. Sci. Natutali* **2015**, *30*, 331–368.
90. Banfi, E.; Galasso, G. *La Flora Esotica Lombarda*; Museo di Storia Naturale di Milano: Milano, Italy, 2010.
91. Stinca, A.; Musarella, C.M.; Rosati, L.; Laface, V.L.A.; Licht, W.; Fanfarillo, E.; Wagensommer, R.P.; Galasso, G.; Fascetti, S.; Esposito, A.; et al. Italian Vascular Flora: New Findings, Updates and Exploration of Floristic Similarities between Regions. *Diversity* **2021**, *13*, 600. [CrossRef]
92. Anzalone, B.; Iberite, M.; Lattanzi, E. La flora vascolare del Lazio. *Inorfm. Bot. Ital.* **2010**, *42*, 187–317.
93. Galasso, G. Notulae to the Italian alien vascular flora: 10. *Ital. Bot.* **2020**, *10*, 57–71. [CrossRef]
94. Iamonico, D. *Polycnemum heuffelii* Láng. *Willdenowia* **2011**, *41*, 314.
95. Stafleu, F.A.; Cowan, R.S. *Taxonomic Literature*, 2nd ed.; The Hague: Utrecht, The Netherlands, 1981; Volume 3.
96. Stafleu, F.A.; Cowan, R.S. *Taxonomic Literature*, 2nd ed.; The Hague/Boston: Utrecht, The Netherlands; Antwerpen, Belgium, 1985; Volume 4.
97. Linnaeus, C. *Species Plantarum*; Impensis Laurentii Salvii: Holmiae, Cilicia, 1753; Volume 2. [CrossRef]
98. Iamonico, D. Linnaean names in *Arum* (Araceae): Typification of *A. arisarum* (≡*Arisarum vulgare*) and *A. tenuifolium* (≡ *Biarum tenuifolium*). *Taxon* **2020**, *69*, 578–581. [CrossRef]
99. Boyce, P.C. A taxonomic revision of Biarum Araceae. *Bot. Mag.* **2008**, *209*, 2–119. [CrossRef]
100. Iamonico, D.; Millozza, A.; Iberite, M. Typification of the names *Epilobium lanceolatum*, *Lotus requienii*, *Orchis romana*, and *Romulea columnae* described from Rome (Italy). *Phytotaxa* **2020**, *454*, 203–212. [CrossRef]
101. Iamonico, D. A conspectus of *Silene* sect. *Lasiocalycinae* (*Caryophyllaceae*). *Phyton* **2018**, *57*, 113–127. [CrossRef]
102. Iamonico, D.; Iberite, M. Nomenclatural remarks on Linnaean names of aquatic plants. *Taxon* **2022**, *71*, 674–678. [CrossRef]
103. Iamonico, D.; Fortini, P.; Hussain, A.N. On the occurrence and naturalization of *Amaranthus hypochondriacus* (*Amaranthaceae*) in some European countries, with notes on its climatic features. *Hacquetia* **2022**, *21*, 211–222. [CrossRef]
104. Iamonico, D. First record of a naturalized population of the tropical *Colocasia esculenta* (*Araceae*) in Italy, and clarifications about its occurrence in southeastern Europe. *Acta Bot. Croat.* **2021**, *80*, 169–175. [CrossRef]
105. Iberite, M.; Abati, S.; Abbate, G.; Iamonico, D.; Pellicciomi, I. *Lemna minuta* Kunth. 1428. Notulae alla checklist della flora vascolare italiana 5 (1420–1474). *Inform. Bot. Ital.* **2008**, *40*, 100–101.
106. Buccomino, G.; Meschini, P.; Monterosso, G.; Verucci, P. Aggiornamento sulla distribuzione di *Lupinus albus* L. subsp. *graecus* (Boiss. & Spruner) Franco & P.Silva (*Fabaceae*) nel Lazio. *Not. Soc. Bot. Ital.* **2019**, *3*, 99–102.
107. Galasso, G.; Domina, G.; Ardenghi, N.M.G.; Arrigoni, P.; Banfi, E.; Bartolucci, F.; Bonari, G.; Buccomino, G.; Ciaschetti, G.; Conti, F.; et al. Notulae to the Italian alien vascular flora: 2. *Ital. Bot.* **2016**, *2*, 55–71. [CrossRef]
108. Bartolucci, F.; Galasso, G.; Peruzzi, L.; Conti, F. Report 2020 on plant biodiversity in Italy: Native and alien vascular flora. *Nat. Hist. Sci. Atti Della Soc. Ital. Di Sci. Nat. Del Mus. Civ. Di Stor. Nat. Di Milano* **2021**, *8*, 41–54. [CrossRef]
109. Preston, C.D.; Sheail, J.; Armitage, P.D.; Davy-Bowker, J. The long-term impact of urbanisation on aquatic plants: Cambridge and the River Cam. *Sci. Total Environ.* **2003**, *314–316*, 67–87. [CrossRef]
110. Hansen, A.J.; Knight, R.L.; Marzluff, J.M.; Powell, S.; Brown, K.; Gude, P.H.; Jones, K. Effects of exurban development on biodiversity: Patterns, mechanisms, and research needs. *Ecol. Appl.* **1893**, *15*, 1893–1905. [CrossRef]
111. Fanelli, G. Analisi fitosociologica dell'area metropolitana di Roma. *Braun-Blanquetia* **2002**, *27*, 1–276.
112. Muratet, A.; Machon, N.; Jiguet, F.; Moret, J.; Porcher, E. The role of urban structures in the distribution of wasteland flora in the greater Paris area, France. *Ecosystems* **2007**, *10*, 661–671. [CrossRef]

113. Lonati, M.; Probo, B.; Gorlier, A.; Pittarello, M.; Scariot, V.; Lombardi, G.; Ravetto Enri, S. Plant diversity and grassland naturalness of differently managed urban areas of Torino (NW Italy). *Acta Hortic.* **2018**, *1215*, 247–253. [CrossRef]
114. Perrino, E.V.; Wagensommer, R.P. Crop wild relatives (CWR) priority in Italy: Distribution, ecology, in situ and ex situ conservation and expected actions. *Sustainability* **2021**, *13*, 1682. [CrossRef]
115. Perrino, E.V.; Wagensommer, R.P. Crop wild relatives (CWRs) threatened and endemic to Italy: Urgent actions for protection and use. *Biology* **2022**, *11*, 193. [CrossRef] [PubMed]
116. Ceschin, S.; Abati, S.; Leacche, I.; Iamonico, D.; Iberite, M.; Zuccarello, V. Does the alien *Lemna minuta* show an invasive behavior outside its original range? Evidence of antagonism with the native *L. minor* in central Italy. *Int. Rev. Hydrobiol.* **2016**, *101*, 173–181. [CrossRef]
117. Abbate, G.; Cicinelli, E.; Iamonico, D.; Iberite, M. Floristic analysis of the weed communities in wheat and corn crops: A case study in western-central Italy. *Ann. Bot.* **2013**, *3*, 97–105.
118. Celesti Grapow, L.; Pignatti, S.; Pignatti, E. Analisi della flora dei siti archeologici di Roma. *Allionia* **1993**, *32*, 113–118.
119. Minissale, P.; Sciandrello, S. The wild vascular flora of the Archaeological Park of Neapolis in Syracuse and surrounding areas (Sicily, Italy). *Biodivers. J.* **2017**, *8*, 87–104.
120. Ceschin, S.; Cutini, M.; Caneva, G. Contributo alla conoscenza della vegetazione delle aree archeologiche romane (Roma). *Fitosociologia* **2006**, *43*, 97–139.
121. Pavesi, A.; Giangreco, L. Osservazioni sulla distribuzione di *Parietaria lusitanica* L. subsp. (*Urticaceae*) nel Lazio. *Infrm. Bot. Ital.* **2006**, *38*, 79–86.
122. Celesti-Grapow, L.; Ricotta, C. Plant invasion as an emerging challenge for the conservation of heritage sites: The spread of ornamental trees on ancient monuments in Rome, Italy. *Biol. Invasions* **2021**, *23*, 1191–1206. [CrossRef]
123. Celesti-Grapow, L.; Abbate, G.; Baccetti, N.; Capizzi, D.; Carli, E.; Copiz, R.; Frondoni, R.; Giunti, M.; Gotti, C.; Iberite, M.; et al. Control of invasive species for the conservation of biodiversity in Mediterranean islands. The LIFE PonDerat project in the Pontine Archipelago, Italy. *Plant Biosyst.* **2017**, *151*, 795–799. [CrossRef]
124. Blasi, C.; Capotorti, G.; Marchese, M.; Marta, M.; Bologna, M.A.; Bombi, P.; Bonaiuto, M.; Bonnes, M.; Carrus, G.; Cifelli, F.; et al. Interdisciplinary research for the proposal of the Urban Biosphere Reserve of Rome Municipality. *Plant Biosyst.* **2008**, *142*, 305–312. [CrossRef]
125. Capotorti, G.; Alós Ortí, M.M.; Copiz, R.; Fusaro, L.; Mollo, B.; Salvatori, E.; Zavattero, L. Biodiversity and ecosystem services in urban green infrastructure planning: A case study from the metropolitan area of Rome (Italy). *Urban For. Urban Green.* **2019**, *37*, 87–96. [CrossRef]
126. Munasinghe, M. *Environmental Economics and Sustainable Development*; Word Bank: Washington, WA, USA, 1993.
127. Capotorti, G.; De Lazzari, V.; Alós Ortí, M.M. Local Scale Prioritisation of Green Infrastructure for Enhancing Biodiversity in Peri-Urban Agroecosystems: A Multi-Step Process Applied in the Metropolitan City of Rome (Italy). *Sustainability* **2019**, *11*, 3322. [CrossRef]
128. Wang, Y.; Chang, Q.; Fan, P.; Shi, X. From urban greenspace to health behaviors: An ecosystem services-mediated perspective. *Environ. Res.* **2022**, *213*, 113664. [CrossRef]

Appendix A

Inventory of the taxa occurring in Appia Antica Regional Park. Names of families in nonbold and uppercase; groups higher than families in bold and uppercase. Abbreviations: A: alien (with CAS: casual taxon; NAT: naturalized taxon; INV: invasive taxon); E: endemic; C: cryptogenic; NC: no longer recorded (not confirmed, but documented by the literature and/or herbaria specimens) S: segetal taxon (with, if occurring, sS: stritcly segetal taxon; cS: characteristic segetal taxon).

FERNS And ALLIES
EQUISETACEAE Michx. ex DC.
Equisetum arvense L., S
Equisetum ramosissimum Desf., S
Equisetum telmateia Ehrh., S

DENNSTAEDTIACEAE Lotsy
Pteridium aquilinum (L.) Kuhn subsp. ***aquilinum***

NEPHROLEPIDACEAE Pic. Serm.
Nerpholepis cordifolia (L.) C.Persl, A NAT

ASPLENIACEAE Newman
Asplenium onopteris L.
Asplenium scolopendrium L. subsp. ***scolopendrium***
Asplenium trichomanes L. subsp. ***quadrivalens*** D.E. Meyer

GYMNOSPERMS
PINACEAE Spreng. ex F.Rudolphi
Pinus halepensis Mill., A CAS
Pinus pinea L., A CAS

CUPRESSACEAE Gray
Cupressus sempervirens L., A CAS

ANGIOSPERMS

LAURACEAE Juss.
Laurus nobilis L.

ARACEAE Juss.
Arum italicum Mill. subsp. ***italicum***, S
Biarum tenuifolium (L.) Schott subsp. ***tenuifolium***
Colocasia esculenta (L.) Schott, A NAT
Lemna gibba L.
Lemna minor L.
Lemna minuta Kunth, A INV
Zantedeschia aethiopica (L.) Spreng., A NAT

ALISMATACEAE Vent.
Alisma lanceolatum With.
Alisma plantago-aquatica L.

DIOSCOREACEAE R.Br.
Dioscorea communis (L.) Caddick & Wilkin

SMILACACEAE Vent.
Smilax aspera L.

ORCHIDACEAE Juss.
Anacamptis coriophora (L.) R.M.Bateman, Pridgeon & M.W.Chase
Anacamptis morio (L.) R.M.Bateman, Pridgeon & M.W.Chase
Anacamptis papilionacea (L.) R.M.Bateman, Pridgeon & M.W.Chase
Ophrys apifera Huds.
Ophrys incubacea Bianca
Ophrys sphegodes Mill. subsp. ***sphegodes***
Ophrys tenthredinifera Willd.
Orchis laxiflora Lam.
Serapias lingua L.
Serapias parviflora Parl.
Serapias vomeracea (Burm. fil.) Briq.
Spiranthes spiralis (L.) Chevall.

IRIDACEAE Juss.
Crocus biflorus Mill.
Gladiolus italicus Mill., cS
Iris germanica L., A NAT
Limniris pseudacorus (L.) Fuss
Romulea bulbocodium (L.) Sebast. et Mauri
Romulea columnae Sebast. et Mauri

ASPHODELACEAE Dumort.
Aloe maculata All., A NAT
Asphodelus ramosus L. subsp. ***ramosus***

AMARYLLIDACEAE J.St.-Hil.
Allium ampeloprasum L.
Allium chamaemoly L. subsp. ***chamaemoly***
Allium neapolitanum Cirillo
Allium roseum L., S
Allium triquetrum L.
Allium vineale L., S
Ipheion uniflorum (Lindl.) Raf., A CAS
Leucojum aestivum L., A CAS
Narcissus pseudonarcissus L., A CAS
Narcissus tazetta L. subsp. ***tazetta***, S
Sternbergia lutea (L.) Ker-Gawler

ASPARAGACEAE Juss.
Agave americana L., A CAS
Asparagus acutifolius L.
Asparagus offininalis L.subsp. ***officinalis***
Bellevalia romana (L.) Reichenb., S
Chlorophytum comosum (Thunb.) Jacques, A CAS
Muscari comosum (L.) Mill., cS
Muscari neglectum Guss. ex Ten.

Loncomelos narbonense (L.) Raf., S
Ornithogalum divergens Boreau, S
Ruscus aculeatus L.
Yucca gloriosa L., A CAS

ARECACEAE Bercht. & J.Presl
Chamaerops humilis L., A CAS
Phoenix canariensis H. Wildpret, A CAS

TYPHACEAE Juss.
Sparganium erectum L.
Typha latifolia L.

CANNACEAE Juss.
Canna indica L., A NAT

JUNCACEAE Juss.
Juncus articulatus L.
Juncus inflexus L. subsp. ***inflexus***
Luzula campestris (L.) DC. subsp. ***campestris***
Luzula forsteri (Sm.) DC.

CYPERACEAE Juss.
Carex distachya Desf.
Carex distans L.
Carex divisa Hudson
Carex flacca Schreb. subsp. ***erythrostachys*** (Hoppe) Holub
Carex hirta L.
Carex otrubae Podp.
Carex pendula Hudson
Cyperus alternifolius L. subsp. ***flabelliformis*** Kük., A NAT
Cyperus badius Desf.
Cyperus longus L.
Cyperus rotundus L., S
Eleocharis palustris (L.) Roem. et Schult., S
Scirpoides holoschoenus (L.) Soják

POACEAE Barnhart
Agrostis stolonifera L. subsp. ***stolonifera***, S
Aira cupaniana Guss., S
Aira elegans Willd. subsp. ***elegans***
Alopecurus myosuroides Hudson subsp. ***myosuroides***, sS
Anisantha diandra (Roth) Tutin ex Tzvelev, cS
Anisantha madritensis (L.) Nevski subsp. ***madritensis***, S
Anisantha rigida (Roth) Hyl., S
Anisantha sterilis (L.) Nevski, S
Arundo donax L., A INV
Arundo plinii Turra, S
Avena barbata Pott. ex Link subsp. ***barbata***
Avena sativa L. subsp. ***sativa***, A NAT
Avena sterilis L. subsp. ***sterilis***, cS
Brachypodium distachyon (L.) P. Beauv., S
Brachypodium rupestre (Host) Roemer et Schultes, S
Brachypodium sylvaticum (Huds.) Beauv. subsp. ***sylvaticum***

Bromopsis erecta (Huds.) Fourr., S
Bromus commutatus Schrad., S
Bromus hordeaceus L. subsp. ***hordeaceus***, S
Catabrosa aquatica (L.) P. Beauv., NC
Catapodium rigidum (L.) C.E. Hubb., S
Cynodon dactylon (L.) Pers., S
Dactylis glomerata L. subsp. ***glomerata***, S
Dactylis glomerata L. subsp. ***hispanica*** (Roth) Nyman
Dasypyrum villosum (L.) P. Candargy, S
Digitaria sanguinalis (L.) Scop., S
Echinochloa crus-galli (L.) Beauv., S
Ehrharta erecta Lam., A CAS
Eleusine indica (L.) Gaertn.
Elymus repens (L.) Gould subsp. ***repens***, S
Festuca danthonii Asch. & Graebn. subsp. ***danthonii***, S
Festuca ligustica (All.) Bertol., S
Festuca myuros L. subsp. **myuros**, S
Gaudinia fragilis (L.) Beauv., S
Holcus lanatus L., S
Hordeum bulbosum L.
Hordeum murinum L. subsp. ***leporinum*** (Link) Arcang., S
Hordeum murinum L. subsp. ***murinum***, S
Hordeum secalinum Schreb.
Hyparrhenia hirta (L.) Stapf subsp. ***hirta***
Imperata cylindrica (L.) Raeusch.
Lagurus ovatus L.
Lolium arundinaceum (Schreb.) Darbysh. subsp. ***arundinaceum***
Lolium multiflorum Lam. subsp. ***multiflorum***, S
Lolium perenne L., S
Lolium rigidum Gaudin subsp. ***rigidum***, S
Macrobriza maxima (L.) Tzvelev, S
Nassella neesiana (Trin. et Rupr.) Barkworth, A NAT
Oloptum miliaceum (L.) Röser & H.R.Hamasha
Parapholis cylindrica (Willd.) Romero Zarco, NC
Paspalum distichum L.
Phalaris aquatica L., S
Phalaris paradoxa L., cS
Phleum pratense L. subsp. ***pratense***, S
Phragmites australis (Cav.)Trin. ex Steud. subsp. ***australis***, S
Phyllostachys reticulata (Rupr.) K.Koch, A NAT
Poa annua L., S
Poa bulbosa L., S
Poa pratensis L., S
Poa trivialis L., S
Polypogon monspeliensis (L.) Desf., S
Rostraria cristata (L.) Tzvelev
Setaria italica (L.) P.Beauv. subsp. ***viridis*** (L.) Thell., A NAT, S
Setaria verticillata (L.) P.Beauv.
Sorghum halepense (L.) Pers., S
Trisetaria panicea (Lam.) Paunero, S
Triticum aestivum L., A CAS
Triticum negletum (Req. ex Bertol.) Greuter, S
Triticum vagans (Jord. & Fourr.) Greuter, S

RANUNCULACEAE Juss.
Anemone hortensis L., S
Clematis vitalba L.
Delphinium consolida L. subsp. ***consolida***, sS
Delphinium halteratum Sm. subsp. ***halteratum***
Ficaria verna Huds. subsp. ***ficariiformis*** (F.W.Schultz) B.Walln., S
Nigella damascena L., S
Ranunculus bulbosus L., S
Ranunculus lanuginosus L., S
Ranunculus parviflorus L., S
Ranunculus peltatus Schrank
Ranunculus repens L., S
Ranunculus sardous Crantz, S
Thalictrum aquilegiifolium L.

PLATANACEAE T.Lestib.
Platanus hispanica Mill. ex Münchh.

PAPAVERACEAE Juss.
Chelidonium majus L.
Fumaria capreolata L. subsp. ***capreolata***, S
Fumaria officinalis L. subsp. ***officinalis***, S
Papaver dubium L., C, sS
Papaver hybridum L., sS
Papaver rhoeas L., C, sS
Papaver somniferum L.

CRASSULACEAE J.St.-Hil.
Kalanchoe daigremontiana Raym., A NAT
Petrosedum sediforme (Jacq.) Grulich subsp. ***sediforme***
Phedimus stellatus (L.) Raf.
Sedum caespitosum (Cav.) DC.
Sedum cepaea L.
Umbilicus horizontalis (Guss.) DC.
Umbilicus rupestris (Salisb.) Dandy

SAXIFRAGACEAE Juss.
Saxifraga trydactilites L., S

VITACEAE Juss.
Parthenocissus quinquefolia (L.) Planch., A NAT
Vitis vinifera L., A NAT

ZYGOPHYLLACEAE R.Br.
Tribulus terrestris L.

FABACEAE Lindl.
Acacia dealbata Link, A NAT
Astragalus glycyphyllos L., S, NC
Astragalus hamosus L., S
Astragalus pelecinus (L.) Barneby
Bituminaria bituminosa (L.) C.H.Stirt.
Cercis siliquastrum L. subsp. ***siliquastrum***
Cytisus villosus Pourr.

Denisophytum bessac (Choiv.) E.Gagnon & G.P.Lewis, A NAT
Emerus major Mill. subsp. ***major***
Ervilia hirsuta (L.) Opiz, sS
Ervum tetraspermum L., sS
Galega officinalis L., S
Gleditsia triacanthos L., A CAS
Hymenocarpos circinnatos (L.) Savi
Lathyrus annuus L., cS
Lathyrus aphaca L. subsp. ***aphaca***, cS
Lathyrus cicera L., S
Lathyrus clymenum L., S
Lathyrus latifolius L., S
Lathyrus oleraceus Lam. subsp. ***biflorus*** (Raf.) H.Schaef., Coulot & Rabaute
Lotus corniculatus L., S
Lotus ornithopodioides L.
Lotus tenuis Waldst. & Kit. ex Willd., S
Lupinus angustifolius L.
Lupinus albus L. subsp. ***graecus*** (Boiss. & Spruner) Franco & P.Silva
Medicago arabica (L.) Huds., S
Medicago falcata L. subsp. ***falcata***, S
Medicago lupulina L., S
Medicago minima (L.) L., S
Medicago orbicularis (L.) Bartal., S
Medicago polymorpha L.
Medicago rigidula (L.) All., S
Medicago sativa L.
Onobrychis viciifolia Scop.
Ononis spinosa L. subsp. ***antiquorum*** (L.) Arcangeli
Ornithopus compressus L., S
Robinia pseudoacacia L., A INV
Scorpiurus muricatus L., S
Securigera cretica (L.) Lassen, S
Securigera securidaca (L.) Degen et Dörfler, S
Spartium junceum L.
Trifolium angustifolium L. subsp. ***angustifolium***, S
Trifolium arvense L., S
Trifolium campestre Schreb., S
Trifolium cherleri L., S
Trifolium echinatum M. Bieb., S
Trifolium incarnatum L. subsp. ***incarnatum***
Trifolium incarnatum L. subsp. ***molinerii*** (Balb. ex Hornem.) Ces.
Trifolium ligusticum Balb. ex Loisel.
Trifolium micranthum Viv.
Trifolium nigrescens Viv. subsp. ***nigrescens***, S
Trifolium pratense L. subsp. ***pratense***, S
Trifolium repens L., S
Trifolium resupinatum L., S
Trifolium scabrum L., S
Trifolium squarrosum L.
Trifolium stellatum L., S
Trifolium subterraneum L., S
Trifolium suffocatum L.
Trifolium tomentosum L., S
Trifolium vesiculosum Savi

Trigonella alba (Medik.) Coulot & Rabaute
Trigonella altissima (Thuill.) Coulot & Rabaute
Trigonella smalii Coulot & Rabaute
Vicia angustifolia L., S
Vicia bithynica (L.) L., cS
Vicia dasycarpa Ten., S
Vicia faba L., A CAS
Vicia hybrida L., S
Vicia incana Gouan
Vicia melanops Sm.
Vicia narbonensis L., S
Vicia sativa L.
Vicia serratifolia Jacq.

ROSACEAE Juss.
Aphanes arvensis L., sS
Agrimonia eupatoria L. subsp. ***eupatoria***, S
Crataegus monogyna Jacq.
Geum urbanum L.
Malus domestica (Suckow) Borkh., A CAS
Malus sylvestris Mill.
Potentilla recta L., S
Potentilla reptans L., S
Poterium sanguisorba L. subsp. ***balearicum*** (Bourg. ex Nyman) Stace, S
Prunus cerasifera Ehrh. var. pissardii (Carrière) C.K.Schneid
Prunus domestica L., A CAS
Prunus spinosa L.
Pyracantha coccinea M.Y.Roemer
Pyrus communis L. subsp. ***pyraster*** (L.) Ehrh.
Pyrus spinosa Forssk.
Rhaphiolepis bibas (Lour.) Galasso & Banfi, A CAS
Rosa canina L.
Rosa chinensis Jacq. var. ***semperflorens*** (Curtis) Koehne
Rosa gallica L.
Rosa sempervirens L.
Rubus caesius L.
Rubus ulmifolius Schott
Sorbus domestica L.

RHAMNACEAE Juss.
Paliurus spina-christi Mill.
Rhamnus alaternus L.
Ziziphus jujuba Mill., A CAS

ULMACEAE Mirb.
Ulmus minor Mill.

CANNABACEAE Martinov
Celtis australis L. subsp. ***australis***
Humulus lupulus L.

MORACEAE Gaudich.
Broussonetia papyrifera (L.) Vent., A NAT
Ficus carica L.

Maclura pomifera (Raf.) C.K. Schneid., A CAS
Morus alba L., A CAS

URTICACEAE Juss.
Parietaria judaica L.
Parietaria lusitanica L. subsp. ***susitanica***
Urtica dioica L., S
Urtica membranacea Poir.
Urtica pilulifera L.
Urtica urens L., S

FAGACEAE Dumort.
Castanea sativa Mill.
Quercus cerris L.
Quercus dalechampii Ten.
Quercus frainetto Ten.
Quercus ilex L. subsp. ***ilex***
Quercus pubescens Willd. subsp. ***pubescens***
Quercus robur L. subsp. ***robur***
Quercus suber L.

JUGLANDACEAE DC. ex Perleb
Juglans regia L., A CAS

CUCURBITACEAE Juss.
Bryonia dioica Jacq.
Cucurbita maxima Duchense subsp. ***maxima***, A CAS
Ecballium elaterium (L.) A. Rich., S

CELASTRACEAE R.Br.
Euonymus europaeus L.

OXALIDACEAE R.Br.
Oxalis articulata Savigny, A NAT, S
Oxalis corniculata L., S
Oxalis dillenii Jacq., A NAT, S
Oxalis pes-caprae L., A NAT, S

VIOLACEAE Batsch
Viola alba Besser subsp. ***denhardtii*** (Ten.) W.Becker, S
Viola arvensis Murray, S
Viola reichenbachiana Jordan ex Boreau

SALICACEAE Mirb.
Populus alba L. subsp. ***alba***
Populus nigra L.
Salix alba L. subsp. ***alba***

PASSIFLORACEAE Juss. ex Roussel
Passiflora caerulea L., A CAS

LINACEAE DC. ex Perleb
Linum usitatissimum L. subsp. ***angustifolium*** (Huds.) Thell., S
Linum strictum L.

HYPERICACEAE Juss.
Hypericum perforatum L., S

EUPHORBIACEAE Juss.
Euphorbia characias L.
Euphorbia exigua L. subsp. ***exigua***, sS
Euphorbia falcata L. subsp. ***falcata***, sS
Euphorbia helioscopia L. subsp. ***helioscopia***, S
Euphorbia peplus L., S
Euphorbia platyphyllos L.
Euphorbia prostrata Aiton, A NAT, S
Euphorbia pulcherrima Willd. ex Klotzsch, A CAS
Euphorbia terracina L., S
Mercurialis annua L., S

GERANIACEAE Juss.
Erodium acaule (L.) Bech. & Thell.
Erodium ciconium (L.) L'Hér., S
Erodium cicutarium (L.) L'Hér., S
Erodium malacoides (L.) L'Hér. subsp. ***malacoides***, S
Erodium moschatum (L.) L'Hér.
Geranium columbinum L., S
Geranium dissectum L., S
Geranium molle L., S
Geranium robertianum L., S
Geranium rotundifolium L., S
Geranium sanguineum L., S

ONAGRACEAE Juss.
Epilobium hirsutum L.
Epilobium lanceolatum Sebast. & Mauri
Epilobium parviflorum Schreber, S
Epilobium tetragonum L. subsp. ***tournefortii*** (Michalet) H.Lév., S

LYTHRACEAE J.St.-Hil.
Lythrum hyssopifolia L., S
Lythrum salicaria L., S
Punica granatum L., A CAS

MYRTACEAE Juss.
Eucalyptus camaldulensis Dehnh. subsp. ***camaldulensis***, A CAS

ANACARDIACEAE R.Br.
Pistacia lentiscus L.
Pistacia terebinthus L. subsp. ***terebinthus***

SAPINDACEAE Juss.
Acer campestre L.
Acer monspessulanum L. subsp. ***monspessulanum***
Acer negundo L., A NAT

RUTACEAE Juss.
Ruta chalepensis L.

SIMAROUBACEAE DC.
Ailanthus altissima (Mill.) Swingle, A INV

MELIACEAE Juss.
Melia azedarach L., A CAS

MALVACEAE Juss.
Althaea cannabina L.
Malva multiflora (Cav.) Soldano, Banfi & Galasso, S
Malva nicaeensis All., S
Malva punctata (All.) Alef.
Malva sylvestris L., S

RESEDACEAE Martinov
Reseda alba L., S
Reseda lutea L. subsp. ***lutea***, S
Reseda phyteuma L. subsp. ***phyteuma***, S
Tarenaya spinosa (Jacq.) Raf., A CAS

CAPPARACEAE Juss.
Capparis orientalis Veill.

BRASSICACEAE Burnett
Alliaria petiolata (Bieb.) Cavara et Grande, S
Arabis hirsuta (L.) Scop.
Berteroa obliqua (Sm.) DC.
Brassica oleracea L., A CAS
Bunias erucago L., S
Calepina irregularis (Asso) Thell., S
Capsella bursa-pastoris (L.) Medik., S
Capsella rubella Reuter, S
Cardamine hirsuta L., S
Diplotaxis erucoides (L.) DC subsp. **erucoides**, S
Diplotaxis tenuifolia (L.) DC., S
Draba verna L.
Lepidium graminifolium L.
Lepidium virginicum L., A CAS
Lunaria annua L., A CAS
Lunaria rediviva L.
Microthlaspi perfoliatum (L.) F.K.Mey., S
Nasturtium officinale W.T. Aiton
Neslia paniculata (L.) Desv. subsp. ***thracica*** (Velen.) Bornm., cS
Raphanus raphanistrum L. subsp. ***landra*** (Moretti ex DC.) Bonnier & Layens, S
Rapistrum rugosum (L.) All., cS
Sinapis alba subsp. ***alba***, cS
Sinapis arvensis L. subsp. ***arvensis***, cS
Sisymbrium officinale (L.) Scop.

SANTALACEAE R.Br.
Osyris alba L.

PLUMBAGINACEAE Juss.
Plumbago auriculata Lam., A CAS

Plumbago europaea L.

POLYGONACEAE Juss.
Fallopia baldschuanica (Regel) Holub, A NAT
Fallopia dumetorum (L.) J. Holub, S
Persicaria amphibia (L.) Delabre
Persicaria lapathifolia (L.) Delarbre
Persicaria maculosa Gray, S
Polygonum aviculare L., S
Polygonum rurivagum Jord. ex Boreau, S
Rumex bucephalophorus L. subsp. ***bucephalophorus***, S
Rumex conglomeratus Murray, S
Rumex crispus L., S
Rumex obtusifolius L. subsp. ***obtusifolius***, S
Rumex pulcher L. subsp. ***pulcher***, S

CARYOPHYLLACEAE Juss.
Arenaria leptoclados (Reichenb.) Guss., S
Arenaria serpyllifolia L., S
Cerastium brachypetalum Desp. ex Pers., S
Cerastium glomeratum Thuill., S
Cerastium ligusticum Viv., S
Dianthus armeria L. subsp. ***armeria***
Herniaria glabra L., cS
Herniaria hirsuta L. subsp. ***hirsuta***, S
Petrorhagia saxifraga (L.) Link subsp. ***saxifraga***, S
Petrorhagia dubia (Raf.) G.López & Romo
Polycarpon tetraphyllum (L.) subsp. ***diphyllum*** (Cav.) O. Bolòs et Font Quer
Polycarpon tetraphyllum L. subsp. ***tetraphyllum***, S
Sagina apetala Ard. subsp. ***apetala***
Saponaria officinalis L.
Silene bellidifolia Jacq.
Silene conica L., S
Silene gallica L., S
Silene gallinyi Rchb., NC
Silene italica (L.) Pers. subsp. ***italica***, S
Silene latifolia Poir., S
Silene nocturna L., S
Silene pendula L.
Silene vulgaris (Moench) Garcke subsp. ***tenoreana*** (Colla) Soldano & F.Conti, S
Silene vulgaris (Moench.) Garcke subsp. ***vulgaris***, S
Spergularia rubra (L.) J.Presl & C.Presl, S
Stellaria media (L.) Vill. subsp. ***media***, S
Stellaria neglecta Weihe, S
Stellaria pallida (Dumort.) Crép.
Spergula arvensis L., sS

CHENOPODIACEAE Vent.
Atriplex patula L., S
Beta vulgaris L. subsp. ***vulgaris***
Chenopodiastrum murale (L.) S.Fuentes, Uotila & Borsch, S
Chenopodium opulifolium Schrader ex Koch et Ziz, S
Chenopodium album L., S
Chenopodium vulvaria L.

Dysphania ambrosioides (L.) Mosyakin & Clemants, A NAT
Dysphania multifida (L.) Mosyakin & Clemants, A CAS
Lipandra polysperma (L.) S.Fuentes, Uotila & Borsch, S
Oxybasis urbica (L.) S.Fuentes, Uotila & Borsch

AMARANTHACEAE Juss.
Amranthus albus L., A NAT
Amaranthus blitoides S. Watson, A NAT
Amaranthus blitum L. subsp. ***blitum***
Amaranthus deflexus L., A INV, S
Amaranthus graecizans subsp. ***silvestris*** (Vill.) Brenan, S
Amaranthus hybridus L., A INV, S
Amaranthus retroflexus L., A INV, S
Amaranthus viridis L., A INV
Polycnemum heuffelii Láng,, C, NC

PHYTOLACCACEAE R.Br.
Phytolacca americana L., A INV

NYCTAGINACEAE Juss.
Mirabilis jalapa L., A NAT

BASELLACEAE Raf.
Anredera cordifolia (Ten.) Steenis, A NAT

PORTULACACEAE Juss.
Portulaca oleracea L., S

CACTACEAE Juss.
Opuntia ficus-indica (L.) Mill., A NAT

CORNACEAE Bercht. & J.Presl
Cornus mas L.
Cornus sanguinea L. subsp. ***sanguinea***

EBENACEAE Gürke
Diospyros kaki Thunb., A CAS

HYDRANGEACEAE Dumort.
Hydrangea macrophylla (Thunb.) Ser., A NAT

PRIMULACEAE Batsch ex Borkh.
Cyclamen hederifolium Aiton
Cyclamen repandum Sm.
Lysimachia arvensis (L.) U.Manns & Anderb. subsp. ***arvensis***, cS
Lysimachia foemina (Mill.) U.Manns & Anderb., S

ERICACEAE Juss.
Arbutus unedo L.

RUBIACEAE Juss.
Cruciata laevipes Opiz, S
Galium album Mill. (*G. erectum* Syme) subsp. ***album***, S
Galium aparine L., S

Galium murale (L.) All.
Galium parisiense L.
Rubia peregrina L.
Sherardia arvensis L., S
Theligonum cynocrambe L.
Valantia muralis L.

GENTIANACEAE Juss.
Blackstonia perfoliata (L.) Huds. subsp. ***perfoliata***
Centaurium erythraea Rafn subsp. ***erythraea***, S
Centaurium maritimum (L.) Fritsch
Centaurium tenuiflorum (Hoffmanns. et Link) Fritsch subsp. ***acutiflorum*** (Schott) Zeltner, S

APOCYNACEAE Juss.
Nerium oleander L., A CAS
Trachelospermum jasminoides (Lindl.) Lem., A CAS
Vinca major L. subsp. ***major***
Vinca minor L.

CONVOLVULACEAE Juss.
Convolvulus arvensis L., S
Convolvulus cantabrica L.
Convolvulus sepium L.
Convolvulus silvaticus Kit.
Cuscuta campestris Yunck, A NAT
Cuscuta cesattiana Bertol., A CAS
Ipomoea indica (Burm.) Merr., A CAS

SOLANACEAE Juss.
Datura stramonium L., S
Lycium chinense Mill., A NAT
Solanum chenopodioides Lam., A NAT
Solanum dulcamara L.
Solanum lycopersicum L., A NAT
Solanum nigrum L., S
Solanum pseudocapsicum L., A CAS
Solanum villosum Mill., S

BORAGINACEAE Juss.
Anchusa azurea Mill., S
Anchusa undulata L. subsp. ***hybrida*** (Ten.) Bég., S
Borago officinalis L., S
Buglossoides arvensis (L.) I.M.Johnst., cS
Cynoglossum creticum Mill., S
Echium italicum L. subsp. ***italicum***
Echium parviflorum M\ch, S
Echium plantagineum L., S
Echium vulgare L., S
Myosotis arvensis (L.) Hill subsp. ***arvensis***, cS
Myosotis ramosissima Rochel subsp. ***ramosissima***, S
Symphytum officinale L., S
Symphytum tuberosum L. subsp. ***angustifolium*** (A.Kerner) Nyman, S

HELIOTROPIACEAE Schrader
Heliotropium amplexicaule Vahl., A CAS
Heliotropium europaeum L., S

OLEACEAE Hoffmanns. & Link
Fraxinus angustifolia Vahl subsp. ***oxycarpa*** (Willd.) Franco et Rocha Afonso
Fraxinus ornus L. subsp. ***ornus***
Ligustrum lucidum W.T. Aiton, A CAS
Ligustrum vulgare L.
Phillyrea latifolia L.

PLANTAGINACEAE Juss.
Antirrhinum majus L.
Antirrhinum tortosuom Bosc ex Lam, S
Callitriche stagnalis Scop.
Cymbalaria muralis G. Gaertn., B. Mey. et Scherb. subsp. ***muralis***
Linaria pelisseriana (L.) Mill., NC
Linaria purpurea (L.) Mill., S
Linaria vulgaris Mill. subsp. ***vulgaris***, S
Misopates calycinum Rothm.
Plantago afra L. subsp. ***afra***, S
Plantago lagopus L.
Plantago lanceolata L., S
Plantago major L., S
Veronica anagallis-aquatica L., S
Veronica arvensis L., cS
Veronica cymbalaria Bodard, S
Veronica hederifolia L., S
Veronica persica Poir., S, A INV
Veronica polita Fr., cS

SCROPHULARIACEAE Juss.
Scrophularia auriculata L.
Scrophularia peregrina L.
Verbascum blattaria L.
Verbascum sinuatum L., S
Verbascum thapsus L. subsp. ***thapsus***

LAMIACEAE Martinov
Ballota nigra L. subsp. ***meridionalis*** (Bég.) Bég., sS
Clinopodium menthifolium (Host) Merino subsp. ***ascendens*** (Jord.) Govaerts, S
Clinopodium vulgare L., S
Lamium amplexicaule L., S
Lamium bifidum Cirillo subsp. ***bifidum***, S
Lamium maculatum L.
Lamium purpureum L., cS
Lycopus europaeus L.
Marrubium vulgare L.
Melissa officinalis L. subsp. ***altissima*** (Sm) Arcang., S
Mentha aquatica L. subsp. ***aquatica***, S
Mentha pulegium L. subsp. ***pulegium***, S
Mentha suaveolens Ehrh. subsp. ***suaveolens***, S
Micromeria graeca (L.) Benth. ex Rchb.
Origanum vulgare L. subsp. ***vulgare***

Prunella laciniata (L.) L.
Salvia clandestina L.
Salvia verbenaca L., S
Thymbra capitata (L.) Cav.
Micromeria graeca (L.) Benth. ex Rchb.
Micromeria juliana (L.) Benth. ex Rchb.
Stachys arvensis (L.) L., S
Stachys germanica L. subsp. ***germanica***, S, NC
Stachys ocymastrum (L.) Briq., S
Stachys romana (L.) E.H.L.Krause, S
Stachys sylvatica L.
Teucrium chamaedrys L. subsp. ***chamaedrys***
Teucrium flavum L. subsp. ***flavum***

OROBANCHACEAE Vent.
Bellardia trixago (L.) All., S
Bellardia viscosa (L.) Fisch. & C.A.Mey., S
Odontites vernus (Bellardi) Dumort. subsp. ***serotinus*** Corb.
Orobanche crenata Forssk., cS
Orobanche hederae Vaucher ex Duby
Parentucellia latifolia (L.) Caruel, S
Phelipanche nana (Reut.) Soják

VERBENACEAE J.St.-Hil.
Lantana camara L., A CAS
Verbena officinalis L., S

ACANTHACEAE Juss.
Acanthus mollis L. subsp. ***mollis***
Ruellia simplex C. Wright, A CAS

BIGNONIACEAE Juss.
Campsis radicans (L.) Bureau, A NAT

CAMPANULACEAE Juss.
Campanula erinus L., S
Campanula rapunculus L., S
Jasione montana L. subsp. ***montana***
Legousia speculum-veneris (L.) Chaix subsp. ***speculum-veneris***, sS

ASTERACEAE Bercht. & J.Presl
Anacyclus radiatus Loisel. subsp. ***radiatus***, S
Andryala integrifolia L., S
Anthemis arvensis L. subsp. ***arvensis***, cS
Anthemis cotula L., cS
Arctium minus (Hill) Bernh.
Artemisia arborescens (Vaill.) L.
Artemisia verlotiorum Lamotte, S
Artemisia vulgaris L., S
Bellis annua L. subsp. ***annua***
Bellis perennis L., S
Bellis sylvestris Cirillo
Bidens subalternans DC., A CAS
Calendula arvensis (Vill.) L., S

Carduus micropterus (Borbàs) Teyber subsp. ***perspinosus*** (Fiori) Arènes
Carduus nutans L. subsp. ***nutans***, S
Carduus pycnocephalus L. subsp. ***pycnocephalus***, S
Carlina corymbosa L., S
Carthamus lanatus L.
Centaurea calcitrapa L., S
Centaurea jacea L. subsp. ***angustifolia*** (DC.) Gremli, S
Centaurea napifolia L., S
Centaurea solstitialis L. subsp. ***solstitialis***, S
Chondrilla juncea L.
Cladanthus mixtus (L.) Chevall.
Cota tinctoria (L.) J.Gay subsp. ***tinctoria***, S
Cichorium intybus L., S
Cirsium arvense (L.) Scop., S
Cirsium creticum (Lam.) d'Urv. subsp. ***triumfetti*** (Lacaita) Werner
Coleostephus myconis (L.) Cass. ex Rchb.f.
Cyanus segetum L., sS
Crepis bursifolia L.
Crepis neglecta L. subsp. ***neglecta***, S
Crepis sancta (L.) Bornm. subsp. ***nemausensis*** (P.Fourn.) Babc., S, A NAT
Crepis setosa Haller f., S
Crepis vesicaria L., S
Dittrichia graveolens (L.) Greuter
Dittrichia viscosa (L.) Greuter subsp. ***viscosa***
Erigeron bonariensis L., A NAT, S
Erigeron canadensis L., A CAS, S
Erigeron sumatrensis Retz., A INV, S
Eupatorium cannabinum L. subsp. ***cannabinum***, S
Filago germanica (L.) Huds., cS
Galactites tomentosus Moench, S
Galinsoga parviflora Cav., A CAS, S
Galinsoga quadriradiata Ruiz et Pav., A NAT, S
Glebionis segetum (L.) Fourr., S
Helianthus tuberosus L., A CAS, S
Hypochaeris achyrophorus L., S
Hypochoeris radicata L.
Jacobaea aquatica (Hill) G. Gaertn., B. Mey. & Scherb.
Jacobaea erratica (Bertol.) Fourr.
Lactuca sativa L. subsp. ***serriola*** (L.) Galasso, Banfi, Bartolucci & Ardenghi, S
Leontodon tuberosus L.
Matricaria chamomilla L., cS
Onopordum acanthium L. subsp. ***acanthium***
Onopordum illyricum L., S
Pallenis spinosa (L.) Cass. subsp. ***spinosa***, S
Pentanema squarrosum (L.) D.Gut.Larr., Santos- Vicente, Anderb., E.Rico & M.M.Mart.Ort., S
Petasites hybridus (L.) G.Gaertn., B.Mey. & Scherb. subsp. ***hybridus***
Helminthotheca echioides (L.) Holub
Picris hieracioides L. subsp. ***hieracioides***, S
Picris hieracioides L. subsp. ***spinulosa*** (Guss.) Arcang.
Pulicaria dysenterica (L.) Bernh., S
Reichardia picroides (L.) Roth, S
Rhagadiolus stellatus (L.) Gaertn., S
Scolymus hispanicus L., S
Senecio leucanthemifolius Poir., S

Senecio vulgaris L., S
Silybum marianum (L.) Gaertn., S
Sonchus asper (L.) Hill subsp. ***asper***, S
Sonchus oleraceus L., S
Sonchus tenerrimus L.
Symphyotrichum squamatum (Spreng.) G.L.Nesom, A NAT
Taraxacum F.H.Wigg. sect. **Taraxacum**, S
Tragopogon porrifolius L., S
Tussilago farfara L., S
Tyrimnus leucographus (L.) Cass.
Urospermum dalechampii (L.) F.W. Schmidt, S
Urospermum picroides (L.) Scop. ex F.W. Schmidt, S
Xanthium spinosum L., A NAT
Xanthium italicum Moretti, S

VIBURNACEAE Rafinesque
Sambucus ebulus L.
Sambucus nigra L.
Viburnum tinus L. subsp. ***tinus***

CAPRIFOLIACEAE Juss.
Lonicera etrusca Santi
Lonicera japonica Thunb., A NAT

DIPSACACEAE Juss.
Dipsacus fullonum L.
Knautia arvensis (L.) Coult., S
Knautia collina Jord.
Knautia integrifolia (L.) Bertol. subsp. ***integrifolia***, S
Sixalix atropurpurea (L.) Greuter et Burdet

PITTOSPORACEAE R.Br.
Pittosporum tobira (Thunb.) W.T. Aiton, A CAS

VALERIANACEAE Batsch
Centranthus ruber (L.) DC. subsp. ***ruber***
Valerianella eriocarpa Desv., cS

ARALIACEAE Juss.
Hedera helix L. subsp. ***helix***

APIACEAE Lindl.
Ammi majus L., S
Ammoides pusilla (Brot.) Breistr.
Anethum piperitum Ucria, S
Angelica sylvestris L.
Anthriscus sylvestris (L.) Hoffm.
Chaerophyllum temulum L., S
Conium maculatum L., S
Daucus carota L. subsp. ***carota***, S
Daucus carota L. subsp. ***maximus*** (Desf.) Ball, S
Eryngium campestre L., S
Ferula communis L. subsp. ***communis***, S
Ferula glauca L.

Helosciadium nodiflorum (L.) W.D.J.Koch subsp. ***nodiflorum***
Oenanthe pimpinelloides L., S
Oenanthe silaifolia M.Bieb.
Opopanax chironium (L.) W.D.J.Koch
Smyrnium olusatrum L.
Tordylium apulum L., S
Tordylium maximum L., S
Torilis arvensis (Hudson) Link subsp. ***arvensis***, S
Torilis japonica (Houtt.) DC., S

Appendix B

Selected specimens collected during the field surveys. The list (alphabetical order of the scientific names) refers to floristic novelties, notable species, no longer recorded taxa, and taxa having *loci classici* and/or nomenclatural types collected in Appia Antica Regional Park (see Section "4.4. Floristic Notes").

Aloe maculata* All. subsp. *maculata
Italy, Latium, Roma, Parco Regionale dell'Appia Antica, Valle della Caffarella, incolto, 27 m a.s.l., 22 April 2012, *G. Nicolella s.n.* (RO); *ibidem*, 23 April 2017, *D. Iamonico* (RO).

***Amaranthus hypochondriacus* L.**
Lazio, Rome Province, Rome city, Appia Antica Regional Park, locality Acquedotti, channels, 56 m a.s.l., 6 October 2016, *D. Iamonico s.n.* (RO!); ibidem, 18 October 2020 (RO!).

***Anredera cordifolia* (Ten.) Steenis**
Italy, Latium, Rome, Appia Antica Regional Park, Acquedotti locality, on *Rubus ulmifolium* dominated community, 41°51′04″ N, 12°33′12″ E, 6 October 2019, *D. Iamonico s.n.* (RO).

***Biarum tenuifolium* (L.) Schott. subsp. *tenuifolium*.**
Italy. Lazio region, Rome Province, Rome city, Appia Antica Regional Park, shrubs vegetation dominated by *Paliurus spina-christi* Mill., 40–42 m a.s.l., 20 August 2015, *D. Iamonico s.n.* (HFLA No. 4861!).

***Bidens subalternans* DC.**
Italy, Latium, Rome, Appia Antica Regional Park, Acquedotti locality, on a little bridge of channel Acqua Mariana, 41°51′02″ N, 12°33′21″ E, 23 November 2020, *D. Iamonico s.n.* (RO).

***Campsis radicans* (L.) Bureau**
Roma, P.R. dell'Appia Antica, Valle della caffarella, marrana, 23 April 2011, *D. Iamonico s.n.* (RO!); ibidem, 1 September 2013 (RO!); Acquedotti locality, 10 August 2021, *D. Iamonico s.n.* (RO!).

***Canna indica* L.**
Italy, Latium, Rome, Appia Antica Regional Park, Acquedotti locality, banks of channel Acqua Mariana, 15 June 2018, *D. Iamonico s.n.* (RO).

***Catabrosa aquatica* (L.) P.Beauv.**
Italy, Latium, Roma, Via Appia Nuova, lungo fossetto tra Ciampino e S. Maria delle Mole. Luoghi aquitrinosi, 20 April 1944, *G. Montelucci s.n.* (RO-Herbarium Montelucci, sub *Aira aquatica* L.).

***Chlorophytum comosum* (Thumbs.) Jacques**
Italy, Latium, Rome, Appia Antica Regional Park, Acquedotti locality, banks of channel Acqua Mariana, 15 May 2019, *D. Iamonico s.n.* (RO).

***Colocasia esculenta* (L.) Schott**
Lazio region, Rome, Appia Antica Regional Park, locality Acquedotti, channels, 58 m a.s.l., 8 August 2015, *D. Iamonico s.n.* (HFLA!); ibidem, 19 m a.s.l., 9 May 2020 (RO!).

***Cyperus alternifolius* L. subsp. *flabelliforme* Kük.**
Lazio region, Rome, Appia Antica Regional Park, Caffarella valley, channels, 25 May 2017 *D. Iamonico s.n.* (RO!); locality Acquedotti, channels, 56 m a.s.l., 15 May 2022, *D. Iamonico s.n.* (RO!).

Denisophytum bessac **(Choiv.) E.Gagnon & G.P.Lewis**
Italy, Latium, Rome, Appia Antica Regional Park, via Appia Pignatelli, shrubs, 24 April 2017, *D. Iamonico s.n.* (RO); *ibidem*, 25 May 2022, *D. Iamonico s.n.* (RO).

Diospyrus kaki **L.**
Italy, Latium, Rome, Appia Antica Regional Park, Acquedotti locality, banks of channel Acqua Mariana, 5 October 2020, *D. Iamonico s.n.* (RO).

Ehrharta erecta **Lam.**
Italy, Latium, Roma, Valle della Ninfa Egeria, 17 March 1876, *G. G.. Cuboni s.n.* (RO-Herbarium Romano no. 51234).

Epilobium lanceolatum **Sebast. & Mauri**
Italy, Lazio, Roma nei prati della Caffarella, 3 June 1812, *F. A. Sebastiani* (RO-Herbarium Romano, n. 19436); Lazio, Via delle Vigne presso Albano, June 1853, *E. Rolli* (RO-Herbarium Romano, n. 19443).

Euphorbia pulcherrima **Willd. ex Klotzsch**
Italy, Latium, Rome, Appia Antica Regional Park, Acquedotti locality, on riverbed of channel Acqua Mariana, 3 June 2022, *D. Iamonico s.n.* (RO).

Heliotropium amplexicaule **Vahl.**
Italy, Latium, Rome, April 1928, *s.c.* (RO-HG, no. 32109); Rome, Appia Antica Regional Park, along on central reservation of the Appia Nuova street, 3 June 2022, *D. Iamonico s.n.* (RO).

Hydrangea macrophylla **(Thunb.) Ser.**
Italy, Latium, Rome, Appia Antica Regional Park, Acquedotti locality, on riverbed of channel Acqua Mariana, 7 August 2016, *D. Iamonico s.n.* (RO); *ibidem*, 29 December 2020, *D. Iamonico s.n.* (RO).

Kalanchoe daigremontiana **Raym.**
Lazio region, Rome, Appia Antica Regional Park, Caffarella valley, *Arando donax* L. dominated community, 20 June 2020, *D. Iamonico s.n.* (RO!).

Lemna minuta **Kunth**
Roma, Parco Regionale dell'Appia Antica, loc. Vaccareccia Caffarella, channel and artificial reservoir, 23 m a.s.l., 7 September 2007, *Iamonico s.n.* (RO!, HFLA!)

Linaria pelisseriana **(L.) Mill.**
Italy, Latium, Roma, Via Ardeatina, 15 May 1892, *L. Salomonsohn s.n.* (RO-Herbarium Romano no. 34429).

Lupinus albus **L. subsp. *graecus* (Boiss. et Spruner) Franco & Pinto da Silva**
Italy, Latium, Rome, Appia Antica Regional Park, Caffarella valley, meadows, 25 April 2021, *D. Iamonico s.n.* (RO).

Melia azedarach **L.**
Lazio region, Rome, Appia Antica Regional Park, Caffarella valley, shrubs, 25 May 2018, 22 m a.s.l., *D. Iamonico s.n.* (RO!); locality Acquedotti, anthropogenic meadows, 57 m a.s.l., 2 July 2022, *D. Iamonico s.n.* (RO!).

Papaver somniferum **L.**
Lazio region, Rome, Appia Antica Regional Park, Caffarella valley, meadows, 7 May 2016, *D. Iamonico s.n.* (RO!).

***Parapholis cylindrica* (Willd.) Romero Zarco**
Italy, Latium, Roma, Alla Caffarella, s.d. (XIX century), *s.c. s.n.* (RO-Herbarium Romano no. 55564, sub *Rottboellia subulata* Savi).

***Passiflora caerulea* L.**
Italy, Latium, Rome, Appia Antica Regional Park, Claudio's aqueduct (Porta Furba), 7 June 2022, *D. Iamonico s.n.* (RO).

***Plumbago auriculata* Lam.**
Italy, Latium, Rome, Appia Antica Regional Park, Caffarella valley, *Rubus ulmifolius* dominated community, 20 June 2020, *D. Iamonico s.n.* (RO).

***Polycnemum heuffelii* Láng**
Italy, Latium, Roma, lungo la via Appia Pignatelli, July 1980, *B. Anzalone s.n.* (RO-Herbarium Romano no. 5993).

***Punica granatum* L.**
Italy, Latium, Rome, Appia Antica Regional Park, Caffarella valley, channels, 23 April 2011, *D. Iamonico s.n.* (RO); *ibidem*, 27 June 2012, *D. Iamonico s.n.* (RO); *ibidem*, 01 September 2013, *D. Iamonico s.n.* (RO), *ibidem*, 29 June 2022, *D. Iamonico s.n.* (RO).

***Rosa chinensis* Jacq. var. *semperflorens* (Curtis) Koehne**
Italy, Latium, Rome, Appia Antica Regional Park, Caffarella valley, shrub community with *Rubus ulmifolius* Schott and *Parthenocissus quinquefolia* Planch, 23 April 2017, *D. Iamonico s.n.* (RO).

***Ruellia simplex* C.Wright**
Italy, Latium, Rome, Appia Antica Regional Park, Acquedotti locality, banks of channel Acqua Mariana, 3 June 2022, *D. Iamonico s.n.* (RO).

Sagina apetala
Italy, Lazio region, Rome city, Appia Antica Region Park, Acquedotti sector, among cobblestones ("basolato") of the ancient Roman road "Latina", 19 May 2020, *Iamonico s.n.* (RO).

***Silene gallinyi* Rchb.**
Italy, Latium, Roma, alla Caffarella (presso catacombe di Pretestato), 6 August 1966, *A. Cacciato s.n.* (RO-Herbarium Anzalone no. 13276), sub *Silene trinervia* Sebast. & Mauri).

Stachys germanica* L. subsp. *germanica
Italy, Latium, Roma, Caffarella, May 1829, *P. Sanguinetti s.n.* (RO-Herbarium Romano no. 38523); incolti aridi della Via Appia Antica pr. La Torre di Cecilia Metella, 3 June 1922, *G. Lusina s.n.* (RO-Herbarium Romano no. 38548); *ibidem* (RO-Herbarium Romano no. 38549); *ibidem* (RO-Herbarium Romano no. 38550); *ibidem* (RO-Herbarium Romano no. 38551).

***Tarenaya spinosa* (Jacq.) Raf.**
Italy, Latium, Roma, macerie a via Appia Pignatelli, 6 September 1966, *B. Anzalone s.n.* (RO-Herbarium Romano s.n., sub *Cleome pungens* Willd); Roma, in via Appia Pignatelli, 12 August 1966, *De Persio s.n.* (RO-Herbarium Romano s.n., sub *Cleome pungens* Willd); Roma, sull'Appia Pignatelli, 14 Semptember 1967, *A. Cacciato s.n.* (RO-Herbarium Romano s.n., sub *Cleome pungens* Willd).

***Trachelospermum jasminoides* (Lindl.) Lem.**

Italy, Latium, Rome, Appia Antica Regional Park, Acquedotti locality, on cliff of channel Acqua Mariana, 15 May 2019, *D. Iamonico s.n.* (RO).

***Typha latifolia* L.**
Italy, Lazio region, Rome administrative province, Appia Antica Regional Park, Caffarella valley, wetlands, 20 m a.s.l., 14 October 2020, *D. Iamonico s.n.* (RO).

***Zantedeschia aetiopica* (L.) Spreng.**
Italy, Latium, Rome, Appia Antica Regional Park, Acquedotti locality, banks of channel Acqua Mariana, 16 May 2019, *D. Iamonico s.n.* (RO).

plants

MDPI

Article

Nomenclatural Synopsis, Revised Distribution and Conservation Status of *Ranunculus gracilis* (Ranunculaceae) in Italy

Fabrizio Bartolucci [1,*], Enzo De Santis [2] and Fabio Conti [1]

[1] Centro Ricerche Floristiche dell'Appennino, Parco Nazionale del Gran Sasso e Monti della Laga—Scuola di Bioscienze e Medicina Veterinaria, Università di Camerino, San Colombo, Barisciano, 67021 L'Aquila, Italy
[2] Independent Researcher, 00189 Roma, Italy
* Correspondence: fabrizio.bartolucci@unicam.it

Abstract: *Ranuculus gracilis* is endemic to the SE Euro-Mediterranean area and its presence in Italy is controversial. Based on analysis of the relevant literature, field surveys and examination of herbarium specimens, a revised distribution of this species in Italy is presented and its conservation status is assessed. *Ranunculus agerii*, described by Antonio Bertoloni from Bologna (Emilia-Romagna, Northern Italy), and *R. schowii*, described by Vincenzo Tineo from Vittoria and Terranova (Sicily), usually regarded as synonyms of *R. gracilis*, are here lectotypified and their taxonomic status discussed. Thanks to our study, the presence of *R. gracilis* in Italy is confirmed and, now, it is reported in a national conservation framework.

Keywords: endemism; floristic research; herbaria; lectotype; Mediterranean flora; nomenclature; taxonomy

Citation: Bartolucci, F.; De Santis, E.; Conti, F. Nomenclatural Synopsis, Revised Distribution and Conservation Status of *Ranunculus gracilis* (Ranunculaceae) in Italy. *Plants* **2022**, *11*, 3094. https://doi.org/10.3390/plants11223094

Academic Editor: Andreas W. Ebert

Received: 20 October 2022
Accepted: 8 November 2022
Published: 14 November 2022

1. Introduction

Ranunculus L. is the largest genus in the family *Ranunculaceae* Juss., with a cosmopolitan distribution, consisting of about 1200 species (including also ca. 600 agamospecies) [1,2]. Based on morphological and molecular data, the genus *Ranunculus* was divided into two subgenera (subg. *Auricomus* and subg. *Ranunculus*) and 17 sections [2]. In Italy, the genus *Ranunculus* comprises 112 taxa (species, subspecies and agamospecies), of which 33 are endemic (four are extinct) and one is an alien species [3–5].

Ranuculus gracilis was described in 1814 by Edward Daniel Clarke from the East Aegean island of Kos (Greece) [6] and belongs to *R.* subg. *Ranunculus* sect. *Ranunculastrum* DC. [2]. This species is endemic to the SE Euro-Mediterranean area and it is distributed in Italy, the Balkan Peninsula, Turkey and Georgia [7,8]. Contrary to what is reported by Euro+Med Plantbase [7] and POWO [8], the presence of *R. gracilis* in Italy is controversial, i.e., [4,9]. According to the latest Italian Flora [9], it is present in a few localities in Sicily and Calabria, no longer recorded in Emilia-Romagna and cultivated in Umbria. On the contrary, following the Italian Checklist [4], it is recorded by mistake in Piemonte, Emilia-Romagna, Umbria, Sicily and doubtfully occurring in Puglia. Two species, described from Italy, are usually regarded as synonyms of *R. gracilis* (e.g., [9–12]): *R. agerii* Bertol. and *R. schowii* Tineo (with doubt). *Ranunculus agerii* was described in 1819 by Antonio Bertoloni from the areas around Bologna (Emilia-Romagna, northern Apennines, Italy) [13] and it is currently regarded as a synonym of *R. gracilis* (i.e., [7–9,12,14–17]). *Ranunculus schowii* was described by Vincenzo Tineo in Gussone [18] from Vittoria and Terranova (Sicily) and regarded as a dubious synonym of *R. gracilis* or *R. agerii* by some authors (i.e., [10–12]) or as a synonym of *R. monspeliacus* L. subsp. *monspeliacus* (i.e., [8]). The purpose of this study is to critically review the presence of *R. gracilis* in Italy and to understand the taxonomic identity of *R. agerii* and *R. schowii*, two names that turned out to be, to the best of our knowledge, not yet typified.

The present contribution is part of an ongoing project promoted by the Italian Botanical Society, aimed at recognizing and typifying all the taxa described from Italy, in order to increase their systematic knowledge and promote further studies [19–22].

2. Materials and Methods

This study is based on an extensive analysis of relevant literature, field surveys and examination of herbarium specimens (including the original material) preserved in APP, BOLO, BR, CAT, CHE, CLF, DR, E, FI, G, GOET, JE, MW, NAP, P, PI, PAL, W and WU (the acronyms follow [23]). We performed a survey for original material for the name *R. agerii* at BOLO where Bertoloni's main collection is housed and at FI, G, NAP and PAL to trace the original material of the name *R. schowii* (see [24,25]). The original material for the name *R. gracilis* was searched at A, BM, CAN, CGE, ECON, GH, FI, K and SWN. The type designations herein follow the Shenzhen Code ([26], hereafter ICN).

The revised Italian distribution of *R. gracilis* is based on examination of herbarium specimens. The distribution data and the occurrence status are given for the Italian administrative regions according to Bartolucci et al. [4].

3. Results

3.1. The Long and Controversial History of R. gracilis in Italy

Over the years, the presence of *R. gracilis* in Italy has been controversial due to the confusion in its identification and to the unclear taxonomic relationships with some currently not accepted species [7,8], such as *R. agerii* Bertol., *R. schowii* Tin. and *R. chaerophyllos* L.

The first herbarium samples collected in Italy that can be referred with certainty to *R. gracilis* date back to the 16th century. There are two specimens without collection locality preserved in the "*En Tibi*" herbarium, made by Francesco Petrollini in Bologna around 1558, kept in Leiden (an image of the specimen is available at https://data.biodiversitydata.nl/naturalis/specimen/L.2110949 (accessed on 10 September 2022)) and in the controversial "Cibo" herbarium kept in Rome in Biblioteca Angelica, recently also attributed to Petrollini [27,28]. According to Stefanaki et al. [28], it is evident that many specimens in the "*En Tibi*" herbarium were collected in the area of Bologna, where Petrollini had his place of residence.

Towards the end of 1500, this plant was also collected by Nicolas Ager (or Agerius) in the Bologna hills. He sent samples to Jean Bauhin, who gave a brief description of this plant in the *Phytopinax* published by Caspar Bauhin as "*Ranunculus racemosa radice Io. Bauhini ... Reperitur in montibus Bononiensibus*" [29]. Later, it was reported by the Bauhin brothers as "*Ranunculus grumosa radice folio ranunculi bulbosi ... Hic Ranunculus agris Bononiensibus familiaribus est, & à D. Agerio collectus*" in the *Prodromos theatri botanici* [30], "*Ranunculus grumosa radice folio ranunculi bulbosi*" in the *Pinax theatri botanici* [31], "*Ranunculus racemosa radice ... Agerio siccam dedit pro Ranunculo Chelidoniae radice*" in the *Historia plantarum universalis* [32] and by Parkinson [33] as "*Ranunculus grumosa radice Bononiensis*". Later, Linnaeus [34], mistakenly included the polynomial published by C. Bauhin in the *Prodromus* [30] and in the *Pinax* [31] in *R. chaerophyllos* L. Antonio Bertoloni was the first to accurately describe [13] the plant from the areas around Bologna (Emilia-Romagna) as "*Ranoncolo Bolognese*", dedicating it to Nicolas Ager, with the name *R. agerii*. After the description, *R. agerii* was treated as synonym of *R. chaerophyllos* L. by Arcangeli [35] and Cesati et al. [36], a name of uncertain application [37]. Later, Fiori et al. [10] and Fiori [11] re-evaluated *R. agerii* as a good species, recording it not only for its *locus classicus*, but also for Sicily and quoting *R. gracilis* and, with doubt, *R. schowii* Tineo, only in [11] as synonyms. Pons [38] recorded *R. agerii* (syn. *R. gracilis*) for several localities around Bologna in Emilia-Romagna and for Catania in Sicily. Tutin [14] and Tutin and Akeroyd [17] in Flora Europaea quoted *R. gracilis* (syn. *R. agerii*) for Italy and Sicily. Zangheri [39] reported *R. gracilis* (syn. *R. agerii*) for Sicily and as naturalized in Northern and Central Italy. Pignatti [15] reported *R. gracilis* (syn. *R. agerii*) in Sicily and no longer recorded for Emilia-Romagna. Greuter et al. [40] in the Med-Checklist quoted *R. gracilis* (syn. *R. agerii*) for Sicily and as doubtfully native in Italy. Jalas and Suominen [16] quoted *R. gracilis* (syn. *R. agerii*) for Calabria and Sicily and with doubt in Emilia-Romagna. Peruzzi and Passalacqua [41] reported *R. gracilis* for the Balkan Peninsula, Turkey, and Crete, while, the Italian records of Calabria and Sicily, should be referred to *R. monspeliacus* L. subsp. *aspromontanus* (Huter, Porta & Rigo) Peruzzi & N.G.Passal. Conti et al. [42] reported *R. gracilis* without synonyms as doubtfully occurring in Italy in Piemonte, Emilia-Romagna and Sicily. In the same year, Scoppola and Spampinato [43] recorded *R. gracilis* (syn. *R. agerii*) for Sicily and as indicated by mistake in Emilia-Romagna. Later, Conti et al. [44] reported *R. gracilis* as indicated by mistake in Italy, updating the occurrence status in Conti et al. [42]. Recently, Pignatti et al. [9] quoted *R. gracilis* (syn. *R. agerii*) as present in a few localities in Sicily, Calabria, no longer recorded in Emilia-Romagna and cultivated in Umbria. On the contrary, according to Bartolucci et al. [4], in the updated checklist of Italian vascular Flora, the species was recorded by mistake in Piemonte, Emilia-Romagna, Umbria, Sicily and as doubtfully occurring in Puglia. Recently, Guarino and La Rosa [45] in the Digital Flora of Italy included in the 4th volume of Flora of Italy [46], recorded *R. gracilis* for Calabria, alien in Umbria and as doubtfully occurring in Sicily and Emilia-Romagna.

3.2. *Typification of the Names*

3.2.1. *Ranunculus agerii* Bertol., Opusc. Sci. 3: 182. 1819

Protologue citation: [Italy, Emilia-Romagna] "Copiosae provenit Bononiae in campis collinis di Monte Donato prope fodinas gypsi inter sata. Floret Aprili. Perenn.".

Lectotype (designated here): [Italy] Reperi copiosum Bononia in/campis collinis prope i Gessi di Montedonato/1818. Aprili. (BOLO [digital photo!], Figure 1B).

Nomenclatural notes: Antonio Bertoloni [13] described *R. agerii*, providing a detailed description, quoting a precise collection locality and citing an illustration "Tab VI". In BOLO, where Bertoloni's main collection is housed, we traced only one herbarium sample with two mounted individuals collected in 1818 on Monte Donato (Figure 1) that can be considered as part of the original material, as well as the illustration "Tab VI" cited in the protologue (Figure 1) (Art. 9.4 of the ICN). The herbarium sample kept in BOLO is complete, well conserved and agrees with the protologue and is selected here as a lectotype for the name *R. agerii*.

Taxonomic notes: based on the original material studied, *R. agerii* should be regarded as a heterotypic synonym of *R. gracilis*.

Figure 1. *Ranunculus agerii*: (**A**) illustration cited in the protologue by Bertoloni as "Tab VI"; (**B**) lectotype of the name *R. agerii* kept in BOLO (reproduced with permission of the *Herbarium Bononiensis*, Alma Mater Studiorum University of Bologna, Italy).

3.2.2. *Ranunculus gracilis* E.D.Clarke, Travels Eur. Asia and Africa 3 part 2(2): 336. 1814

Protologue citation: [Greece] "island of Cos [Kos]".

Type: not traced.

Nomenclatural notes: According to Miller [47], Clarke's herbarium was included in the Herbarium of A.B Lambert. Later, Clarke's specimens collected in Greece and Asia Minor were bought by G.S. Gibson and subsequently acquired by the herbarium BM. We searched for the original material in BM and in other herbaria where parts of the Lambert and Gibson collections are kept (e.g., CAN, CGE, BM, FI, FH, K and SWN), without finding original material.

3.2.3. *Ranunculus schowii* Tineo, in Gussone Fl. Sicul. Syn. 2(2): 889. 1845. [1 January–September 1845]

Protologue citation: [Italy, Sicily] "In arvis arenosis: fra Vittoria e Terranova (Tin.)".

Lectotype (designated here): [Italy, Sicily [Illustration: *Ranunculus schowii* Tin./V. Cartoccio dis./1845 (NAP barcode NAP0000512 [digital photo!], Figure 2).

Nomenclatural notes: Vincenzo Tineo in Gussone [18] described *R. schowii,* providing a detailed description, quoting a precise collection locality and citing an unpublished illustration "Tin. ined.". In order to trace the original material, we checked the PAL herbarium, where Tineo's main collection is housed. We also searched in FI, G, NAP and P, where duplicates by Tineo's collections are kept. Lojacono Pojero [48] wrote that he saw the only authentic specimen of *R. schowii* in H. Pan. (i.e., *Herbarium Panormitanum*), today PAL. In the *Herbarium Mediterraneum Panormitanum* (PAL), this sample is no longer present (G. Domina, pers. comm.). In NAP (ex-Herbarium Gussone Sicilia), where duplicates by Tineo's specimens are usually hosted, we did not trace any samples but only the unpublished illustration ("*Ranunculus schowii* Tin./V. Cartoccio dis./1845"; NAP barcode NAP0000512) cited in the protologue. The illustration is also labelled with a representation label [49] "12a. *Ranunculus schowii* Tin./Aprile, Majo", where "12a" is a reference to the position of the species within the genus in Gussone's Synopsis. We were not able to trace original material in FI, G and P. The unpublished illustration in NAP (NAP0000512) is the only element belonging to the original material (Art. 9.4 of the ICN), agrees with the protologue and is here designated as the lectotype for the name *R. schowii* (Figure 2).

Taxonomic notes: based on the protologue and the lectotype, *R. schowii* seems to have unique characteristics, only marginally close to particular forms of *R. isthmicus* Boiss. The individual depicted in the illustration (lectotype) shows fusiform root tubers, basal leaves tripartite with entire or lobed (only at the apex) segments and deflexed sepals at flowering. Further studies to assess the morphological variability in this species and to clarify its taxonomic status will be needed. In the case of synonymy of *R. isthmicus* Boiss. (published in 1846, [50]) and *R. schowii* (published in 1845, [18]), the latter would have priority and it should be advisable to proceed with a formal conservation proposal for the name *R. isthmicus*.

Figure 2. Lectotype of the name *R. schowii* kept in NAP (reproduced with permission of the *Herbarium Neapolitanum*, University of Naples Federico II, Italy).

3.3. Taxonomic Treatment

Ranunculus gracilis E.D.Clarke, Travels Eur. Asia and Africa 3 part 2(2): 336. 1814

= *Ranunculus agerii* Bertol., Opusc. Sci. 3: 182. 1819

Lectotype (designate here): [Italy] Reperi copiosum Bononia in/campis collinis prope i Gessi di Montedonato/1818. Aprili. (BOLO [digital photo!], Figure 1B)

= *Ranunculus granulatus* Griseb., Spic. Fl. Rumel. 1: 306. 1843

Lectotype (designated by Strid [51] (p. 306)): [Turkey] *In m. Bulgurlu, Grisebach 24* (GOET barcode GOET009797 [digital photo!]).

= *Ranunculus peloponnesiacus* Boiss., Diagn. Pl. Orient. 1: 63. 1843

Lectotype (designated by Strid [52] (p. 49)): [Greece] *colles elati Argolidis et Arkadiae, Apr. 1842, Boissier s.n.* (P, isolectotype MW barcode MW0592427 [digital photo !]; an image of the isolectotyoe is available at https://plant.depo.msu.ru/public/scan.jpg?pcode=MW0592427 (accessed on 10 September 2022)).

–*Ranunculus chaerophyllos* auct. fl. ital. p.p., non L.

Description (based on [52,53] and personal observations on the studied material): perennial, softly hirsute, 5–40 cm. Root tubers broadly ovate, mixed with long filiform roots. Stems erect, 1 to 3-flowered, appressed-pubescent in upper part, glabrous in the lower part. Basal leaves glabrous to subglabrous, long-petiolate, with the petiole sparsely pilose, dilated into a scarious sheath below; often heteromorphic, outer ones 3-lobed or 3-partite with cuneate-spreading segments divided into obtuse, ovate or oblong lobes, the inner 3-partite to or near base, with lobes variously dissected into obtuse segments. Cauline leaves few and reduced, subsessile, with linear segments. Sepals 6–8 mm, broadly lanceolate, strongly deflexed, sparsely pilose to subglabrous. Petals 6–14 mm, obovate, obtuse, bright yellow. Receptacle glabrous. Achenes numerous, in a dense, narrowly ellipsoid to ovate head (6)8–10 × 4–5 mm Achenes ovate-oblong, wingless, compressed, smooth, c. 1.5–2 mm, attenuate into a straight or somewhat hooked beak 0.5–1.0 mm.

Distribution: endemic to the SE Euro-Mediterranean area, distributed in Italy, Balkan Peninsula, Turkey and Georgia. The presence of *R. gracilis* in the latter country was reported by Grossheim [54], under the name *R. agerii* Bertol., for a single locality (Akhaltsikhe district; see Map No. 82 included in [54]), but, in our opinion, this report requires further checks. In Italy, in the current state of knowledge, it is present only in Lazio based on our finding reported here, no longer recorded in Emilia-Romagna and Sicily, recorded by mistake in Calabria and Puglia and formerly cultivated in botanical gardens in Toscana and Umbria (Figure 3). The presence of the species in each Italian administrative region is discussed below:

- Piemonte: the species was cited as doubtfully occurring in Piemonte by Conti et al. [42] and as recorded by mistake by Bartolucci et al. [4]; it was never recorded for the region (D. Bouvet and A. Selvaggi, pers. comm.).
- Emilia-Romagna: the presence of the species is confirmed by several old herbarium specimens kept in BOLO, CHE, FI, G, P, PI and RO (see specimens examined). Furthermore, the 16th century samples preserved in the *"En Tibi"* and *"Cibo"* herbaria were also collected in Emilia-Romagna near Bologna (Stefanaki et al. 2018, 2019). No recent herbarium samples or bibliographic records have been found (see, [55]); therefore, the species should be considered as no longer recorded. Targeted field research will be needed before considering the species as locally extinct.
- Toscana: the species was never recorded for the region. We traced an old herbarium specimen in FI, collected in the Botanical Garden of the University of Firenze, where the species was probably cultivated.
- Umbria: the species was recorded for the region in the past [9,11,15] as naturalized in the Botanical Garden of the University of Perugia. We traced the herbarium specimen linked to the old report by Fiori [11] in FI (see Specimens examined).
- Lazio: during field investigations carried out in the territory of Anagni (Frosinone, Central Italy) in March and April 2022, we discovered the species on Mt. Campitelli (Figure 3). Our finding corroborates the old report for this area by Sibilia ([56], under the name *R. agerii*) and confirms the presence of this species in Italy. The Sibilia record has never been incorporated in the regional floras [4,57–59]. In the current state of knowledge, this is the only population present in Italy.
- Puglia: the species was recorded by Di Pietro and Misano [60]. This record was later regarded as doubtful by Bartolucci et al. [4]. We were not able to trace herbarium specimens linked to this record and the species should be regarded as probably indicated by mistake (R. Di Pietro, pers. comm.).
- Calabria: according to Peruzzi and Passalacqua [41], the Calabrian records should be referred to *R. monspeliacus* subsp. *aspromontanus*.
- Sicily: the species was reported in Sicily from different localities by Giardina et al. [12]: between Vittoria and Terranova based on the description of *R. schowii* [18], between Catania and Misterbianco based on Strobl [61] and from Polizzi Generosa [62]. We have shown that *R. schowii* is not related to *R. gracilis;* therefore, the report of the latter between Vittoria and Terranova is erroneous. We traced, in PAL, the sample collected in Polizzi Generosa ("sotto il paese di Polizzi Generosa vicino all'acquedotto, 30 April 1990, Raimondo and Certa"), which should be referred to *R. paludosus*. A specimen cited in Wikiplantbase Sicilia [63] as *R. gracilis* and stored in PAL (No. 43515) collected at Busambra belongs to *R. paludosus* as well as a specimen in CAT No. 048272 (Monte Lauro, 9/V/1991, Brullo et al.). The only datum that we were able to confirm is the indication by Strobl [61], for the Amenano between Catania and Misterbianco at the foot of Etna, thanks to the tracing of an old herbarium sample stored in FI, collected in 1874 by Heidenreich at Misterbianco (quoted also by Fiori et al. [10]). Based on our

data, *R. gracilis* should be considered as no longer recorded in Sicily. Targeted field research will be needed before considering the species as locally extinct.

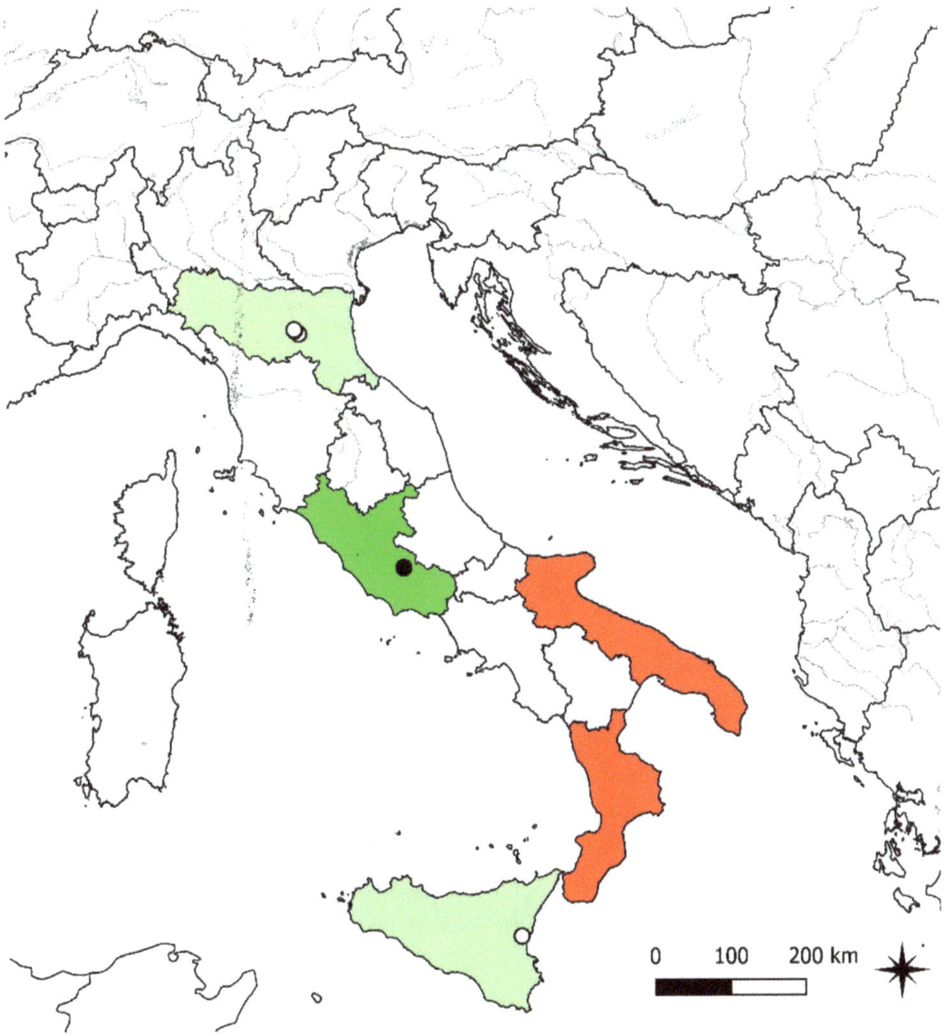

Figure 3. Distribution map of *Ranunculus gracilis* in Italy according to the herbarium material studied and field investigations: black symbols indicate the population currently present based on field investigations and empty symbols refer to the old herbarium specimens seen. Green background: occurring; pale green background: no longer recorded (reliable historical record); red background: recorded by mistake.

Figure 4. *Ranunculus gracilis* (Lazio, Mt. Campitelli, photo E. De Santis): (**A**) habit; (**B**) habitat; (**C**) flower with deflexed sepals; (**D**) inner basal leaves; (**E**) basal portion of the plant; (**F**) achenes; (**G**) outer basal leaf; (**H**) particular of the broadly ovate tubers.

Phenology: flowering late March to April; fruiting in April and May.

Habitat: humid habitats, meadows, fields and open woodland at an elevation of 0–1400 m a.s.l. In Italy, in the only currently known locality (Mt. Campitelli, Anagni, Lazio), it grows on the edge of *Quercus cerris* L. wood, on fresh and moist sandy soil, at an elevation of 750 m a.s.l.

Chromosome number: 2n = 16 [64].

Conservation status: *Ranunculus gracilis* currently occurs outside the NATURA 2000 network on Mt. Campitelli (Anagni, Frosinone) in Lazio (Central Italy). The populations in Emilia-Romagna (Northern Italy) and Sicily, confirmed by old herbarium specimens, have not been observed for over 120 years. The area of occupancy (AOO) is 4 km^2, calculated with GeoCAT (Geospatial Conservation Assessment Tool) software [65]. The species actually occurs in one location and a decline in the AOO was observed, considering the possible extinction of some populations. According to IUCN [66] criterion B2ab(i,ii,iv), the species is assessed as Critically Endangered (CR) at the regional level (Italy).

Taxonomic remarks: *Ranuculus gracilis* belongs to *R.* sect. *Ranunculastrum*. The section is characterized by species with a beak equal to or longer than the achene body, a receptacle glabrous and with dimorphic roots partly tuberous [2]. In Italy, there are nine native taxa belonging to this section [4]: *R. garganicus* Ten., *R. illyricus* L., *R. isthmicus* Boiss., *R. millefoliatus* Vahl, *R. monspeliacus* L. subsp. *aspromontanus* (Huter, Porta & Rigo) Peruzzi & N.G.Passal. (endemic), *R. monspeliacus* L. subsp. *monspeliacus*, *R. monspeliacus* L. subsp. *saxatilis* Nyman (extinct), *R. paludosus* Poir. and *R. spicatus* Desf. subsp. *rupestris* (Guss.) Maire (endemic). *Ranunculus gracilis* is easily distinguished from other species for a specific combination of characters, such as broadly ovoid tubers, basal leaf subglabrous shallowly 3-lobed (outer), 3-partite to or near base, with lobes variously dissected into obtuse segments (inner) and sepals deflexed at flowering. In Table 1, the qualitative morphological diagnostic features [9,17,41] of the Italian native taxa belonging to *R.* sect. *Ranunculastrum* are reported.

Table 1. Comparisons of morphological qualitative characters among native species of *R.* sect. *Ranunculastrum* in Italy.

	R. gracilis	*R. garganicus*	*R. illyricus*	*R. isthmicus*	*R. millefoliatus*	*R. monspeliacus* **subsp.** *monspeliacus*	*R. monspeliacus* **subsp.** *aspromontanus*	*R. monspeliacus* **subsp.** *saxatilis*	*R. paludosus*	*R. spicatus* **subsp.** *rupestris*
Basal leaf shape	3-lobed to 3-partite divided into obtuse, ovate or oblong lobes (outer), 3-partite to or near base, with lobes variously dissected into linear, obtuse segments (inner)	3-partite to base, lobes 2–3 pinnatisect into linear-lanceolate subacute segments	3-partite to base into linear-lanceolate lobes, entire or tripartite, rarely simply pinnatisect	3-partite to base or pinnatisect with lobes dissected in broadly linear, obtuse segments	3-partite, lobes 2–3 pinnatisect into linear-lanceolate subacute segments	3-partite to base, with lobes 3-fid (the middle stipitate)	orbicular to3-lobed	orbicular to3-lobed	flabellate to suborbicular (outer), 3-partite (inner) to base and usually divided into broadly linear segments	orbicular to 3-lobed
Basal leaf indumentum	glabrous to subglabrous	subglabrous	lanate-subsericeous	sparsely appressed-pubescent	glabrous	pubescent	pubescent	sericeous	appressed pubescent to subglabrous	appressed pubescent
Tubers shape	broadly ovoid	cylindrical	ellipsoid to broadly fusiform	fusiform	ovoid	fusiform	fusiform	fusiform	ellipsoid or broadly cylindrical	fusiform
Sepals	deflexed at flowering	appressed to corolla at flowering	deflexed at flowering	deflexed at flowering	appressed to corolla at flowering	deflexed at flowering	deflexed at flowering	deflexed at flowering	appressed to corolla at flowering	appressed to corolla at flowering
Achene indumentum	glabrous	glabrous	glabrous	glabrous	glabrous	pubescent	pubescent	pubescent	glabrous or pubescent	pubescent

Specimens examined of *Ranunculus gracilis* E.D.Clarke: **Italy**. EMILIA-ROMAGNA: *Reperi copiosum Bononia in campis collinis prope* i Gessi di Montedonato, April 1818, *s.coll.* (BOLO, lectotype of *R. agerii*); legi *Bononia in campis collinis di Gaibolla* inter sata, May 1823, *s. coll.* (BOLO, under the name *R. agerii*); legi Bononia a Gaibola, s.d., *s. coll.* (BOLO, under the name *R. agerii*); in collibus prope Bononiam, 27 April 1885, A. *Fiori s.n.* (CHE barcode CHE007984, P barcode P02403935, under the name *R. agerii*); *ex collibus Bononiensibus*, 1840, *M. de Martem s.n.* (P barcode P02817111, under the name *R. agerii*); environs de Bologne, Apennins, 1824, *Schleicher s.n.* (P barcode P02819312, under the name *R. agerii*); *circa Bononiam*, s.d., *s.coll.* (P barcode P06233836, under the name *R. agerii*); *legi in silvis prope Gaibola Bononiae*, April 1898, *G. Betti s.n.* (PI No. 015140); nei prati presso Gaibola, May 1884, *Mattei s.n.* (PI No. 061810, under the name *R. agerii*); colline bolognesi a M. Donato, April 1864, *s.coll.* (PI No. 061811, under the name *R. agerii*); Bologna, M. Donato, s.d., *Rosellini s.n.?* (PI No. 061812, under the name *R. agerii*); *R. agerii* Bertol., s.loc., s.d., *s.coll.* (PI No. 003289); colline bolognesi a M. Donato, April 1864, *O. Beccari s.n.* (FI barcode FI066998, under the name *R. agerii*); lungo il Ravone presso Bologna, 18 April 1894, *A. Fiori s.n.* (FI barcodes FI066999, FI067000, FI067001, under the name *R. agerii*); vallata del Ravone presso Bologna, April 1882, *Gibelli s.n.* (FI barcodes FI067002, FI067017, PI Nos 061808, 061809, under the name *R. agerii*); Gaibola presso Bologna, nei prati, April 1891, *Mattei s.n.* (FI barcode FI FI067003, under the name *R. agerii*); Bologna, June 1889, *Mattei s.n.* (FI barcode FI067004, under the name *R. agerii*); vicinanze di Bologna, June 1883, *Mattei s.n.* (FI barcode FI067005, under the name R. agerii); prati presso Gaibola nelle vicinanze di Bologna, May 1886, *Mattei s.n.* (FI barcode FI067006, under the name *R. agerii*); nei campi alla Croara, April 1873, *G. Cugini s.n.* (FI barcode FI067007, under the name *R. agerii*); Bologna, Monte S. Donato, 25 May 1873, *G. Bertoloni s.n.* (FI barcode FI067008, under the name *R. agerii*); in pratis Gaibola prope Bononiam, 27 April 1886, A. Fiori s.n. (FI barcode FI067009, under the name *R. agerii*); nei colli Bolognesi, s.d., *G. Bertoloni s.n.* (FI barcode FI067010, under the name *R. agerii*); prati e vigne presso Gaibola Bologna, April 1885, *A. Baldacci s.n.* (FI barcode FI067011, under the name *R. agerii*); *ex collibus Bononiensibus* a Montedonato, 1842, *A. Bertoloni s.n.* (FI barcode FI067012, under the name *R. agerii*); *legi in parte meridionale montis* Castello della Croara (in cultis dia 4 May 1837), *s.coll.* (FI barcode FI067013, under the name *R. agerii*); contorni di Bologna, 1834, *Bubani s.n.* (FI barcode FI067014, under the name *R. agerii*); nei colli Bolognesi, s.d., *s.coll.* (FI barcode FI067018, under the name *R. agerii*); lungo il Ravone presso Bologna, 18 April 1892, *A. Fiori s.n.* (FI barcode FI067022, under the name *R. agerii*); Bologna, April 1892, *A. Fiori s.n.* (FI barcode FI067000, under the name *R. agerii*); Bononiae, s.d., *Moricand s.n.* (G barcode G00145280, under the name *R. agerii*); Monte Donato presso Bologna, s.d., *Bertoloni s.n.* (RO, under the name *R. agerii*); Bologna, May 1839, *s.coll.* (RO, under the name *R. agerii*); Barbianello, Bolognese, terreno argilloso, May 1899, *Betti s.n.* (RO, under the name *R. agerii*); Rio Ravone presso Bologna, 27 April 1885, *A. Fiori s.n.* (RO, under the name *R. agerii*); in un campo presso Pontecchio e nei prati a Gaibola, pre. di Bologna, April-May 1882, *G. Pirzini s.n.* (RO); TOSCANA. Florentia, 14 April 1866 (ex horto botanico Musei) (FI barcode FI067019, under the name *R. agerii*); UMBRIA. Perugia: largamente inselvatichito nell'Orto Botanico dell'Università, ma proveniente dal Bologense, 6 April 1899, *L. Palomba s.n.* (FI barcode FI067020, under the name *R. agerii*); LAZIO. Monti Campitelli (Anagni, Frosinone), radure e margine boschivo, 730 m, 20 April 2022, *E. De Santis s.n.* (APP No. 66136); SICILIA. Catania, in humidis pr. Misterbianco, 24 March 1874, *Heindereich s.n.* (FI barcode FI067021, under the name *R. agerii*). **Greece.** Samaria (Criti), 25 April 1976, *W. Greuter s.n.* (P barcode P00040497); Bassoré (Grèce), 1200 m, 23 April 1986, *F. Billy s.n.* (CLF barcode CLF006633); Montes Argolidis, s.d., *Boissier s.n.* (JE barcode JE 00021604) L; Insel Lefkádha (Nom. Lefkádhos). Kollivάta: Vorhof und Umgebung der Friedhofs östlich oberhalb der Ortschaft (ca. 1 km E Aléxandhros) (UTM 34S DH 7388), frische (wechseltrockene) Rudertalfluren, 445 m, 24 April 2011, *Gutermann W.* et al. *Iter Ionicum XIX 39880* (WU barcode WU0085740); Greece. Cephalonia. Grizata (Γριζάτα). Surrounding of gorge approx. 600 m NW of Grizata church, Maquis, (MGRS 34S DH 6830), 60–90 m, 30 March 2005, *Gilli C.* et al. *Iter Ionicum XXI 41217* (WU barcode WU0091948); Insel Kérkira (Nom. Kerkíras). Pandokrátoras: Karstplateau der Westseite, bei den großen Dolinen südlich des Fahrwegs (0.7–1.2 km westlich des Gipfels). [UTM: 34S DK 0200], 750–770 m, Rasenfragmente der Dolinen-Sonnseite, 16 May 2000,

Gutermann, W. et al. *Iter ionicum XVIII 34982* (WU barcode WU 0097318); Griechenland: Jonische Inseln: Kefallinia. Umgebung von Sámi: Gebiet zwischen der Stadt Sámi und der Ortschaft (Halbinsel) Dhihália sowie der Moní Iperajías Theotóku Agrilíon (ca. 2 km NE oberhalb von Sámi), 0–ca. 200 m, 11 April 1974, *Fischer M.A. & Fischer G. s.n.* (WU barcode WU0097319); Corfù, Monte Deca, 25 April 1887, *Gelmi s.n.* (WU barcode WU0097320); Corcyra (Corfu) Im Gerölle am Mte unter Büschen von Quercus coccifera, 11 April 1877, *Spreitzenhofer G.C. Iter jonicum a. 1877 s.n.* (WU No. 0097321); Corcyra (Corfu), am Plateau des Mte. San Salvatore, 13 April 1877, *Spreitzenhofer G.C. Iter jonicum a. 1877 127* (WU No. 0097322); Corcyra (Corfu) Mte. Deca im Gerölle in circa 15-1600 Fuß Höhe, in der Richtung gegen San Deca, 11 April 1877, *Spreitzenhofer G.C. Iter jonicum a. 1877 62* (WU No. 0097323); Zakynthos. Kulturland bei der Weggabelung (am Fahrweg zum Vrachiónas) 1 km NE Mariés (UTM 34S DG 7285) ca. 460 m, Flach terrassierte, steinige Olivenhaine, angrenzende Cistus-Heide, 25 April 2012, *Gutermann W.* et al. *Iter Ionicum XX 40612* (WU No. 0102503); Zante, Nördl. von Volimäs, 26 March 1936, *K. Ronniger s.n.* (W qrcode W0148865); S. Deka; prope ecclesiam Pantokrator, 529 m, 13 April 1896, Baenitz C. s.n. (DR barcode DR057363, under the name *R. agerii*; mixed with *R. paludosus*); colles elati Argolidis et Arkadiae, Apr. 1842, *Boissier s.n.* (MW barcode MW0592427); **Turkey**. Sinop, Ince Burum at Gerne, 1 May 1967, *Tobey C. 1616A* (E barcode E00442245, E00442247); Sinop, Ince Burum, 30 m, from clearings in woods, moist, 23 April 1966, *Tobey C. 1616A* (E barcode E00442245); *ibidem*, volcanicconglomerate, seaside. Also in forest clearings, 23 April 1966, *Tobey C. 1616* (E barcode E00442244); Bolu, between Istanbul and Ankara, E. of Gerede, short grassy turf, with Pinus, Juniperus and Quercus scrub, 28 April 1970, *Rix E. M. 1514* (E barcode E00442246); Istanbul, Silivri: Near Beyciler, edge of shrubby community, dry meadow, 16 April 1961, *Demiriz H. 4388* (E barcode E00442249); Istanbul (As.), Yakacik-Aydos road, Fountain, maquis, 6 May 1946, *Demiriz H. 2498* (E barcode E00442250); Aziatisch Turkije, Zonguldak (a.d. Zwarte Zee), 1 May 1947, Dijkstra, SJ s.n. (L qrcode L.1745470); Istanbul, Proti, 23 April 1938, *Post B.v.D. s.n.* (E barcode E00442253); Istanbul, Camlica, 26 April 1919, *Post B.v.D. s.n.* (E barcode E00442405); Istanbul, Cypress Hill grave, 25 April 1918, *Post B.v.D. s.n.* (E barcode E00442243); Bithynia, 1839, *Grisebach s.n.* (BR barcode BR0000005295913); in m. Bulgurlu, *Grisebach 24* (GOET barcode GOET009797).

4. Conclusions

Nomenclature plays a central role in the description of the diversity of life on our planet and the typification process is essential for any taxonomic study. At the same time, floristic research and the study of herbarium collections are of crucial importance in biodiversity conservation of vascular plants and are necessary to collect data for planning the correct conservation strategies. Our study on *R. gracilis* in Italy allowed us, primarily, to evaluate the taxonomic identity of *R. agerii* and *R. schowii*, both described from Italian territory. After typification, *R. agerii* should be regarded as a heterotypic synonym of *R. gracilis*, while *R. schowii* showed a combination of unique characters, close in some ways to atypical forms of *R. isthmicus*, and needs further studies to assess the morphological variability in the species and to clarify its taxonomic status. Thanks to our contribution, the presence of *R. gracilis* in Italy is confirmed, expanding the distribution range of this endemic species to the SE Euro-Mediterranean towards the west. In Italy, *R. gracilis* is present, in the current state of knowledge, with a single population at risk of extinction found in Lazio (Central Italy). Furthermore, we were able to confirm the historical presence of the species, based on the study of herbarium collections, in the Emilia-Romagna (Northern Italy) and Sicily, where it has not been observed for over 120 years. It will now be possible to plan specific field surveys to verify whether *R. gracilis* is still present in these areas or is to be considered extinct. Furthermore, the species is now reported in the national conservation framework.

Author Contributions: Conceptualization and methodology, F.B. and F.C.; field investigations, E.D.S.; nomenclature, F.B.; writing—original draft preparation, F.B.; writing—review and editing, F.B., E.D.S. and F.C.; supervision, F.B. All authors have read and agreed to the published version of the manuscript.

Funding: This research received no external funding.

Institutional Review Board Statement: Not applicable.

Informed Consent Statement: Not applicable.

Data Availability Statement: The data presented in the current study are available within the article.

Acknowledgments: The authors wish to thank A. Brach (A, GH, ECON), M. Carine, N. Holstein and J. Wajer (BM), J. M. de Vos (BAS), G. Domina (PAL), J. Doubt (CAN), R. Galesi (CAT), S. Kenyon (SWN), U. Mossetti (BOLO), C. Nepi and L. Lastrucci (FI), F. Roma-Marzo and G. Astuti (PI), D. Magri, A. Tilia and M. Pierfranceschi (RO), P. Caputo and R. Vallariello (NAP) and other curators of the herbaria listed in the Section 2 for their help with searching for the original material and for providing us digital images of herbarium specimens. We also thank A. Alessandrini, L. Bernardo, D. Bouvet, G. Galasso, N. Passalacqua, L. Peruzzi, R. Di Pietro, A. Soldano, A. Selvaggi and R.P. Wagensommer, who provided us with distributive data.

Conflicts of Interest: The authors declare no conflict of interest.

References

1. Tamura, M. Angiospermae. Ordnung *Ranunculales*. Fam. *Ranunculaceae*. II. Systematic part. IV 17A. In *Die natürlichen Pflanzenfamilien*, 2nd ed.; Hiepko, P., Ed.; Duncker Humblot: Berlin, Germany, 1995; pp. 223–519.
2. Hörandl, E.; Emadzade, K. Evolutionary classification: A case study on the diverse plant genus *Ranunculus* L. (Ranunculaceae). *Perspect. Plant Ecol. Evol. Syst.* **2012**, *14*, 310–324. [CrossRef]
3. Peruzzi, L.; Conti, F.; Bartolucci, F. An inventory of vascular plants endemic to Italy. *Phytotaxa* **2014**, *168*, 1–75. Available online: https://docs.google.com/spreadsheets/d/18SCKlDJ8TQHNm58td5j94UCZpymUPlwDZ5Yx9718MXE/edit#gid=4 (accessed on 10 September 2022). [CrossRef]
4. Bartolucci, F.; Peruzzi, L.; Galasso, G.; Albano, A.; Alessandrini, A.; Ardenghi, N.; Astuti, G.; Bacchetta, G.; Ballelli, S.; Banfi, E.; et al. An updated checklist of the vascular flora native to Italy. *Plant Biosyst.* **2018**, *152*, 179–303. [CrossRef]
5. Galasso, G.; Conti, F.; Peruzzi, L.; Ardenghi, N.M.G.; Banfi, E.; Celesti-Grapow, L.; Albano, A.; Alessandrini, A.; Bacchetta, G.; Ballelli, S.; et al. An updated checklist of the vascular flora alien to Italy. *Plant Biosyst.* **2018**, *152*, 556–592. [CrossRef]
6. Clarke, E.D. *Travels in Various Countries of Europe, Asia and Africa: Greece, Egypt, and the Holy Land*; R. Watts for Cadell and Davies: London, UK, 1814; Volume 3, Part 2.
7. Hörandl, E.; von Raab-Straube, E.; Ranunculeae. Euro+Med Plantbase—The Information Resource for Euro-Mediterranean Plant Diversity. 2015. Available online: https://europlusmed.org/cdm_dataportal/taxon/d6c91320-a11e-4069-b29a-51b6073596dd (accessed on 30 September 2022).
8. POWO—Plants of the World Online, Facilitated by the Royal Botanic Gardens, Kew. Available online: http://www.plantsoftheworldonline.org/ (accessed on 20 July 2022).
9. Pignatti, S.; Guarino, R.; La Rosa, M. (Eds.) *Flora d'Italia*, 2nd ed.; New Business Media: Milan, Italy, 2017; Volume 1.
10. Fiori, A.; Béguinot, A.; Paoletti, G. *Flora Analitica d'Italia*; Tipografi del Seminario: Padova, Italy, 1898; Volume 1, pp. 257–607.
11. Fiori, A. *Nuova Flora Analitica d'Italia*; Tipografia di M. Ricci: Firenze, Italy, 1924; Volume 1, pp. 641–800.
12. Giardina, G.; Raimondo, F.M.; Spadaro, V. A catalogue of plants growing in Sicily. *Bocconea* **2007**, *20*, 5–582.
13. Bertoloni, A. Memoria di Antonio Bertoloni professore di botanica nell'università di Bologna sopra due specie nuove di piante italiane. *Opusc. Sci.* **1819**, *3*, 180–186.
14. Tutin, T.G. Ranunculus L. In *Flora Europaea*; Tutin, T.G., Burges, N.A., Chater, A.O., Edmondson, J.R., Heywood, V.H., Moore, D.M., Valentine, D.H., Walters, S.M., Webb, D.A., Eds.; Cambridge University Press: Cambridge, UK, 1964; Volume 1, pp. 223–238.
15. Pignatti, S. *Flora d'Italia*; Edagricole: Bologna, Italy, 1982; Volume 1.
16. Jalas, J.; Suominen, J. (Eds.) *Atlas Florae Europaeae. Distribution of Vascular Plants in Europe. Nymphaeaceae to Ranunculaceae*; The Committee for Mapping the Flora of Europe and Societas Biologica Fennica Vanamo: Helsinki, Finland, 1989; Volume 8, 261p.
17. Tutin, T.G.; Akeroyd, J.R. Ranunculus L. In *Flora Europaea*, 2nd ed.; Tutin, T.G., Burges, N.A., Chater, A.O., Edmondson, J.R., Heywood, V.H., Moore, D.M., Valentine, D.H., Walters, S.M., Webb, D.A., Eds.; Cambridge University Press: Cambridge, UK, 1993; Volume 1, pp. 269–287.
18. Gussone, G. *Florae Syculae Synopsis*; Tramater: Napoli, Italy, 1845; Volume 2.
19. Domina, G.; Giusso del Galdo, G.; Gargano, D.; Labra, M.; Peccenini, S.; Peruzzi, L.; Raimondo, F.M. The Italian Loci Classici census. *Taxon* **2012**, *61*, 1351–1354. [CrossRef]
20. Peruzzi, L.; Domina, G.; Bartolucci, F.; Galasso, G.; Peccenini, S.; Raimondo, F.M.; Albano, A.; Alessandrini, A.; Banfi, E.; Barberis, G.; et al. An inventory of the names of vascular plants endemic to Italy, their loci classici and types. *Phytotaxa* **2015**, *196*, 1–217. [CrossRef]
21. Peruzzi, L.; Galasso, G.; Domina, G.; Bartolucci, F.; Santangelo, A.; Alessandrini, A.; Astuti, G.; D'Antraccoli, M.; Roma-Marzio, F.; Ardenghi, N.M.G.; et al. An inventory of the names of native, non-endemic vascular plants described from Italy, their loci classici and types. *Phytotaxa* **2019**, *410*, 1–215. [CrossRef]

22. Brundu, G.; Peruzzi, L.; Domina, G.; Bartolucci, F.; Galasso, G.; Peccenini, S.; Raimondo, F.M.; Albano, A.; Alessandrini, A.; Banfi, E.; et al. At the intersection of cultural and natural heritage: Distribution and conservation of the type localities of the Italian endemic vascular plants. *Biol. Conserv.* **2017**, *214*, 109–118. [CrossRef]
23. Thiers, B. Index Herbariorum: A Global Directory of Public Herbaria and Associated Staff. Available online: http://sweetgum.nybg.org/science/ih/ (accessed on 20 July 2022).
24. Stafleu, F.A.; Cowan, R.S. *Taxonomic Literature: A Selective Guide to Botanical Publications and Collections with Dates, Commentaries and Types*, 2nd ed.; Bohn, Scheltema & Holkema: Utrecht, The Netherlands, 1976; Volume 1.
25. Stafleu, F.A.; Cowan, R.S. *Taxonomic Literature: A Selective Guide to Botanical Publications and Collections with Dates, Commentaries and Types*, 2nd ed.; Bohn, Scheltema & Holkema: Utrecht, The Netherlands, 1986; Volume 6.
26. Turland, N.J.; Wiersema, J.H.; Barrie, F.R.; Greuter, W.; Hawksworth, D.L.; Herendeen, P.S.; Knapp, S.; Kusber, W.-H.; Li, D.-Z.; Marhold, K.; et al. *International Code of Nomenclature for Algae, Fungi, and Plants (Shenzhen Code) Adopted by the Nineteenth International Botanical Congress Shenzhen, China, July 2017*; Regnum Vegetabile; Koeltz Botanical Books: Glashütten, Germany, 2018. [CrossRef]
27. Stefanaki, A.; van Thijsse, G.; Ufellen, G.; van Eurlings, M.; Andel, T. The En Tibi herbarium, a 16th century Italian treasure. *Bot. J. Linn. Soc.* **2018**, *187*, 397–427. [CrossRef]
28. Stefanaki, A.; Porck, H.; Grimaldi, I.M.; Thurn, N.; Pugliano, V.; Kardinaal, A.; Salemink, J.; Thijsse, G.; Chavannes-Mazel, C.; van Kwakkel, E.; et al. Breaking the silence of the 500-year-old smiling garden of everlasting flowers: The En Tibi book herbarium. *PLoS ONE* **2019**, *14*, e0217779. [CrossRef] [PubMed]
29. Bauhin, C. *Phytopinax*; Per Sebastianum Henricpetri: Basel, Switzerland, 1596.
30. Bauhin, C. *Prodromos Theatri Botanici Caspari Bauhini*; Typis P. Iacobi, impensis I. Treudelii: Frankfurt, Germany, 1620.
31. Bauhin, C. *Pinax Theatri Botanici*; Sumptibus & Typis Ludovici Regis: Basel, Switzerland, 1623.
32. Bauhin, J.; Cherler, J.H. *Historia Plantarum Universalis*; Fr. Lud. von Graffenried; Ebroduni: Yverdon, Switzerland, 1650.
33. Parkinson, J. *Theatrum Botanicum*; Tho. Cotes: London, UK, 1640.
34. Linnaeus, C. *Species Plantarum*, 2nd ed.; Salvii, L., Ed.; Holmiae: Stockholm, Sweden, 1762; Volume 1.
35. Arcangeli, G. *Compendio della Flora Italiana*; Ermanno Loescher Editore: Turin, Italy, 1882.
36. Cesati, V.; Passerini, G.; Gibelli, G. *Compendio della Flora Italiana*; Vallardi: Milan, Italy, 1885; Volume 2.
37. Jarvis, C. *Order Out of Chaos*; Linnaean Society of London: London, UK, 2007.
38. Pons, G. Saggio di una rivista critica delle specie italiane del genere *Ranunculus* L. (continuazione e fine). *Nuovo G. Bot. Ital.* **1901**, *8*, 5–27.
39. Zangheri, P.A. *Flora Italica*; CEDAM: Padova, Italy, 1976.
40. Greuter, W.; Burdet, H.M.; Long, G. (Eds.) *Med-Checklist. A Critical Inventory of Vascular Plants of the Circum-Mediterranean Countries: Dicotyledones (Lauraceae–Rhamnaceae)*; Editions des Conservatoire et Jardin botaniques de la Ville de Genève: Genève, Switzerland, 1984; Volume 4.
41. Peruzzi, L.; Passalacqua, N.G. On *Ranunculus aspromontanus* Huter (Ranunculaceae) and its taxonomic relationship. *Willdenowia* **2003**, *33*, 255–264. [CrossRef]
42. Conti, F.; Abbate, G.; Alessandrini, A.; Blasi, C. (Eds.) *An Annotated Checklist of the Italian Vascular Flora*; Palombi Editori: Roma, Italy, 2005.
43. Scoppola, A.; Spampinato, G. Atlante delle specie a rischio di estinzione. In *Stato delle Conoscenze Sulla Flora Vascolare d'Italia*; Scoppola, A., Blasi, C., Eds.; Versione 1.0. CD-Rom; Palombi Editori: Roma, Italy, 2005.
44. Conti, F.; Alessandrini, A.; Bacchetta, G.; Banfi, E.; Barberis, G.; Bartolucci, F.; Bernardo, L.; Bonacquisti, S.; Bouvet, D.; Bovio, M.; et al. Integrazioni alla checklist della flora vascolare italiana. *Nat. Vicentina* **2007**, *10*, 5–74.
45. Guarino, R.; La Rosa, M. Flora d'Italia Digitale. In *Flora d'Italia*, 2nd ed.; Pignatti, S., Guarino, R., La Rosa, M., Eds.; Edagricole: Milan, Italy, 2019; Volume 4.
46. Pignatti, S.; Guarino, R.; La Rosa, M. (Eds.) *Flora d'Italia*, 2nd ed.; Edagricole: Milan, Italy, 2019; Volume 4.
47. Miller, H.S. The herbarium of Aylmer Bourke Lambert: Notes on its acquisition, dispersal, and present whereabouts. *Taxon* **1970**, *19*, 489–553. [CrossRef]
48. Lojacono Pojero, M. *Flora Sicula*; Tipo-Litografia Salvatore Bizzarrilli: Palermo, Italy, 1889; Volume 1.
49. La Valva, V. La Collezione Gussone Sicilia. *Webbia* **1993**, *48*, 515–537.
50. Boissier, P.E. *Diagnoses Plantarum Orientalium Novarum*; Series 1; B. Hermann: Leipzig, Germany, 1846; Volume 6.
51. Strid, A. New taxa described in Grisebach's "Spicilegium Florae Rumelicae et Bithynicae" (1843–1846). *Preslia* **2000**, *72*, 241–321.
52. Strid, A. Ranunculus L. In *Flora Hellenica*; Strid, A., Kit, T., Eds.; Gantner Verlag: Ruggell, Liechtenstein, 2002; Volume 2, pp. 38–69.
53. Davis, P.H. (Ed.) Ranunculus L. In *Flora of Turkey and the East Aegean Islands*; Edinburgh University Press: Edinburgh, UK, 1965; Volume 1, pp. 146–197.
54. Grossheim, A.A. *Flora Kavkaza*, 2nd ed.; Academy of Sciences of USSR: Moskov, Russia, 1950; Volume 4.
55. Pezzi, G.; Buldrini, F.; Alessandrini, A. (Eds.) *Dal Libro Alla Mappa, la Flora Storica della Provincia di Bologna*; Centro Stampa RER: Bologna, Italy, 2021.
56. Sibilia, C. Ricerche Floristiche sul Territorio di Anagni. Nota 2. *Nuovo G. Bot. Ital.* **1924**, *31*, 161–176.
57. Anzalone, B. *Prodromo della Flora Romana (Elenco Preliminare delle Piante Vascolari Spontanee del Lazio)*; Quad. Lazionatura 5; Tip. Centenari: Roma, Italy, 1984; 250p.

58. Anzalone, B. Prodromo della Flora Romana (elenco preliminare delle piante vascolari spontanee del Lazio), Aggiornamento. Parte 1a. *Ann. Bot.* **1996**, *52* (Suppl. 11), 1–81.
59. Anzalone, B.; Iberite, M.; Lattanzi, E. La Flora vascolare del Lazio. *Inform. Bot. Ital.* **2010**, *42*, 187–317.
60. Di Pietro, R.; Misano, G. Analisi fitosociologica e considerazioni sintassonomiche sulla vegetazione forestale delle Gravine occidentali dell'Arco Ionico (Murge pugliesi e lucane, Italia meridionale). *Inform. Bot. Ital.* **2009**, *41*, 215–246.
61. Strobl, P.G. Flora des Etna. *Oesterr. Bot. Z.* **1884**, *34*, 329–332. [CrossRef]
62. Raimondo, F.M.; Certa, G.; Gianguzzi, L.; Ilardi, V.; Notara, G. Materiali per una nuova flora palermitana. *Quad. Bot. Amb. Appl.* **1995**, *6*, 125–130.
63. Domina, G.; Peruzzi, L.; Bedini, G. (Eds.) Wikiplantbase #Sicilia v3.0. 2019. Available online: http://bot.biologia.unipi.it/wpb/sicilia/index.html (accessed on 27 September 2022).
64. Popova, M. Cytotaxonomic investigation in genus *Ranunculus* L. *Izv. Na Bot. Inst.* **1973**, *24*, 233–239.
65. Bachman, S.; Moat, J.; Hill, A.W.; de la Torre, J.; Scott, B. Supporting Red List threat assessments with GeoCAT: Geospatial conservation assessment tool. *ZooKeys* **2011**, *150*, 117–126. [CrossRef]
66. IUCN Guidelines for Using the IUCN Red List Categories and Criteria. Version 15.1. Prepared by the Standards and Petitions Subcommittee, Cambridge. Available online: https://www.iucnredlist.org/resources/redlistguidelines (accessed on 23 September 2022).

plants

MDPI

Article

New Morphological, Distribution, and Ecological Data on *Scabiosa garganica* (Caprifoliaceae), a Poorly Known Species of the Italian Flora, with Evaluation of Its Conservation Status and Typification of the Name

Daniele Bonsanto [1], Nello Biscotti [1,*] and Robert Philipp Wagensommer [2,*]

[1] Department of Agricultural, Food and Environmental Sciences (D3A), Marche Polytechnic University, I-60131 Ancona, Italy

[2] Faculty of Education, Free University of Bozen-Bolzano, I-39042 Brixen-Bressanone, Italy

* Correspondence: nellobisco@gmail.com (N.B.); robertphilipp.wagensommer@unibz.it (R.P.W.)

Abstract: This paper presents the results of a research performed on Gargano Promontory (SE-Italy) on the populations of *Scabiosa garganica*, a species with little herbarium records and whose few morphological descriptions are outdated. *S. garganica* belongs to the *S. holosericea* aggr., a group including very similar taxa that still have different taxonomic classifications. Its typical location is Monte Sant'Angelo in the Gargano area. Surveys have ascertained the existence of many populations, whose stational data help to understand the distribution and ecological conditions *S. garganica* is linked to. The morphological analysis of a large sample (75 plants from 9 sites) allows for the description of the qualitative and quantitative characteristics of this species. The new morphological framework highlights the species autonomy of *S. garganica* and can contribute to clarifying the relationship with *S. holosericea* and *S. taygetea* to which it is closer. In addition, for the correct delimitation of the species, the name *Scabiosa garganica* is lectotypified. Finally, the species was assessed against the IUCN criteria for the evaluation of its conservation status.

Keywords: ecology; Italy; IUCN; lectotype; morphology; taxonomy

Citation: Bonsanto, D.; Biscotti, N.; Wagensommer, R.P. New Morphological, Distribution, and Ecological Data on *Scabiosa garganica* (Caprifoliaceae), a Poorly Known Species of the Italian Flora, with Evaluation of Its Conservation Status and Typification of the Name. *Plants* **2023**, *12*, 1915. https://doi.org/10.3390/plants12091915

Academic Editor: Zafar Ahmad Handoo

Received: 21 April 2023
Revised: 2 May 2023
Accepted: 5 May 2023
Published: 8 May 2023

1. Introduction

This research focuses on *Scabiosa garganica* Porta & Rigo ex Wettst. [=*S. taygetea* Boiss. & Heldr. subsp. *garganica* (Porta & Rigo ex Wettst.) Hayek] (Caprifoliaceae), a species alternatively considered endemic to Italy [1], occurring in Italy and Greece [2], or as a synonym of *S. taygetea* in Albania [3]. In Italy, it is known to grow with certainty only in the Gargano area (Apulia) [4]. This species belongs to an extremely polymorphous group (*Scabiosa holosericea* aggr.), thus having a still temporary taxonomic classification [1] that is also considered still unresolved [5]. The species was described on the basis of a collection of Porta and Rigo [6], whose tag says "near Monte Sant'Angelo" in the Gargano area. Actually, a "*Scabiosa garganica*" had been reported a century earlier in southern Gargano by Micheli [7], but unexpectedly it was not included in the Catalogus plantarum Horti Caesarei Florentini [8], where several other plants from Gargano are listed. Some years later, *Scabiosa garganica* was mentioned by Tilli [9], describing it as "*frutescens, villosa et incana, foliis laciniatis, flore ex caeruleo purpuracente*" (shrublike, hairy, and grayish, with fringed edges leaves and flowers of a purplish cerulean). After many years, in the spring of 1874 and without making any reference to what Micheli [7] and Tilli [9] had stated, Porta and Rigo [6] report to have found in "Monte Sant'Angelo and not anywhere else", a *Scabiosa* they state being new, and, they wrote, "we decided to call it 'garganica' after its location". In their second travel in Gargano in 1875 (for more information on the two travels in 1874 and 1875, see [10]), Porta and Rigo [6] provided further details about the location ("*apricis Montis S. Angelo, pascuis saxosis*"—sunny Monte Sant'Angelo, in rocky pastures) and the plant's phenology ("summer flowering"). The species is afterwards reported in Compendio della Flora Italiana [11] with the binomial *Scabiosa garganica* Arc., but in "Flora Italiana" by Parlatore and Caruel [12], it is included as a variety of *Scabiosa pyrenaica* All., together with *S. holosericea* Bertol., whose presence in the Gargano area is reported by Pasquale and Licopoli [13]. However, the authors state that "'*Scabiosa garganica*' gathered by Porta and Rigo in Gargano area is a very beautiful and more tomentose form"; moreover, this work highlights for the first time its similarity with *S. taygetea* described from Greece [14]. In Wettstein [15], it is accompanied by an illustrated table and described as a species, and new morphological details are provided: " ... a wonderful plant ... 30 to 50 cm high ... with a rhizome full of sterile leaves and 1 to 4 flowering stems" (*Rhizoma caespites foliorum steriles et caules floriferos 1–4 edens*). Wettstein too thinks that such characteristics make it undoubtedly close to *S. taygetea* (found in Peloponnese only), but it is "easily recognisable due to its red-violet/purple colour, a shorter corolla and above all its significantly shorter lobes of the calyx"; in addition, it has "protruding, unusually thick shiny white hairs". In his annotations, Wettstein concludes that "the presence of the same species in Monte Gargano and in the high Albanian mountains is of great geographical interest". In the first years of the 20th century, the "Herbarstudien" by Huter [16] reports that "nr. 72" (Porta and Rigo's herbarium sheet) "can only be used as a form of *S. holosericea* Bert.". Huter too thinks that its main feature are the thick hairs on the leaves ("soft wool, with a colour ranging from ash grey to velvety whitish"); like in *S. holosericea*, the hairs are simple, not branchy, Huter says, and (except for the basal ones) they are "pinnatifid, or lyrate pinnatifid but with a very large terminal lacinia, at times almost round". In 1921, it is included among the critical and rare species of the Italian flora analyzed by Lacaita [17], who writes as having found it in 1919 in Gargano, "copious in the cliffs of the western limestone ridge, not far from the village but towards Manfredonia, about 450 metres a.s.l.". According to Lacaita, the woolly appearance of the leaves "is much more pronounced in Gargano samples than in Wettstein's sketch of the plant in Albania". Lacaita adds that *S. garganica* has only been found near the village of Monte Sant'Angelo where Porta and Rigo have found and documented it during their two trips in Gargano area. The morphological notes of Lacaita are such that he cannot distinguish it from *S. holosericea*. To resolve his doubt, he concludes that "several samples from the different Gargano's stations are needed". Unfortunately, the partial knowledge of the plant remains unchanged in the following years, but *Scabiosa garganica* undergoes continuous reviews that modify its taxonomic status many times. In Fiori's "Flora Analitica" [18], Porta and Rigo's "*Scabiosa garganica*" is mentioned as a species [*S. garganica* (Porta) Fiori]. In "Prodromo della Flora Garganica" [19], it is treated as a subspecies of *S. holosericea* (*S. holosericea* Bert. subsp. *garganica* Huter, Porta & Rigo). In this new combination, Fenaroli [19] gathers the different reports of *Scabiosa columbaria* L. var. *holosericea* (Bert.) Fiori by Pasquale and Licopoli (collected in 1872), Martelli (loc. Testa del Gargano, collected in 1893 (FI!)), Fiori (loc. Monte Sant'Angelo, collected in 1898 (FI!)), Trotter and Forti (loc. Manfredonia and Monte Sant'Angelo, collected in 1907 (FI!)), Lacaita (collected in 1919). Fenaroli attributes to this same combination also the samples collected by himself with Grilli (loc. Santuario di Pulsano in 1960 (Herb. Fenaroli, TR)) and by Agostini ("between Monte Sant'Angelo and Mattinata", in 1961 (Herb. Fenaroli, TR)). According to Fenaroli, in the Gargano area, either *Scabiosa holosericea* and its "*garganica*" subspecies must be present, and they are distinguishable also for their distribution (since the "*garganica*" subspecies was known only in Monte Sant'Angelo's station). Later, *Scabiosa garganica* was proposed by Zangheri [20] as a subspecies of *Scabiosa vestita* Jord. (*S. vestita* subsp. *garganica* (Porta & Rigo ex Wettst.) Zangh., *comb. inval.*). In "Flora d'Italia" by Pignatti [21], the "Scabiosa of Gargano" is included in the variability of *Scabiosa holosericea*, highlighting the presence in Gargano Promontory of characteristic populations with "thickly woolly leaves with elongated white-greyish hairs". In this work, Fenaroli's combination is deemed "illegitimate", even though it is stressed that "the rank of this taxon needs to be defined through an appropriate research" that has not been carried out. In the first years of 2000, in fact, the main study materials for this plant still are the samples gathered by Porta and

Rigo in the Gargano area, and Licht [22] indicates *S. taygetea* subsp. *garganica* as no longer recorded from Gargano area. In the Checklist of the Italian vascular flora by Conti et al. [23], *S. garganica* is considered as a variety of *Scabiosa taygetea* (*Scabiosa taygetea* var. *garganica* (Porta & Rigo ex Wettstein) Hayek). This combination highlights the greater similarities with the populations described for the fir woods on Mount Taygetus in Peloponnese, as Parlatore and Caruel [12] already emphasized. In the more recent publications about the Italian flora, the *Scabiosa* from Gargano area is considered at species rank (*Scabiosa garganica* Porta & Rigo ex Wettstein) [24], or as a subspecies (*Scabiosa tygetea* Boiss. & Heldr. subsp. *garganica* (Porta and Rigo) Hayek) [1,4]; it is reported as such also in the recent "Flora vascolare del Gargano e delle Isole Tremiti" [25]. Pignatti [1] also mentions *S. holosericea* as present in the Gargano region, while Bartolucci et al. [4] consider it as not present in Apulia. Moreover, Pignatti [1] still keeps the awareness of the existence of a little-known group (*holosericea*) whose arrangement is temporary. After more than a century, even now the same gaps pointed out by Lacaita [17] are still the same, due to the complete lack of studies on this species, which would be absolutely needed to clarify the actual distribution in Italy, also given its relationship with *Scabiosa holosericea* and *S. taygetea*, and considering that a report from Basilicata (Monte Alpi), previously attributed to *Scabiosa taygetea* [26], was recently doubtfully referred to *S. garganica* [27]. Monte Sant´Angelo is the reference site including the latest reports [28–30]. At the same time, *Scabiosa holosericea* has been nonetheless reported also in southern areas of Gargano Promontory, but with no herbarium material [31]. *Scabiosa holosericea* s. str. is moreover reported in the analytical key by Licht [22]. Licht (Herb. Garg.-MJG, available online at http://jacq.org/#database, accessed on 16 March 2023) also mentions six herbarium sheets of samples of *Scabiosa taygetea* subsp. *garganica* collected in Vallone di Pulsano, Monte Saraceno, Torre Pucci, Monte Spigno, SS 89 (from km 2.2 to km 2.4; at km 10.2). In the light of these reports, the presence in Gargano Promontory of *Scabiosa garganica* is well beyond Monte Sant'Angelo, distinguishing the historical reports from Monte Sant'Angelo [6,17], Vallone di Pulsano [19], and Testa del Gargano [19], from the recent ones (Monte Pucci, Monte Spigno, km 2–10 of SP 89) with herbarium samples (collected in 1986–2006 by Licht) (Table 1). At the same time, reports of *Scabiosa holosericea* from Gargano need to be (re-)considered.

Table 1. Historical reports of *Scabiosa garganica* from Gargano.

Author(s)	Scientific Name Used by the Author(s)
Micheli [7]	*Scabiosa garganica*
Tilli [9]	*Scabiosa garganica*
Pasquale and Licopoli [13]	*Scabiosa holosericea—S. taygetea*
Porta and Rigo [6]	*Scabiosa garganica*
Arcangeli [11]	*Scabiosa garganica*
Parlatore and Caruel [12]	*Scabiosa pyrenaica* var. *garganica*
Wettstein [15]	*Scabiosa garganica*
Huter [16]	*Scabiosa holosericea*
Lacaita [17]	*Scabiosa garganica*
Fiori [18]	*Scabiosa garganica*
Fenaroli [19]	*Scabiosa holosericea* subsp. *garganica*
Zangheri [20]	*Scabiosa vestita* subsp. *garganica*
Pignatti [21]	Included in the variability of *Scabiosa holosericea*
Conti et al. [23]	*Scabiosa taygetea* var. *garganica*
Di Pietro & Wagensommer [30]	*Scabiosa taygetea* subsp. *garganica*
Peruzzi et al. [24]	*Scabiosa garganica*
Bartolucci et al. [4]	*Scabiosa taygetea* subsp. *garganica*
Pignatti [1]	*Scabiosa taygetea* subsp. *garganica*
Licht [29]	*Scabiosa taygetea* subsp. *garganica*
Licht & Wagensommer [25]	*Scabiosa taygetea* subsp. *garganica*

Therefore, the aims of this paper are as follows: (i) the description of the correct morphology of *Scabiosa garganica*; (ii) clarifying the current distribution in Gargano area; (iii) investigating the ecological conditions to which the species is linked to; (iv) clarifying its relationships within the *S. holosericea* aggr.; (v) the evaluation of the conservation status of the species, according to the IUCN categories and criteria [32]; (vi) the typification of the name *Scabiosa garganica* for the correct interpretation of this name, given that it appears to be still untypified [24].

2. Materials and Methods

We analyzed 75 individuals of *Scabiosa garganica* collected in autumn 2020 in 9 sites, 5 in the southern area of the Gargano Promontory, and 4 in the northern area of the promontory (Table S1). Every site was georeferenced by a GPS (Garmin Etrex 100, average accuracy of 3 m).

For each individual, morphological qualitative and quantitative characters were analyzed. Leaves (185 basal leaves and 187 cauline leaves) were scanned with a 600 dpi resolution, and following this, they were measured with ImageJ [33] (Figure 1).

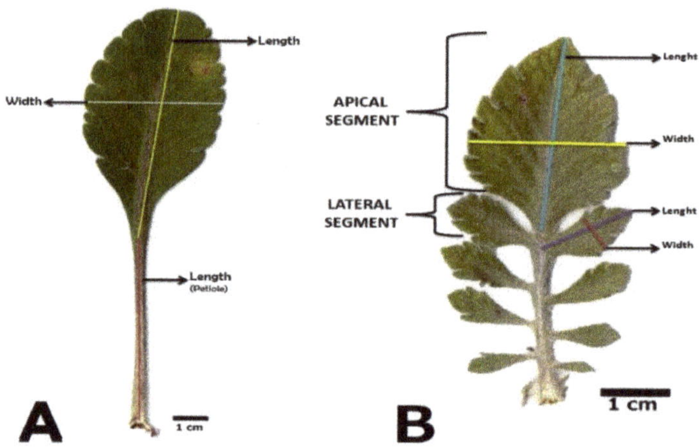

Figure 1. Measurements carried out on leaves: (**A**) basal leaves, (**B**) cauline leaves.

Scabiosa garganica were compared through the descriptions in the literature [1,6,7,14–17,21,25] and from the analysis of digitized herbarium sheets of *S. taygetea* and *S. holosericea*, listed in the specimen visa.

For each individual, morphological qualitative and quantitative characters were analyzed.

The measurements carried out were as follows: Plant height, number of scapes, number and diameter of inflorescences, length of flower peduncles, for the basal leaves lamina length and width, length/width, petiole length, and for the cauline leaves total length and width, apical segment (length and width, length/width), lateral segment (length and width, length/width), apical segment width/lateral segment width, and number of lateral lobes. Using SPSS [34], we then performed the standard statistical analysis (mean, standard deviation, min. and max.) of the dataset composed by all measurements, gathered in an Excel spreadsheet (Table S2).

Principal component analysis (PCA), non-metric multidimensional scaling (NMDS), and cluster analysis (CA) using the average linkage method (UPGMA) were performed with PAST package v4.09 software (Natural History Museum, Oslo, Norway) [35]. The similarity matrix was calculated using the Gower coefficient, suitable for mixed data [36].

With regard to the bioclimatic framework, we referenced the recent classification of Pesaresi et al. [37].

The acronyms of the herbaria are according to Thiers [38].

The conservation status of the species was assessed according to the IUCN categories and criteria [32]. The extent of occurrence (EOO) was calculated as convex hull, whereas the area of occupancy (AOO) was calculated using 2×2 km grid cells.

3. Results

3.1. Distribution of *Scabiosa garganica*

The survey confirmed the presence of many populations both in the northern area of Gargano Promontory (around Peschici) and in the southern area (Monte Sant'Angelo, Manfredonia, Mattinata). In the northern zones, the populations are distributed in a coastal strip going from Torre di Monte Pucci until the last rocky ridges close to Palude di Sfinale, with an altitude range of 20 to 130 m a.s.l. In the southern zones, the populations found are on the other hand distributed in a quite continuous way only in the north-west ridge of Monte Saraceno (Mattinata), and from there alongside the provincial road 55 that leads to Monte Sant'Angelo, with an altitude range between 250 and 670 m a.s.l.

3.2. Ecological Characterisation of *Scabiosa garganica*

The stational data supply useful ecological information: The populations can be found at altitudes ranging between 20 m a.s.l. (Sfinale) and 1000 m a.s.l. (Monte Spigno); the exposure varies from north-west to south-west; and the substrates are rocky dry substrates with flat or even deep soils, often coves of land on small terraces or mountain sides, or at the foot of rocky slopes. The bioclimatic framework [37] belongs to the Mediterranean macrobioclimate, pluviseasonal-oceanic bioclimate, and lower mesomediterranean thermotype with lower subhumid ombrotype. With regard to the vegetation, both northern and southern areas are dominated by dry rocky *Pinus halepensis* Mill. woodlands, where the populations of *Scabiosa garganica* are linked to clearings or can be found at the borders of the forest. An important aspect is that the species can be found mostly in areas recovering after a wildfire, and thus the zones are characterized by the presence of species of the genus *Cistus* (*C. salviifolius* L., *C. monspeliensis* L.). In such conditions, it is obvious that the species can play an important role in the successions of recovery after a wildfire, prior to the beginning of the bushes (*Pistacia lentiscus* L.) stage, or to the germination resumption of the woodland itself. Inside the forests, plants of *S. garganica* develop an impressive growth, with very long procumbent stems. In the same environment, populations of *S. garganica* can be observed along the sides of paved roads (provincial roads), as part of a road's vegetation (Peschici area). In this case, the plant community *S. garganica* belongs to is usually characterized by *Brachypodium retusum* (Pers.) P.Beauv. There is typically the coexistence on rocky substrates in southern Gargano of plant species characterizing the chasmophytic flora of Gargano [28,39], such as *Lomelosia crenata* (Cirillo) Greuter & Burdet subsp. *dallaportae* (Boiss.) Greuter & Burdet and *Inula verbascifolia* (Willd.) Hausskn. (sites of Monte Sant'Angelo/Manfredonia), or of the latter together with *Centaurea subtilis* Bertol. (Monte Saraceno, Monte Sant'Angelo/Mattinata). Together with *Campanula garganica* Ten. and *Coronilla juncea* L., *Centaurea subtilis* is on the other hand typical of the plant communities of *S. garganica* in the northern areas of the promontory.

3.3. Biological Characterisation of *Scabiosa garganica*

The several analyzed samples confirmed the nature of a scapose hemicryptophyte, with a vegetative phase and a quite long reproductive activity, with floral scapes developing from late spring to the end of autumn. Full flowerings are typical, even during autumn. The lateral ascending rooting shoots help the horizontal growth of the plant, which often takes a circular shape.

3.4. Quantitative Morphometric Characters of *Scabiosa garganica*

The analyzed individuals (Table S1) present the structure of very flashy plants, 23 cm to 100 cm high. Every individual had 1 to 15 scapes; many flower peduncles developed from them with a length between 3 and 40 cm. The number of flower heads per individual ranged from 2 to 146, while the flower heads' diameters were between 1.5 and 4.3 cm. The plants had many basal leaves, and the cauline ones grew up to one-third of the whole plant. The basal leaves' length was between 1.07 and 8 cm, and the width between 1.04 and 9.21 cm; the petiole length ranged from 1.02 to 8 cm. The length/width ratio of the leaf was on average 1.22. The cauline leaves on the other hand resulted in being 0.9 to 9.47 cm long, 0.49 to 4.32 cm wide, with a number of lobes between 2 and 16. In every cauline leaf, it was always possible to distinguish an apical segment whose length ranged between 0.58 and 5.76 cm, and whose width was between 0.09 and 3.36 cm; the length/width ratio of the apical segment was on average 2.3. The lateral segments resulted in being long, between 0.21 and 2.25 cm, and wide between 0.14 and 4.65 cm, with a ratio of 3.45. Moreover, the apical segment width/lateral segment width ratio was 4.85.

3.5. Qualitative Characters of *Scabiosa garganica*

The individuals had a very ramified thick root system, from whose base many leafy scapes grew. The leaves had a green/greyish velvety indumentum with patent whitish hairs on both their lower and upper pages. Hairs characterized scapes and flower peduncles. The difference between the cauline and the basal leaves was pronounced: The first ones were opposite and laciniate, imparipinnatesect with a crenate edge, generally linear towards the higher point of the scape; the basal leaves were simple and elliptic-spathulate (generally with rounded apex). The inflorescence had herbaceous bracts on two rows and wrapping a hairy receptacle with bracteoles. The corolla was usually blueish, close to lilac and purplish. The central (pentamerous) flowers were actinomorphic, and the peripheral ones were zygomorphic. The calyx had bristles and setae; each flower had an eight-furrowed tubular involucel. All the four stamens were bilobed.

3.6. Typification of the Name *Scabiosa garganica*

The name *Scabiosa garganica* appears to be still untypified [24]. Given that a type is essential for the correct interpretation of the name, we here proceeded with the designation of the type.

The species was firstly correctly described by Wettstein [15], who described it indicating "Monte Gargano in Italien" (Mt. Gargano in Italy) and referring to the specimens collected by Porta and Rigo, hosted in the herbarium by P. Porta [6]. According to Stafleu and Cowan [40], the Porta collections are preserved in more than thirty herbaria. We were able to trace original material for the name *S. garganica* in BM, FI, JE, K, and TR. All the specimens fit the protologue, in particular those of the *Itinere II Italico*, given that Wettstein reports that the specimens were marked with the number 72 and corresponded to the current taxonomic circumscription of *S. garganica*. According to all data stated, we here designated the specimen preserved in FI barcode FI065250 (left-hand specimen) as a lectotype of the name *Scabiosa garganica*.

Scabiosa garganica Porta & Rigo ex Wettst. 1892: 67(–68, 97, pl. 4, figs. 1–2)

Type (lectotype, designated here): ITALY. *Ex itinere II italico*, Italia austral. Apulia: Gargano in pasc. saxos. apricis ad montem St. Angelo, sol. calcar., 600–2000′, 3 July 1875, Porta et Rigo 72 (FI barcode FI065250 left-hand specimen [digital image!], available online at http://parlatore.msn.unifi.it/types/search.php; isolectotypes: BM barcode BM001134486 [digital image!], FI barcode FI065250 right-hand specimen [digital image!], FI barcode FI065251 [digital image!], FI barcode FI065252 [digital image!], FI barcode FI065253 [digital image!], FI barcode FI065255 [digital image!], FI barcode FI065257 [digital image!], FI barcode FI065258 [digital image!], JE barcode JE00016102 [digital image!], JE barcode JE00016103 [digital image!], K barcode K000762977 [digital image!], K barcode K000762978 [digital image!], K barcode K000762979 [digital image!]).

Further original material traced: ITALY. *Ex itinere I. italico Portae et Rigoi*, Italia austr. Apulia, in pascuis Mt. S. Angelo in Gargano, 1(000)-2000′ s.m., sol. calcar., 4 July 1874, Porta et Rigo s.n. (FI barcode FI065254 [digital image!], FI barcode FI065256 [digital image!],

K barcode K000762976 [digital image!]); Apulia, in saxosis apricis Gargani circa M.te S. Angelo, sol. cal., alt 1(000)-2000′, 03/07/1875, Porta s.n. (TR 030714 [digital image!]).

3.7. Conservation Status of *S. garganica*

The conservation status of *S. garganica* was evaluated according to the IUCN categories and criteria [32]. The only criterion that could be applied was criterion B (i.e., geographic range). With an extent of occurrence (EOO) of 970 km^2, calculated as convex hull; an area of occupancy (AOO) of 52 km^2, calculated using 2 × 2 km grid cells; a number of locations (sensu IUCN) of six; and an estimated continuing decline of EOO and quality of habitat, mainly due to the construction of buildings and the recovery of the pine forest, we considered the species Vulnerable, VU B1ab(i,iii) + 2ab(i,iii).

3.8. Comparison with *S. holosericea* and *S. taygetea*

The NMDS analysis, performed with three dimensions, yielded an ordination with a stress value of 0.09122. The scatterplot shows on the first two axes a clear distinction between *S. garganica* and *S. holosericea* and *S. taygetea*, and no overlapping areas among *S. garganica* and the other individuals (Figure 2). The UPGMA dendrogram (Figure 3) yielded two well-defined clusters, one including all individuals of *S. garganica* and the other all individuals of *S. holosericea* and *S. taygetea*. The PCA plot (Figure 4) shows how plant height and the number of flower heads per plant were the most important discriminatory variables between *S. garganica* and *S. holosericea* and *S. taygetea*; furthermore, the PCA plot indicates the clear separation between *S. garganica* and the other taxa, with the samples of the latter to the left and the *S. garganica* samples to the right of axis 1 of the Component1 principal coordinates, representing 25.08% of the total variation of the dataset.

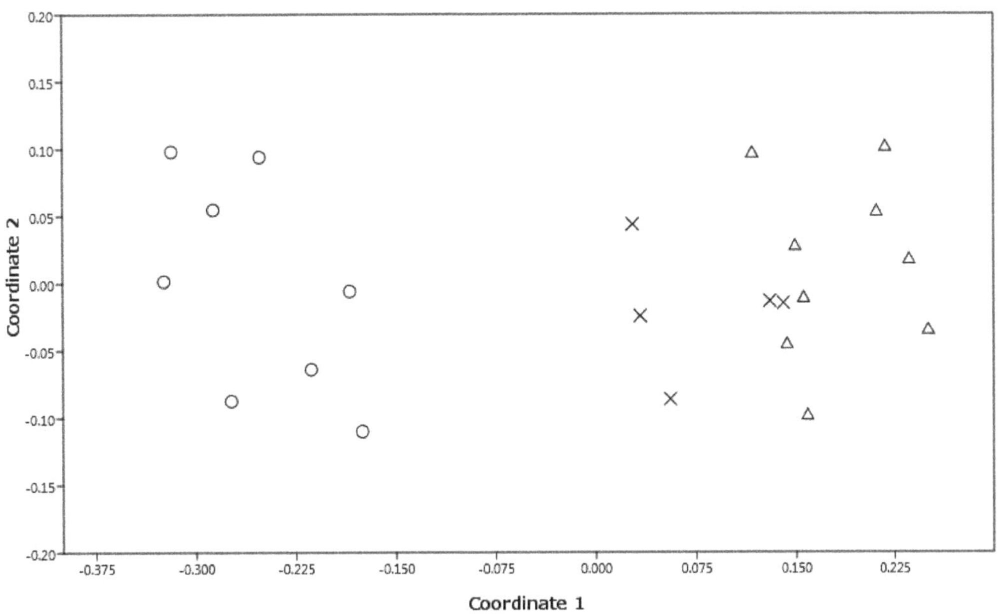

Figure 2. Non—metric multidimensional scaling scatterplot showing the first two dimensions of the analysis. Legend: circle (*S. garganica*), triangle (*S. taygetea*), X (*S. holosericea*).

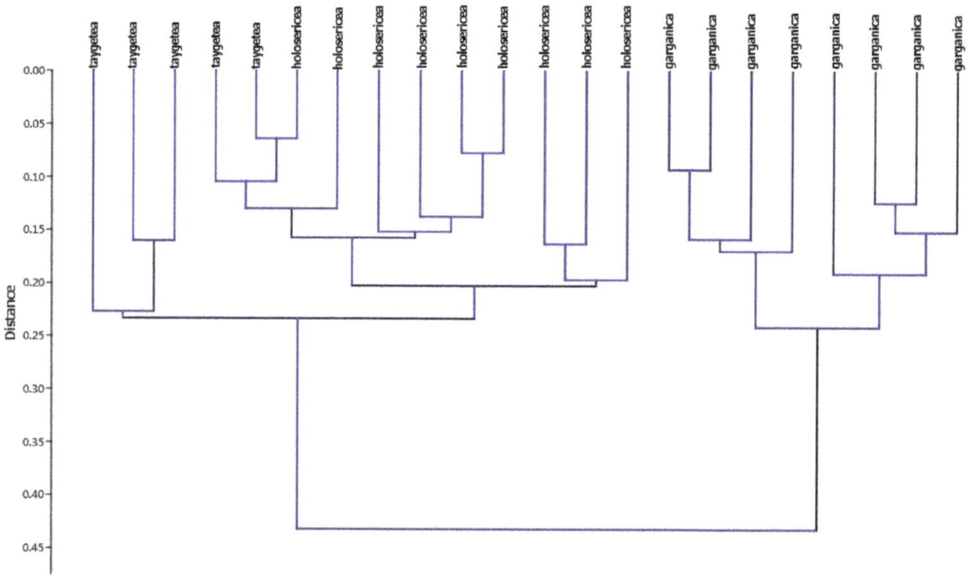

Figure 3. Hierarchical clustering of individuals of *S. garganica*, *S. holosericea*, and S. *taygetea* using a paired group algorithm (UPGMA) and Gower Similarity Index.

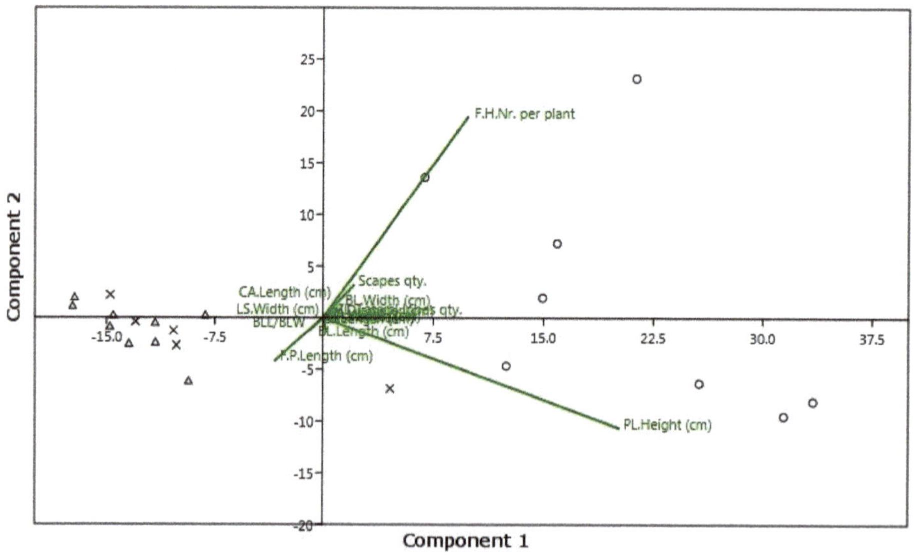

Figure 4. PCA plot. Component1 (25.08%), Component2 (19.55%). Legend: circle (*S. garganica*), triangle (*S. taygetea*), X (*S. holosericea*).

Comparisons of morphological characters between *S. garganica*, *S. holosericea*, and *S. taygetea* are summarized in Table 2.

Table 2. Quantitative characters, reported as mean ± standard deviation, minimum and maximum (extreme values in brackets).

	S. holosericea	*S. garganica*	*S. taygetea*
Plant height (cm)	(26.54) 31.72 ± 3.84 (37.79)	(23) 59.24 ± 18.77 (100)	(15.31) 34 ± 10.6 (49.79)
No. of inflorescences	(1) 1.37 ± 0.51 (2)	(2) 18.59 ± 23 (146)	(1)1.71 ± 0.75 (3)
Inflorescence diameter (cm)	(0.61) 1.42 ± 0.61 (2.47)	(1.5) 2.79 ± 0.45 (4.3)	(0.7) 1.41 ± 0.7 (2.44)
Flower peduncles (cm)	(14.74) 21.55 ± 5.41 (31.35)	(3) 14.18 ± 6.39 (40)	(3.24) 15.84 ± 8.68 (30.83)
No. of scapes	(1) 1.33 ± 0.5 (2)	(1) 4.18 ± 3.33 (15)	(1) 1.71 ± 0.75 (3)
BASAL LEAVES			
Lamina length (cm)	(2.25) 3.48 ± 0.71 (4.93)	(1.07) 3.49 ± 1.47 (8)	(2.25) 3.26 ± 0.77 (4.77)
Width (cm)	(0.64) 1.22 ± 0.33 (1.77)	(1.04) 3.31 ± 1.26 (9.21)	(0.96) 1.42 ± 0.34 (1.86)
Petiole length (cm)	(1.38) 1.94 ± 0.45 (2.82)	(1.02) 3.32 ± 1.29 (8)	(1.39) 2.83 ± 1.76 (5.90)
Lamina length/Width (average)	3	1.22	2.32
CAULINE LEAVES			
Apical segment			
Length (cm)	(0.99) 2.12 ± 0.83 (3.22)	(0.58) 2.82 ± 1.03 (5.76)	(1.73) 2.15 ± 0.58 (3.24)
Width (cm)	(0.23) 0.77 ± 0.39 (1.31)	(0.09) 1.48 ± 0.67 (3.36)	(0.83) 1.06 ± 0.18 (1.25)
Length/width (average)	3.11	2.3	2.03
Lateral segment			
Length (cm)	(0.39) 0.88 ± 0.34 (1.22)	(0.21) 1.09 ± 0.38 (2.25)	(0.57) 0.83 ± 0.16 (1.02)
Width (cm)	(0.14) 0.28 ± 0.12 (0.41)	(0.14) 0.41 ± 0.37 (4.65)	(0.14) 0.28 ± 0.12 (0.41)
Length/width (average)	3.16	3.45	3.02
Total leaf			
Length (cm)	(2.62) 2.96 ± 0.34 (3.34)	(0.9) 4.75 ± 1.52 (9.47)	(2.26) 3.22 ± 0.77 (4.32)
Width (cm)	(0.14) 0.28 ± 0.12 (0.41)	(0.49) 2.21 ± 0.73 (4.32)	(1.02) 1.47 ± 0.23 (1.65)
No. of leaf lobes	(2) 4.4 ± 1.51 (6)	(2) 6.69 ± 2.78 (16)	(4) 5 ± 1.09 (6)
Apical segment width/lateral segment width (average)	2.7	4.85	2.19

The comparison between *S. garganica*, *S. holosericea*, and *S. taygetea* shows that *S. garganica* is an independent species, endemic to southern Italy (Gargano and doubtfully occurring in Basilicata).

4. Discussion

In a triangular territory whose farther points are Monte Sant'Angelo, Manfredonia, and Mattinata, we found most of the locations reported by Fenaroli [19], i.e., Micheli [7] (ascending Mount Gargano), Trotter and Forti [41] (between Manfredonia and Monte Sant'Angelo), Porta and Rigo [6] (near Monte Sant'Angelo), and Lacaita [17] (400 m a.s.l., Monte Sant'Angelo/Manfredonia road). These stations refer also to two out of the nine studied populations in the present research, located along the road leading from Monte Sant'Angelo to Manfredonia. In this same area, we might recognize Fiori's station (Valle delle Macchie in Monte Sant'Angelo territory) and Agostini's station (between Monte Sant'Angelo and Mattinata—collected in 1961 and preserved in Herb. Fenaroli-TR), erroneously attributed by Fenaroli to *Scabiosa holosericea*, a species that is not present in Gargano. In this part of the territory, in fact, are three of our collecting stations, all characterized by the occurrence of *Scabiosa garganica* (without *Scabiosa holosericea*). Moreover, the samples collected at Pulsano Abbey (between Monte Sant'Angelo and Manfredonia) have been attributed by Fenaroli to *S. holosericea*, but they belong to *S. garganica* as well.

Outside this area, the station reported by Martelli in Testa del Gargano (FI!, collected in 1893), a few kilometers far from Vieste (not ascertained by our survey), is interesting and requires more field research to confirm or exclude it from the current distribution of the species.

Our collecting stations in the northern Gargano area significantly expanded the knowledge on the distribution of *Scabiosa garganica* in this part of the promontory, where the species was reported by Licht [29] in Monte Pucci (west of Peschici).

With regard to the distribution, further reflection is needed about the different altitude conditions in northern Gargano stations (exposed to winds from the north/north-east) and the southern ones. A similar disjunction on the Gargano Promontory has been observed in the distribution of another plant species of conservation interest, namely, *Centaurea subtilis* [42]. Such distributional differences between the north and south parts of Gargano can be due to climatic (exposure to the winds from north-east) and/or pedological reasons (different soil's conditions). In northern Gargano, the bioclimatic zone of *Scabiosa garganica* does not span beyond 100 m a.s.l., while in southern Gargano, it appears until 1000 m a.s.l.; the pedological conditions the plant is found in in northern Gargano starts from 20 m a.s.l., yet it can be found in the southern zone only above 300 m a.s.l.

The analysis performed on the abundant samples collected define important new quantitative parameters that might help to recognize the species today: According to Wettstein [15] and Pignatti [1], the plants' height is between 30 and 50 cm, but our data reveal a much wider range, i.e., 23 to 100 cm. The number of scapes also was proven to be much greater (four on average), being included in a range between 1 and 15. Apart from the scape number, another difference we noted between our data and those from the literature is the number of flower heads: Pignatti [1] reports a maximum of three flower heads, but plants with up to 146 heads have been found, and they never have a unique flower head. Other interesting data regard the flowering period: According to Porta and Rigo [6], it is in summer, while our survey found flashy and rich blossoming that continued until December, and thus we might say it has a summer/autumn blossoming. Similar as reported in "Flora d'Italia" [1], the color of the corolla is typically purple/blueish.

Lastly, the data collected are somehow enough to clarify the morphological relationship with species close to *Scabiosa garganica*. The comparison highlights significant differences with regard to the flower heads' number (generally unique in *S. holosericea*, up to 3 in *S. taygetea* and up to 146 in *S. garganica)*, plant height (*S. garganica* is higher), and their indumentum (soft green/yellowish hairs in *S. taygetea*, whitish and thick in *S. holosericea* and *S. garganica*). Other differences can be observed in terms of the color of the corolla, being purple/blueish in *S. garganica* and *S. holosericea*, but different from Tilli's [9] cerulean/purplish and Wettstein's [15] red/violet of *S. taygetea*. As for *S. holosericea*, our research also examined wide zones of Gargano Promontory without finding any presence of plants attributable to *S. holosericea*, so that the reports of Pasquale and Licopoli [13], Fenaroli [19], Fanelli et al. [31], and Licht [22] can be attributed to *Scabiosa garganica*, as also indicated by Licht and Wagensommer [25].

According to the IUCN categories and criteria [32], *S. garganica* is Vulnerable. Even if most of the territory occupied by the species is protected by the Gargano National Park, specific conservation actions, both in situ and ex situ, should be implemented in order to ensure the long-term conservation of this rare species endemic to southern Italy.

5. Conclusions

Thanks to the research performed in the Gargano area, *Scabiosa garganica* is now a species defined under the morphological aspect, and hence it gets an undeniable autonomy from *S. holosericea* and *S. taygetea*; therefore, the taxon *Scabiosa garganica* Porta & Rigo ex Wettst. can be considered fully valid. Between these taxa, there are a few unquestionable similarities, such as the typical hairs of all the three species; on the other hand, the morphological differences—starting from the color of the corolla and ending with the plant structure—are significant, as demonstrated by the statistical analysis. *S. garganica* can be recognized thanks to its flashy structure (the plant is up to 1 m high and has many scapes), to the large number of flower heads (up to 146), and from its summer/autumn blossoming. Our research also defined the habitat of *S. garganica* (altitude, exposure, nature of the substrate). Even if *S. garganica* should be considered an Italian endemism (Gargano and doubtful in Basilicata), its relationships with *S. taygetea* make it interesting from a phytogeographical point of view. In fact, both if *S. garganica* is considered occurring in Italy and Greece [2] or as endemic to Italy [1], i.e., as a vicariant of the related *S. taygetea* growing in Greece, from a phytogeopraphical point of view, it represents one of the many taxa occurring in the Gargano Promontory and linked to the Balkan flora, such as *Bromus parvispiculatus* H.Scholz [43], *Cerinthe retorta* Sm. [44,45], *Linum elegans* Spruner ex Boiss. [46], and *Ophrys oestrifera* M.Bieb. aggr. [47]. In Italy, all these species are rather rare.

In conclusion, phytosociological surveys in the plant communities where *S. garganica* has been found would be beneficial to better understand the synecological role of this interesting species of the Italian flora.

***Specimina visa* of *Scabiosa holosericea* and *S. taygetea*:**

Scabiosa holosericea Bertol.: ITALY. S.l., August 1908, Fiori A. (LY barcode LY0312723); Toscana: Alpes Apuanes à Fornole, July 1862, Savi P. (LY barcode LY0312729); Alpi Apuane, 1867, Bertoloni, A. (K barcode K000762986); Alpi Apuane, 1818, Schleicher, J.C. (K barcode K000762985); Alpi Apuane, M. Altissimo versante settentr., 5 October 1951, Pichi Sermolli R.; van Steenis C.G.G.J., Contardo A.; det. R. Corradi (L barcode L2979831); Calabria: M. Pollino, in pascuis saxosis elatis, solo calc., 1800–2000, 29 July 1898, Rigo G. (LY barcode LY0312718).

Scabiosa taygetea Boiss. & Heldr.: Greece. Kalamata, Montes Taygetos, ad semitam inter pagum Tòryza et montem Profictis Elias, 1200–1500 m, in regione abietina, 26 June 1978, Cernoch F. (BR barcodes BR00000027547878V, BR0000025719024V); M. Taygetus, 1885, Haussknecht C. (JE barcode JE00016176); In reg. sylvatica M. Taygeti, July 1844, Heldreich T.H.H. (W barcode WU077546); In rupestris regionum superiorum Taygeti, August 1844, Heldreich T.H.H. (W barcode W0050831).

Supplementary Materials: The following supporting information can be downloaded at https://www.mdpi.com/article/10.3390/plants12091915/s1, Table S1: Collecting sites of *Scabiosa garganica* samples used for the present research. Table S2: Dataset composed of all the measurements carried out on *Scabiosa garganica*, *S. holosericea*, and *S. taygetea*.

Author Contributions: Conceptualization, D.B. and N.B.; methodology, D.B., N.B. and R.P.W.; validation, R.P.W.; investigation, D.B., N.B. and R.P.W.; writing—original draft preparation, D.B., N.B. and R.P.W.; writing—review and editing, R.P.W.; supervision, R.P.W. All authors have read and agreed to the published version of the manuscript.

Funding: This research received no external funding.

Data Availability Statement: Data is contained in the article, including the Supplementary Material.

Acknowledgments: The authors thank Matteo Falcone and Michele Matassa for some reports in the territory of Monte Sant'Angelo, as well as the curators of the herbaria cited in the text for freely granting the digital herbarium sheets.

Conflicts of Interest: The authors declare no conflict of interest.

References

1. Pignatti, S. *Flora d'Italia*, 2nd ed.; Edagricole-New Business Media: Bologna, Italy, 2017.
2. Domina, G. 2017: Dipsacaceae. In: Euro+Med Plantbase—The Information Resource for Euro-Mediterranean Plant Diversity. Available online: http://ww2.bgbm.org/EuroPlusMed/PTaxonDetail.asp?NameId=7502407&PTRefFk=7500000 (accessed on 27 November 2021).
3. Barina, Z.; Somogyi, G.; Pifkó, D.; Rakay, M. Checklist of vascular plants of Albania. *Phytotaxa* **2018**, *378*, 1–339. [CrossRef]
4. Bartolucci, F.; Peruzzi, L.; Galasso, G.; Albano, A.; Alessandrini, A.; Ardenghi, N.M.G.; Astuti, G.; Bacchetta, G.; Ballelli, S.; Banfi, E.; et al. An updated checklist of the vascular flora native to Italy. *Plant Biosyst.* **2018**, *152*, 179–303. [CrossRef]
5. The Plant List: Scabiosa garganica Porta & Rigo ex Wettst. Available online: http://www.theplantlist.org/tpl1.1/record/kew-2593345 (accessed on 19 May 2021).
6. Rigo, G. Relazione botanica del viaggio eseguito da Porta e Rigo nelle provincie meridionali d'Italia dalla fine di marzo fino a tutto il 10 agosto 1875. *N. Giorn. Bot. Ital.* **1877**, *9*, 282–317.
7. Micheli, P.A. *Itinera botanica, Tomus II*; Biblioteca del Reale Istituto Botanico Fiorentino, Università degli Studi, Biblioteca di Scienze: Firenze, Italy, 1710; Manoscritti Ms 27; pp. 1704–1730.
8. Tozzetti, G.T. *Catalogus Plantarum Horti Caesarei Florentini, Opus Postumum;* Ex typographia Bernardi Paperinii: Florentia, Italy, 1748.
9. Tilli, M.A. *Catalogus Horti Pisani, Typis Regie Celfitudinis;* Apud Tartinium & Franchium: Florentia, Italy, 1723.

10. Fenaroli, F. Florae Garganicae Mantissa. *Delpinoa* **1975**, *14–15*, 61–123.
11. Arcangeli, G. *Compendio della Flora Italiana*; Vincenzo Bona Tip.: Torino, Italy, 1882; p. 889.
12. Parlatore, F.; Caruel, T. *Flora Italiana Continuata da Teodoro Caruel, Vol. 7. Asteriflore*; Le Monnier: Firenze, Italy, 1887.
13. Pasquale, G.A.; Licopoli, G. *Di un Viaggio Botanico al Gargano*; Accademia Delle Scienze Fisiche e Matematiche; Stamperia del Fibreno: Naples, Italy, 1873; Volume 5, pp. 1–31.
14. Boissier, P.E. *Diagnoses Plantarum Orientalium Novarum, sér. 1, vol. 1 (6)*; Apud B. Herrmann: Lipsiae, Germany, 1846; pp. 73–74.
15. Wettstein, R. Beitrag zur Flora Albaniens. *Biblioth. Bot.* **1892**, *5*, 1–103.
16. Huter, R. Herbarstudien. *Öster. Bot. Zeitschr.* **1905**, *55*, 400–406. [CrossRef]
17. Lacaita, C. Piante italiane critiche o rare. *N. Giorn. Bot. Ital.* **1921**, *28*, 150–152.
18. Fiori, A. *Nuova Flora Analitica d'Italia*; Tipografia di M. Ricci: Firenze, Italy, 1923; Volume 1,2.
19. Fenaroli, L. Florae Garganicae Prodromus: Pars quarta. *Webbia* **1974**, *29*, 123–301. [CrossRef]
20. Zangheri, P. *Flora Italica*; Cedam: Padova, Italy, 1976.
21. Pignatti, S. *Flora d'Italia*; Edagricole: Bologna, Italy, 1982.
22. Licht, W. *Bestimmungsschlüssel zur Flora des Gargano (Süd-Italien)*; Shaker Verlag: Aachen, Germany, 2008; p. 384.
23. Conti, F.; Abbate, G.; Alessandrini, A.; Blasi, C. (Eds.) *An Annotated Checklist of the Italian Vascular Flora*; Palombi Ed.: Roma, Italy, 2005; p. 420.
24. Peruzzi, L.; Galasso, G.; Domina, G.; Bartolucci, F.; Santangelo, A.; Alessandrini, A.; Astuti, G.; D'Antraccoli, M.; Roma-Marzio, F.; Ardenghi, N.M.G.; et al. An inventory of the names of native, non-endemic vascular plants described from Italy, their loci classici and types. *Phytotaxa* **2019**, *410*, 1–215. [CrossRef]
25. Licht, W.; Wagensommer, R.P. *Flora Vascolare del Gargano e Delle Isole Tremiti. Chiavi Analitiche per la Determinazione*; Edizioni del Parco, Claudio Grenzi Editore: Foggia, Italy, 2020; p. 955.
26. Peruzzi, L.; Gargano, D. Reports 1357–1361. In *Mediterranean Chromosome Report 13*; Kamari, G., Blanchè, C., Garbari, F., Eds.; Flora Mediterranea: Palermo, Italy, 2003; Volume 13, pp. 347–404.
27. Bartolucci, F.; Domina, G.; Andreatta, S.; Argenti, C.; Bacchetta, G.; Ballelli, S.; Enrico, B.; Davide, B.; Giuseppina, B.; Gianni, B.; et al. Supplementary data to Notulae to the Italian native vascular flora: 11. *Ital. Bot.* **2021**, *11*, 77–92. [CrossRef]
28. Bianco, P.; Brullo, S.; Pignatti, E.; Pignatti, S. La vegetazione delle rupi calcaree della Puglia. *Braun Blanquetia* **1988**, *2*, 133–151.
29. Licht, W. *Conspectus Florae Garganicae Praecursor*; University of Mainz: Mainz, Germany, 2018; Available online: https://plants1.iome.uni-mainz.de/plant-evolution/associates/dr-wolfgang-licht/ (accessed on 16 March 2023).
30. Di Pietro, R.; Wagensommer, R.P. A new *Sesleria juncifolia* association from south-eastern Italy and its position in the amphi-Adriatic biogeographical context. *Acta Bot. Croat.* **2014**, *73*, 171–207.
31. Fanelli, G.; Lucchese, F.; Paura, B. Le praterie a *Stipa austroitalica* di due settori adriatici meridionali (basso Molise e Gargano). *Fitosociologia* **2001**, *38*, 25–36.
32. IUCN Standards and Petitions Committee. Guidelines for Using the IUCN Red List Categories and Criteria. *Version 15.1. Prepared by the Standards and Petitions Committee.* 2022. Available online: https://www.iucnredlist.org/resources/redlistguidelines (accessed on 30 December 2022).
33. ImageJ. Available online: https://imagej.nih.gov/ij/download.html (accessed on 19 May 2021).
34. SPSS. Available online: https://www.spss.it/statistics-for-data-analysis (accessed on 19 May 2021).
35. Hammer, Ø.; Harper, D.A.; Ryan, P.D. PAST: Paleontological statistics software package for education and data analysis. *Paleontol. Electron.* **2001**, *4*, 9.
36. Gower, J.C. A general coefficient of similarity and some of its properties. *Biometrics* **1971**, *27*, 857–874. [CrossRef]
37. Pesaresi, S.; Biondi, E.; Casavecchia, S. Bioclimates of Italy. *J. Maps* **2017**, *13*, 955–960. [CrossRef]
38. Thiers, B.; Index Herbariorum: A Global Directory of Public Herbaria and Associated Staff. New York Botanical Garden's Virtual Herbarium. 2022. Available online: https://sweetgum.nybg.org/ih/ (accessed on 9 October 2022).
39. Di Pietro, R.; Wagensommer, R.P. Analisi fitosociologica su alcune specie rare e/o minacciate del Parco Nazionale del Gargano (Italia centro-meridionale) e considerazioni sintassonomiche sulle comunità casmofitiche della Puglia. *Fitosociologia* **2008**, *45*, 177–200.
40. Stafleu, F.A.; Cowan, R.S. *Taxonomic Literature. A Selective Guide to Botanical Publications and Collections with Dates, Commentaries and Types. Volume IV: P-Sak*, 2nd ed.; [Regnum Vegetabile 110]; Koeltz Botanical Books: Schmitten, Germany, 1983; p. 1214.
41. Trotter, A. A traverso il Gargano. *Boll. Orto Bot. Napoli* **1911**, *3*, 233–249.
42. Wagensommer, R.P. Indagini sulla distribuzione puntuale delle specie rare: Alcuni casi studio della flora del Gargano (Puglia). *Inform. Bot. Ital.* **2010**, *42*, 451–463.
43. Karl, R.; Scholz, H. *Bromus parvispiculatus* H. Scholz. [In: Greuter, W., Raus, Th. (Eds.) Med-Checklist Notulae 28]. *Willdenowia* **2009**, *39*, 340.
44. Wagensommer, R.P.; Fröhlich, T.; Fröhlich, M. First record of the southeast European species *Cerinthe retorta* Sibth. & Sm. (Boraginaceae) in Italy and considerations on its distribution and conservation status. *Acta Bot. Gall. Bot. Lett.* **2014**, *161*, 111–115.
45. Stinca, A.; Musarella, C.M.; Rosati, L.; Laface, V.L.A.; Licht, W.; Fanfarillo, E.; Wagensommer, R.P.; Galasso, G.; Fascetti, S.; Esposito, A.; et al. Italian vascular flora: New findings, updates and exploration of floristic similarities between regions. *Diversity* **2021**, *13*, 600. [CrossRef]

46. Wagensommer, R.P.; Bartolucci, F.; Fiorentino, M.; Licht, W.; Peccenini, S.; Perrino, E.V.; Venanzoni, R. First record for the flora of Italy and lectotypification of the name *Linum elegans* (Linaceae). *Phytotaxa* **2017**, *296*, 161–170. [CrossRef]
47. Van de Vijver, B.; Van Looken, W.; Thiers, G.; Cuypers, A. *Ophrys oestrifera* subsp. montis-gargani, a new subspecies from the Gargano (Italy). *J. Eur. Orch.* **2010**, *42*, 167–180.

plants

MDPI

Review

The Biology of the Genus *Ceiba*, a Potential Source for Sustainable Production of Natural Fiber

Ximena Gómez-Maqueo * and Alicia Gamboa-deBuen *

Instituto de Ecología, Universidad Nacional Autónoma de México, Mexico City 04510, Mexico
* Correspondence: ximenagmb@ciencias.unam.mx (X.G.-M.); agamboa@ecologia.unam.mx (A.G.-d.)

Abstract: The species of the genus *Ceiba* produces fruits with fibers with a high content of cellulose. The fiber is used for textiles, cushion filling and for industrial purposes and its characteristics have been studied in some species including *Ceiba pentandra* (kapok), *C. speciosa* and *C. aesculifolia*. The use of the trunk and seeds of *Ceiba* has also been described for different species. This article presents a review on the biological diversity of the genus Ceiba (Malvaceae). The genus *Ceiba* has 18 recognized species that are distributed naturally in America and Africa. However, some *Ceiba* trees have been introduced to various countries, especially in Asia, due to their ornamental interest and potential uses for their fiber. Ecophysiological studies of different *Ceiba* species have shown that resistance to adverse environmental conditions varies from species to species. Therefore, *Ceiba* species are considered potentially useful in restoring ecosystems impacted by human activity. The information related to the classification, morphological characteristics, phenology, ecophysiology and distribution of the different species will be extremely relevant for the sustainable production of kapok fiber. Finally, the recent genomic and transcriptomic studies also provide a valuable resource for further genetic improvement and effective use of *Ceiba* trees.

Keywords: kapok; *Ceiba*; fiber; sustainability; Malvaceae; silk-cotton

Citation: Gómez-Maqueo, X.; Gamboa-deBuen, A. The Biology of the Genus *Ceiba*, a Potential Source for Sustainable Production of Natural Fiber. *Plants* **2022**, *11*, 521. https://doi.org/10.3390/plants11040521

Academic Editor: Robert Philipp Wagensommer

Received: 26 January 2022
Accepted: 12 February 2022
Published: 15 February 2022

1. Introduction

Natural fibers are obtained from different plants and animals and have many uses, both locally and industrially. Although the popularization of oil-based fibers has reduced the historic demand of natural fibers, with synthetic fibers holding about 58% of total fiber use by 2013, it is estimated that production of natural fibers worldwide corresponded to approximately 33 million tons by the same year. About 96% of all natural fibers produced are derived from plants, with cotton accounting for 79% of total production, while other plant-based fibers, such as jute, hemp, sisal, coir or kapok, contribute to about 18% [1].

In recent years, there has been increasing concern on the sustainability and negative impacts on the environment posed not only by oil-based fibers, but also by fibers naturally sourced or regenerated from cellulose [2]. For instance, cotton is the most widespread plant-based fiber; about 80 countries produce cotton commercially, occupying approximately 2.5% of the farmable land around the world [1]. Still, cotton production requires about 25% of the total insecticide and 10% of total pesticides produced worldwide, as well as intensive irrigation with an estimated 7000–29,000 L of water required to produce 1 kg of cotton [3]. Meanwhile, the other fiber-producing species are exploited at smaller scales around the globe and could open an opportunity to cope with the need of sustainable and environmentally friendly production of natural fibers. Some of these alternative fibers, such as kapok, need to be mixed with other fibers to be properly spun for textile production. However, by incorporating kapok in the production of other synthetic or cotton fibers, it could be possible to reduce the total environmental footprint of the finished products.

"Kapok" is a common term used to identify a type of "seed fiber" (e.g., cotton) produced mainly by two plant species known as "kapok trees": *Bombax ceiba* L. (Malvaceae

Juss., also known as "red cotton tree" or "red silk-cotton"), and *Ceiba pentandra* (L.) Gaertn. (Malvaceae, also known as "silk-cotton" or "Java cotton"). There is usually confusion in the literature as to which of the two species is used as the source of "kapok", though most of the information regarding the physical and chemical properties of kapok fibers comes from *C. pentandra*. It is noteworthy to mention that cotton, jute and kapok, three of the main fiber-producing plants, belong to the Malvaceae family. As observed in taxonomic revisions of the subfamilies within the Malvaceae, there are many other genera aside from *Bombax* L. and *Ceiba* Mill. that belong to the "kapok clade" and produce floss-bearing fruits, such as the genera *Pachira* Aubl., *Eriotheca* Schott and Endl. or *Pseudobombax* Dugand [4], though the viability of producing fiber or cellulose from these species remains to be explored. *C. pentandra* is cultivated and found widely spread in several plantations in Southeast Asia, where most of the countries that produce, and export kapok fibers are located. Before WWII, kapok trees were an important cash crop because the fiber was extensively used in life jackets and aviation clothing, among other things. In the 1960s, as a result of the massive production of synthetic fibers, the kapok trade declined substantially [5]. By 2015, Indonesia was one of the largest producers and exporters of kapok in the world, followed by Thailand, which are the two countries on the FAO database [6,7]. Over the past 10 years, there has been an increasing interest in studying the properties of kapok. This fiber has been found to be environmentally friendly and biodegradable as well as having anti-bacterial and anti-mite properties. Most of the research that has been carried out is related to materials science and engineering. For example, potential applications of kapok fiber are related to oil sorbents [6]. Still, the production of kapok fiber has had a slight declining trend, as seen in Figure 1 [7].

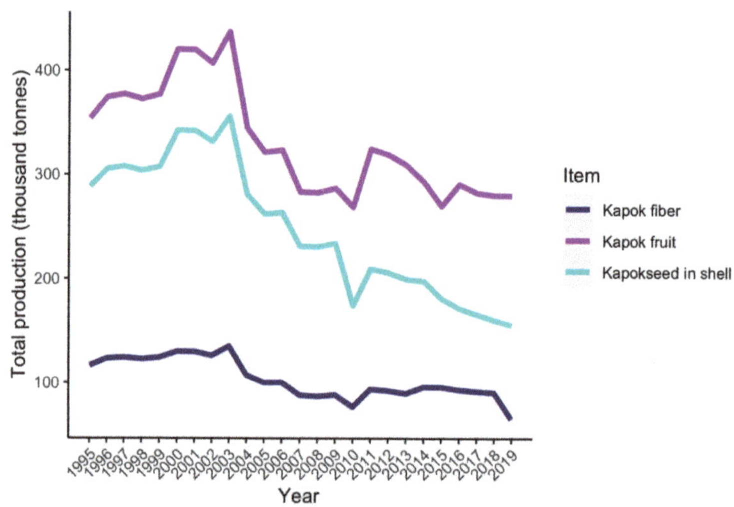

Figure 1. FAO estimates of worldwide production of kapok fiber, fruits and seeds since 1995. The data correspond to the combined production of Indonesia and Thailand, the only countries indicated in the database.

Aside from the economic importance of *C. pentandra*, several species from the genus are regarded as culturally and environmentally important species; they are found incorporated in local rituals almost everywhere [8]. In Mexico, the Mayan and Aztec pre-Hispanic cultures considered *C. pentandra* and *C. aesculifolia* (Kunth) Britten and Baker f. respectively, as sacred trees that connected the human world with the place where the Gods lived [9–11]. In Guatemala, *C. pentandra* is considered the national tree [12]. In west Africa, *C. pentandra* is also regarded as a sacred tree [8], while in Asia, cultivated kapok offers an important source of seasonal jobs and income for the kapok-producing countries [1]. In this review we will focus on *C. pentandra* and other species of the genus *Ceiba* as potential sources of sustainable kapok, as well as other ecological and cultural benefits.

2. Characteristics of the Genus *Ceiba*

The genus *Ceiba* comprises 18 species, 17 of which are naturally distributed in the Neotropics. *Ceiba* species are trees usually between 10 to 25 m tall, with some species reaching above 50 m (*C. pentandra*, *C. lupuna* P.E. Gibbs and Semir), or as small treelets of about 2 m tall, such as like *C. jasminodora* (A. St.-Hil.) K. Schum. [13]. They present digitate composite leaves with serrated or plain borders, and characteristic aculeate trunks and branches [14]. Some species, such as *C. chodatii* (Hassl.) Ravenna, *C. speciosa* (A. St.-Hil.) Ravenna, *C. glaziovii* (Kuntze) K. Schum., and *C. pubiflora* (A. St.-Hil.) K. Schum., present ventricose or "swollen" trunks, which explain some of the common names given to these trees in South America, such as "palo borracho" due to the bottle-shaped trunk or "barriguda" as in having a swollen belly [14]. The flowers present a corolla conformed by five petals, which are diverse in size and color, ranging from pale tones, such as white, ivory, yellow or light pink, to vibrant colors, such as pink or red, and sometimes with yellow tones towards the base. In some cases, the petals present dark colored striations. The fruits are woody capsules that contain a modified endocarp into long, tubular trichomes that constitutes the kapok fiber, in which the seeds are imbedded. This fiber facilitates seed dispersion through wind. Due to the diversity of flower morphology, *Ceiba* species are pollinated by several species of bats, butterflies, bees and hummingbirds [14]. In Figure 2, we present some representative illustrations of the different flowers, leaves and trunk shapes found within the genus. Most species are deciduous and flower when leafless; the flowers are usually short-lived and present crepuscular anthesis. These characteristics have hampered the taxonomic efforts of identification and classification of the clade [14]. In Table 1, we present a short compilation of morphological characteristics and habitat distribution for each species.

Table 1. Morphological characteristics, distribution and phenology of the 18 recognized species by [14,15]. Descriptions have been obtained from the aforementioned authors; information obtained from other sources is indicated in brackets. SDTF: seasonally dry tropical forest.

Species	Height	Flower Features	Trunk	Habitat	Distribution	Flowering Time	Pollinators
C. aesculifolia (Kunth) Britten and Baker f.	tree, 8–15 m [16]	cream, ivory	slender trunk with spines	SDTF	Mexico and Central America	March–July (September–January)	bats
C. boliviana Britten and Baker f.	tree, 10 m	pink with intense dark red striations	usually ventricose with spines	SDTF	Bolivia, Peru	March–April(January)	possibly bats
C. chodatii (Hassl.) Ravenna	tree, 12 m	ivory to pale yellow, sometimes with crimson flecks	ventricose, usually with spines	SDTF	Argentina, Bolivia and Paraguay	February–May	possibly sphingid moths
C. crispiflora (Kunth) Ravenna	tree, over 10 m	dark pink magenta with few striations distally, yellowish at the base	usually with spines	SDTF and humid forests	Brazil	February–March	possibly diurnal butterflies
C. erianthos (Cav.) K. Schum.	tree, 10 m	white with sparse carmine striations distally, becoming uniform towards the base	presents spines	SDTF	Brazil	March–July	bats
C. glaziovii (Kuntze) K. Schum.	tree, 10–15 m	white, sometimes with magenta striations towards the base	ventricose, with spines	SDTF	Brazil	July-September	possibly bats

Table 1. *Cont.*

Species	Height	Flower Features	Trunk	Habitat	Distribution	Flowering Time	Pollinators
C. insignis (Kunth) P.E. Gibbs and Semir	tree, 10 m	white to light pink with a yellowish base, occasionally with red striations	ventricose, usually with spines	SDTF	Ecuador and Peru	May–July (October)	possibly bats
C. jasminodora (A. St.-Hil.) K. Schum.	treelet, 1–2 m	cream, markedly reflexed	spiny branches	SDTF rocky outcrops [13]	Brazil	April–July	possibly moths
C. lupuna P.E. Gibbs and Semir	giant tree, up to 50 m	deep red distally, pale yellow with red speckles towards the base	usually with spines	humid forests	Ecuador, Peru and Brazil	May–June	unknown
C. pentandra (L.) Gaertn.	giant tree, up to 60 m; the savannah ecotype about 10 m; cultivated types 25 m [5]	white to distinctive light pink	presents spines and large buttresses (cultivated types can lack both) [5]	SDTF and humid forests [13]	pantropical, introduced in Asia and Europe	August–September	bats, possibly bees
C. pubiflora (A. St.-Hil.)	Tree, over 20 m	pale pink with sparse dark flecks, or pink-lilac with carmine striations	sometimes ventricose, with spines	SDTF	Paraguay, Argentina and Brazil	February–May	possibly humming-birds
C. rubriflora Carv.-Sobr. and L.P. Queiroz	tree, 20 m	deep red	ventricose, with spines	SDTF, calcareous outcrops	Brazil	July–August	unknown
C. samauma (Mart.) K. Schum.	tree, 15 m	white but with dense golden brown trichomes	may present buttresses, spiny branches	SDTF and humid forests	Ecuador, Peru, Brazil and Bolivia	December–March (May)	unknown
C. schottii Britten and Baker f.	tree, 8 m	white	presents spines	SDTF, mangroves and flood zones [17]	Mexico and Central America	June–October [17]	diurnal butterflies
C. soluta (Donn. Sm.) Ravenna	Not reported	white	presents spines	SDTF	Guatemala	February	unknown
C. speciosa (A. St.-Hil.) Ravenna	tree, 10–20 m	dark pink magenta distally, base white to yellow, usually with dark striations	ventricose, usually with spines	SDTF and humid forests	Argentina, Bolivia and Brazil	January–May	possibly diurnal butterflies
C. trischistandra (A. Gray) Bakh.	tree, 15–30 m	white, externally tinted green, somewhat reflexed	presents spines	SDTF	Ecuador	April-July	unknown
C. ventricosa (Nees and Mart.) Ravenna	tree, 10 m or more	white to cream with dark redish flecks towards the base	ventricose, usually with spines	SDTF	Brazil	February-April	possibly bats

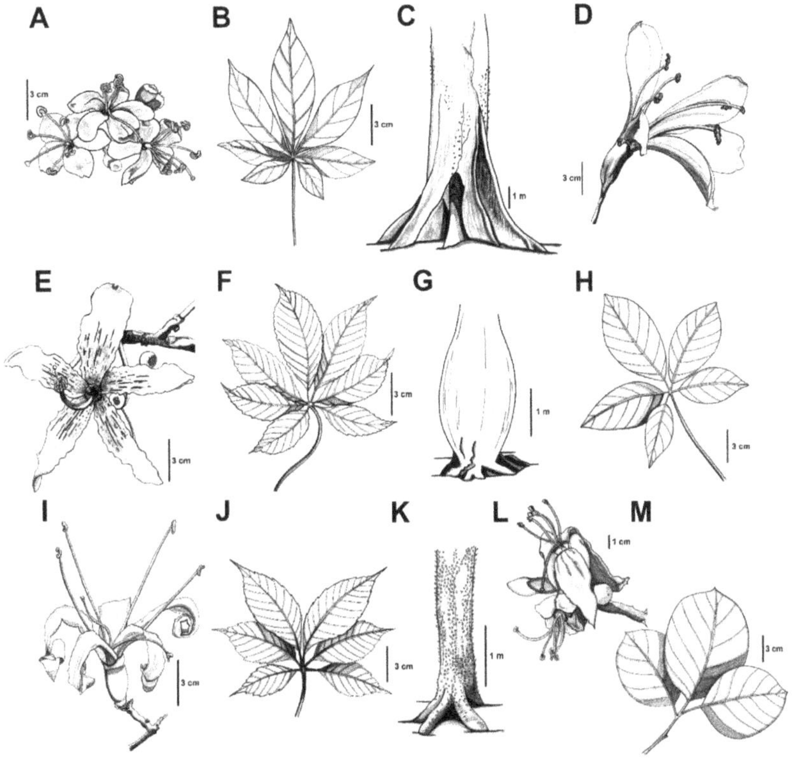

Figure 2. Representative illustrations of the diversity of flowers, leaves and trunks seen within the genus *Ceiba*. (**A–C**): *C. pentandra*; (**D**,**H**): *C. samauma*; (**E–G**): *C. speciosa*; (**I–K**): *C. aesculifolia*; (**L**,**M**): *C. jasminodora*.

3. Ecology of *Ceiba* Species

3.1. The Ecosystem and the Challenges Faced by Ceiba Species Due to Habitat Loss and Degradation

The seasonally dry tropical forest (SDTF), in which most *Ceiba* species are present (Table 1, 13 species), is characterized by several months of severe drought (with rainfall less than 100 mm) [18,19]. These types of tropical forests usually have a mean annual temperature above 17 °C and annual rainfall ranging from 250 to 2000 mm, which occurs mostly during six to eight months [19]. Unlike savannahs, which generate under the same climatic conditions and are dominated by a xeromorphic, fire-tolerant grass layer, the SDTF are tree-dominated and have an almost continuous canopy [18]. These ecosystems occur in disjunct floristic nuclei that show high levels of beta diversity [20]. Although covering about 42% of the tropical ecosystems worldwide [19], it is currently one of the most endangered ecosystems by deforestation in order to allocate arable land due to the fertility of the soil, as well as other land uses [18]. Some species, such as *C. rubriflora* and *C. jasminodora*, are also endangered due to habitat loss; the former been endemic to calcareous outcrops in the Serra do Ramalho (Brazil) [15], and the latter being restricted to rocky outcrops in the Espinhaço mountain range (Brazil) [13].

The water-availability cycles experienced by plants in the SDTF drives fundamental phenological transitions that limits growth and reproduction to the wet season [19,21]. Additionally, seed germination, seedling establishment and regeneration respond to these water cycles [22]. Most of the species that inhabit in the SDTF shed their leaves during the dry months in order to deal with water limitations [22] and abiotic factors, including certain levels of drought or increases in temperature, which may trigger important transitions such as flowering [21]. Thus, disruption or alteration of these factors by human activity and habitat loss may bring about shifts in phenological transitions, disrupting pollination and reproduction, an event already documented in *C. aesculifolia* [23]. Temperature increases have caused altitudinal shifts in seedling establishment and survival in *C. aesculifolia*, especially in zones near urban settlements in Morelia, Michoacán (Mexico) [24].

According to Pezzini et al. [13], four *Ceiba* species (*C. pentandra*, *C. crispiflora*, *C. samauma* and *C. speciosa*) can also inhabit humid tropical forests, and are associated with river valleys and flood zones or grow within gallery forests. Other sources indicate that *C. schottii* can be found in semi-evergreen forests and mangroves within the Yucatán peninsula in Mexico [17]. Finally, *C. lupuna* is the one species restricted to humid tropical forests [13]. In the Peruvian and Brazilian Amazon, wild populations of *C. pentandra* are threatened by intensive exploitation by the plywood industry [25]. Currently, *C. soluta* and *C. crispiflora* are present in the IUCN red list of threatened species [26]. As in the case of SDTF, humid tropical forests face important challenges associated with climate change that impact on fundamental ecosystem properties, including nutrient cycling, carbon storage, shifts in temperature and rain regimes, that will ultimately have negative impacts on diversity and ecosystem services [27].

3.2. *Conservation and Ecosystem Restoration Potential of Ceiba Species*

Several *Ceiba* species have been shown to play an important role within their respective ecosystems, which in turn make them suitable for conservation and restoration programs. In this section we will present *C. aesculifolia* and *C. pentandra* as case studies, since most *Ceiba* species are largely understudied.

3.2.1. *Ceiba aesculifolia*

The research performed in several populations of *C. aesculifolia* throughout the Mexican SDFTs recognizes the species as a pioneer-secondary species, which can grow in shallow soils and higher hydric stress [28,29], as well as in low-soil P concentrations [30]. The seedlings have been shown to resist drought through high sapwood water storage capacity, and although this trait renders them vulnerable to xylem embolism, the buffering role of water storage allows the seedlings to maintain their water potential above the soil's potential as water stress intensifies [31]. Adult trees of this species seem to be resistant to anthropogenic disturbances related to gap formation, such as removal of branches or neighboring vegetation, since there are no differences in growth or density in contrast to individuals in undisturbed areas [32]. The seedlings have better survival to high temperatures in contrast to other local species [24,33], attributes which might play an important role in the implementation of strategies towards conservation, considering the challenges posed by climate change. Still, the concerns raised by [24] Valle-Díaz et al. (2009) regarding the need to assist plant regeneration in disturbed areas near the city of Morelia (Mexico) implies the need of in situ and ex situ plant propagation. To this end, several studies have been conducted on seed germination in wild populations found in Veracruz (Mexico). Recent field work and greenhouse experiments by Velázquez-Rosas et al. [28], Martínez-González et al. [29] and Martínez-González et al. [34] were centered on the importance of seed size variation during germination, during seedling establishment and seedling survival to foliar damage, respectively. Seed size seems to have an effect on germination of individuals emerging in pastures (disturbed areas), whereas there was no effect of seed size in conserved SDTF patches [29] or in greenhouse germination tests [28]. Seedling survival after six months post-germination did not show an association to seed size either [29]. Still, greenhouse experiments showed that seed size does have an effect on total dry weight increase of the seedlings by means of shifts in the root–shoot ratios and changes in leaf area. Total chlorophyll seems to also respond to both seed size and foliar damage [34]. Furthermore, a study performed by Olvera-Mendoza et al. [35] evaluated the genetic diversity within the introduced individuals in the restoration effort conducted by Valle-Díaz et al. [24]. The genetic diversity estimated in the introduced individuals was higher than either of the provenances from which seeds were sourced. This higher diversity could offer the opportunity for reintroduced populations to adapt to an ever-changing environment. Moreover, germination tests performed under field and controlled conditions have demonstrated the positive effects of natural and matrix priming treatments on *C. aesculifolia* seeds (Gómez-Maqueo et al. [36], and references therein). This germination and seedling survival research will help to develop better restoration and conservation strategies of this species as well as provide insight into a sustainable exploitation of this tree by in/ex situ propagation.

3.2.2. *Ceiba pentandra*

Conservation and restoration efforts in Africa and Asia propose the species as a suitable tree despite being an introduced species. In Madagascar, one proposal stems from field observations of several vertebrate species that feed from or within the trees. These vertebrates eat the flowers, or the insects found in the trees, while some others use the tree as refuge. Several of these vertebrates are effective seed dispersers of the native flora, thus the strategic planting of *C. pentandra* trees could aid in seed dispersion by allowing the movement of animal dispersers over the tree canopy and between forest fragments [37]. Meanwhile in India, research has been conducted on the tolerance to salt stress during germination [38]. The authors indicate that *C. pentandra* is moderately tolerant to salt stress during germination, with mild effects on development and growth. However, soil salinity decreased final germination as well as root and shoot development when the electrical conductivity of the soil exceeded 9 dSm^{-1}, with severe effects experienced from 12 to 15 dSm^{-1}. Moreover, in India, *C. pentandra* is suitable for sustainable management of agro-forest systems and afforestation field experiments aiming to reclaim degraded coastal farmlands and to increase the productivity of these degraded soils. This field work has offered insight into viable strategies for management of these degraded soils, while also using a multi-purpose species that offers an important source of pollen for local beekeepers, as well as fibers, oils and cattle feed that sustain local livelihood [39].

4. A Brief History on the Origin of the Clade *Ceiba* and the Arrival of Cultivated *C. pentandra* to Asia

Nowadays, several *Ceiba* species have been introduced worldwide (Figure 2). In east Africa and Asia, the main purpose was for exploitation of different products derived from these plants [5,8], while in the northern hemisphere, they are mostly regarded as ornamental plants in gardens and botanical collections [40,41]. There has been historical uncertainty on whether the genus *Ceiba* originated in the Neotropics or in west Africa [8,42], and even some claims of an Asian origin due to its long history of exploitation [43,44]. As presented in Figure 2, all recognized species are present in America, supporting a Neotropical origin [14,15]. The uncertainty arises from the fact that *C. pentandra* is naturally distributed in both America and Africa, with fossil records of pollen grains about 13,000 years old present in Ghana, suggesting that the species was present prior to any evidence of human-facilitated dispersion in Africa [42]. Although the usual explanation for disjunct distribution of flora between South America and Africa involves an origin of the clade prior to the separation of Gondwana about 96 million years ago (Ma), the study by Dick et al. [42] tested several vicariance hypotheses to explain the disjunct distribution of *C. pentandra*, finding evidence for one of the few cases of extreme long-distance dispersion from the Neotropics to equatorial Africa after the separation of both continents. This was further supported by the phylogenetic analysis performed by Pezzini et al. [13], where they analyzed 14 *Ceiba* species in the Neotropics, indicating a Neotropical origin for the clade, with an estimated emergence during the mid-Miocene, about 21 Ma, and the divergence of *C. pentandra* at 12.7 Ma. Thus, the current knowledge indicates that *C. pentandra* migrated to west Africa and several characteristics, such as its rapid growth, tolerance to water stress, and a self-compatible mating system, could have contributed to its successful colonization of west Africa. Once established in the African moist semi-deciduous forests, *C. pentandra* was able to colonize the savannah, generating a smaller tree ecotype (about 10 m high) [5]. Some authors have proposed several subspecies in order to distinguish the American and African from the cultivated forms in Asia. However, Baker (1965) [8] (p. 6), as well as Gibbs and Semir [14] consider that the species should be considered as a single highly polymorphic species. In Africa, there is evidence that both the semi-deciduous forest and the savannah ecotypes can generate hybrids with an intermediate phenotype [8]. One of these hybrids is most likely the origin of cultivated kapok (usually referred to as *C. pentandra* var. *indica*) [5].

The introduction of *C. pentandra* into Asia is still a debated issue, with no definitive answer. However, authors, such as Baker (1965) [8] (p. 6) and Zeven [5], support the notion that the cultivated forms of *C. pentandra* come from a reduced pool of parental trees, due to the low diversity observed among the different populations present in Asia. In Figure 2, we present the migration routes towards Asia proposed by Baker (1965) [8] (p. 6) and Blench [8]. Some accounts have hypothesized that the Portuguese might have brought the species from America to Africa, and later introduced it to Asia; this was quickly dismissed due to the presence of some pictorial records depicting the species east of the Indian Ocean about 1500 years ago (Steinman 1934, in Blench [8] p. 5), before any possible incursion could be made by the Portuguese [8]. However, Steinman's claims of the pictorial representations have also been questioned, considering that the paintings might represent some other local species bearing similarities with *C. pentandra* [8]. Other accounts propose that the species was introduced to India first and then to southwest Asia between 500 BCE and 500 CE and, supported by the pictorial representations presented by Steinman (1934) and Toxopeus (1941, in [5] pp. 271–272), has already been cultivated by the start of the 10th century [5].

5. Kapok Fiber Characteristics and Uses

The kapok fiber, a light fiber with a hollow tubular structure, is about 1 to 2 cm long. The fibers are comprised of microtubes with a mean external diameter of about 10 μm and a wall thickness of 0.1 μm; meanwhile cotton fibers present mean external diameter of about 16.8 μm and a wall thickness of 3.9 μm [45,46]. These characteristics, that provide less strength, were also reported for other species of *Ceiba* [47]. The cellular origin of the kapok fiber, the cells of the endocarp, facilitates fiber collection, such as cotton lint, as it is not attached to the seed. Cotton fiber originates from the epidermal cells of the seed coat [5,48]. The most common use of the fiber produced by any of the *Ceiba* species, reported in different regions of the world, is its use as fillers for pillows and cushions. However, due to their hollow structure, kapok fiber aggregates have key properties, including superhydrophobicity and porosity, ideally suited for life-saving supplies due to their maneuverability and increased buoyancy, as well as other attributes that artificial buoyancy materials lack, such as biodegradability, acid/alkali resistibility and natural abundance [45]. For textile uses, kapok fibers are short and light so kapok fiber used for fabrics or yarns must be blended with other cellulosic fibers, such as cotton or rayon, in order to improve its stability [49]. However, blending kapok with cotton or other fibers to make fabrics or yarns could reduce the amount of water and resources used during manufacturing, reducing the overall carbon footprint and environmental impact of the end product, in contrast to a similar product produced entirely of cotton or synthetic fibers. The clothing and textile industries are two of the most environmentally costly industries and face several challenges towards sustainability at every level of production [50]. Thus, diversifying prime materials and eco-friendly manufacturing alternatives will aid towards ameliorating current and future impacts.

Kapok fibers are a potential source of cellulose and nanocellulose, comprised of up to 69% cellulose [51]. The high cellulose content has also been described in fibers of *C. speciosa* [52] and *C. aesculifolia* [47], suggesting that the fibers of the different *Ceiba* species could be an important source of cellulose and nanocellulose, polymers extensively used in biotechnological industries.

Kapok fiber is an excellent oil absorbent due to its hydrophobic nature; it has a high proportion of acetyl groups (approximately 13%). It has been suggested that this fiber could be used to recover oil spilled in water [53]. Moreover, kapok fiber as a natural material that has relatively lower cost and better biodegradability could be a better option compared to usual synthetic products [54].

6. Other Exploitable Resources from *Ceiba* Species

C. pentandra is also cultivated commercially for its seeds. Each kapok tree bears 1000 to 2000 pods annually that yield about 15 to 25 kg seeds. Chemical analyses of seeds has demonstrated that they are composed of 31–33% protein, 19–22% sugar and 27–28% lipids [55]. Kapok oil, which is extracted from the seeds, is used for the manufacture of the soap and as a substitute for cotton-seed oil. The use this oil as biofuel has also been proposed [44]. The most abundant fatty acids are linoleic acid, palmitic acid and oleic acid, and malvalic and sterculic acids have been also identified in *C. pentandra* and *C. speciosa* seeds [56,57].

The seeds and roots of *Ceiba aesculifolia* are commercialized as food in central Mexico, but mostly in the traditional markets of communities where these trees are found [58,59]. The seeds, bark and roots are also used traditionally to treat several illnesses, including gastritis, kidney disorders and skin infections, and to reduce blood sugar levels [58]. The bark also possesses antioxidant properties. A recent study demonstrated that the tubers of *C. aesculifolia* are edible with a good potential. These tubers contain protein (3.64%), lipids (3.18%) and carbohydrates (68.27%) [60].

7. High Throughput Technologies and Molecular Approaches towards Plant-Resource Management

Over the past 20 years, there has been an important increase in plant genome assemblies. However, half of the 137 land–plant orders lack a representative genome, while 6 orders are over-represented. Malvales, with 32 genomes, is one of these orders with 30 assemblies from species in the Malvaceae family. The *Gossypium* L. genus is over-represented with 22 genomes and the *G. raimondi* L. assembly was the first reported for this family in 2012 [61]. In 2018, the genome of *Bombax ceiba* L. was reported, along with the complete chloroplast and mitochondria genome sequences. The phylogenetic analysis using these genomes showed that *B. ceiba* has a close relationship with the genus *Gossypium* [62–64]. In 2020, the database MaGenDB was published, which included the genomic information of 13 Malvaceae species. This database could be a useful tool for comparative genomics between Malvaceae species [65].

Meanwhile, genetic information for *Ceiba* species is limited. Microsatellite markers were developed for *C. pentandra* in 2003, with aims to explore the mating system, genetic diversity and flow as well as other population dynamics in the Peruvian and Brazilian Amazon [25]. In 2019, the chloroplast genome of *C. speciosa* was sequenced and characterized; the phylogenetic analysis showed that *C. speciosa* was closest to *B. ceiba* [66]. In 2020, an extensive transcriptomic study from germinating seeds of *C. aesculifolia* was reported. About 54,000 transcripts were assembled, representing 12,683 complete coding transcripts with similarity to *Arabidopsis thaliana* (L.) Heynh. These transcripts represent most of the putative genes for protein synthesis that participate in the germination process as described for other species, which have been estimated in about 12,000 to 18,000 genes [36,67,68]. These germinating seed transcriptomes included information related to genes that are involved in either abiotic or biotic stress. The genes reported include LEA proteins and heat shock proteins as well as proteins involved in pathogen resistance. These data could be a valuable resource for different molecular, biochemical and cellular studies of *Ceiba* species related to drought resistance, thermotolerance and fiber development, among other processes.

The genetic bases involved in the regulation of cotton fiber development have been extensively studied [69]. The two genes from the MYB family of transcription factors

GhMML3 and *GhMML4* have been identified to act as master regulators of cotton fiber initiation. Evolutionary analysis of this gene family revealed that these genes are grouped in two Malvaceae-specific clades and have been detected in *Theobroma cacao* L., *Durio zibethinus* Rumph. ex Murray and *B. ceiba* [70]. The germinating seed transcriptome from *C. aesculifolia* present information of MYB transcription factors. As expected, all the detected transcripts seem to be orthologous of the *A. thaliana* genes because the specific MYB genes of Malvaceae are involved in epidermal cell differentiation during fruit and seed development [71].

8. The Future of the Genus *Ceiba* and Perspectives

Despite the relative cultural importance of the genus in America, there is still insufficient information on the ecology and management towards sustainable exploitation of the different resources they can offer. As seen in previous sections, most of the ecological research has been carried out in the two species widely distributed in north and central America, although the vast majority of species are located in south America (Figure 3). Moreover, in contrast to Asia, exploitation of *Ceiba* species is either in the form of being ornamental (particularly of *C. speciosa* and *C. chodatti*) or it occurs as exploitation of resources at a small scale by local communities. Therefore, there is still much work needed in order to develop sustainable exploitation in the Americas, while also implementing proper conservation and management of wild populations. In the case of *C. aesculifolia*, due to the exploitation of their reproductive structures and roots, Arellanes-Cancino et al. [72] conducted a study in the valley of Tehuacán-Cuicatlán (Mexico) in order to assess current status of wild populations and offer insight into sustainable management strategies of the species. The inhabitants of the valley have utilized and managed the species over many generations, creating a strong cultural bond with the species, although no formal cultivation occurs [58]. This instead has put some pressure over several populations found within the valley, indicating that some of them could be at risk if no proper measures are implemented [72]. Further demographic studies similar to this study are needed to monitor wild populations, as well as a closer involvement of different social actors, decision-makers and academia in order to protect both the species and the livelihood of the inhabitants of the valley.

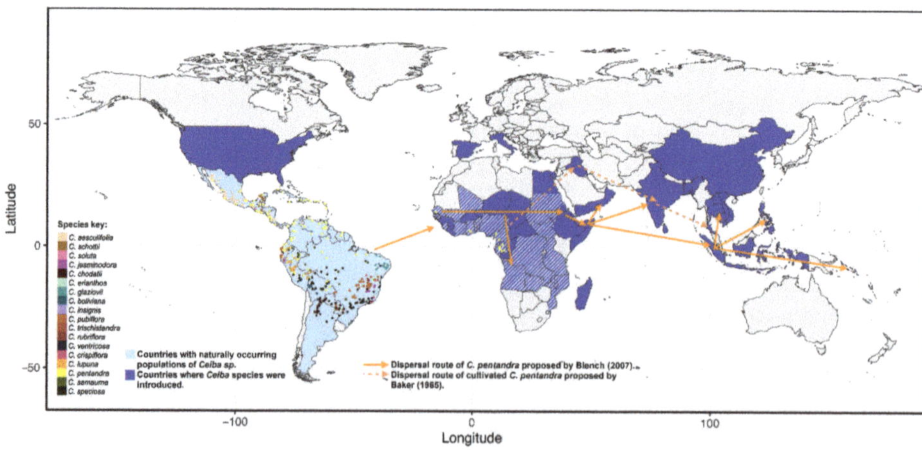

Figure 3. Distribution of the 18 recognized *Ceiba* species, and proposed routes of human-facilitated dispersion of *C. pentandra*. The dots represent the data of naturally-occurring specimens, adapted from Dick et al. [42] and Pezzini et al. [13]. Countries with both light and dark blue lines in Africa represent those countries where there are mixed reports of both natural populations of *C. pentandra* and assisted introductions. Dispersal routes towards Asia were adapted from Blench [8]. Data on assisted introductions was compiled from several sources, primarily [73,74].

9. Conclusions

Although many ecological, physiological and sustainability-oriented aspects on the biology of *Ceiba* species are still insufficient for full-scale exploitation of the genus, it is evident that there is work in process towards that end. Most of the research presented here must be replicated in understudied species, especially considering the endemic status and the risk of habitat loss that most species face. Still, the species in this genus offer an opportunity to cope with these same threats, due to their ability to withstand environmental stress and human-induced disturbances. They could also offer alternative sources of natural fibers, by designing rational and data-driven strategies for conservation and sustainable exploitation of resources. Including kapok-based fibers into the initial stages of textile manufacturing could significantly reduce the carbon footprint of the final products and reduce our dependency of cotton and oil-based fibers, two prime materials with the highest carbon footprints. However, kapok exploitation should also consider the relevance of *ad-hoc* strategies based on the natural diversity of *Ceiba* species and the ecological context in which they thrive, while also involving local communities, investors and governments.

Author Contributions: A.G.-d. and X.G.-M. conceptualized the study, reviewed the literature, and wrote the manuscript. All authors have read and agreed to the published version of the manuscript.

Funding: This work was supported by Programa de Apoyo a Proyectos de Investigación e Innovación Tecnológica (PAPIIT UNAM) grant IN210621 to A.G.-d, and by the Project OE001 "*Ceiba aesculifolia* como especie modelo para el estudio molecular del envejecimiento de las semillas en relación al desempeño germinativo" financed by the Dirección General del Sector Primario y Recursos Naturales Renovables within Secretaría de Medio Ambiente y Recursos Naturales (SEMARNAT) and by the Comisión Nacional para la Biodiversidad (CONABIO).

Institutional Review Board Statement: Not applicable.

Informed Consent Statement: Not applicable.

Data Availability Statement: Not applicable.

Acknowledgments: We thank María Güemez for the illustrations presented in Figure 2.

Conflicts of Interest: The authors declare no conflict of interest.

References

1. Townsend, T.; Sette, J. Natural Fibres and the World Economy. In *Natural Fibres: Advances in Science and Technology towards Industrial Applications*; Fangueiro, R., Rana, S., Eds.; Springer: Dordrecht, The Netherland, 2016; Volume 12, pp. 381–390. [CrossRef]
2. Chen, H.L.; Burns, L.D. Environmental analysis of textile products. *CTRJ* **2006**, *24*, 248–261. [CrossRef]
3. Gedik, G.; Avinc, O. Hemp fiber as a sustainable raw material source for textile industry: Can we use its potential for more eco-friendly production? In *Sustainability in the Textile and Apparel Industries*; Muthu, S.S., Gardetti, M.A., Eds.; Springer Nature: Chaum, Switzerland, 2020; pp. 87–109. [CrossRef]
4. Carvalho-Sobrinho, J.G.; Alverson, W.S.; Alcantara, S.; Queiroz, L.P.; Mota, A.C.; Baum, D.A. Revisiting the phylogeny of Bombacoideae (Malvaceae): Novel relationships, morphologically cohesive clades, and a new tribal classification based on multilocus phylogenetic analyses. *Mol. Phylogenet. Evol.* **2016**, *101*, 56–74. [CrossRef]
5. Zeven, A.C. Kapok tree, *Ceiba pentandra* Gaertn. In *Outlines of Perennial Crop Breeding in the Tropics*; Ferwerda, F.P., Wit, F., Eds.; Landbouwhogeschool: Wageningen, The Netherlands, 1969; pp. 269–287.
6. Zheng, Y.; Wang, J.; Wang, A. Recent advances in the potential applications of hollow kapok fiber-based functional materials. *Cellulose* **2021**, *28*, 5262–5292. [CrossRef]
7. Food and Agriculture Organization of the United Nations. FAOSTAT Statistical Database. Rome, Italy. Available online: https://www.fao.org/faostat/en/#home (accessed on 5 January 2022).
8. Blench, R.M. The intertwined history of the silk-cotton and baobab. In *Fields of Change: Progress in African Archaeobotany*; Cappers, R.T.J., Ed.; Barkhuis: Eelde, The Netherlands, 2007; Volume 5, pp. 1–19.
9. Barrera-Vázquez, A. La ceiba-cocodrilo. *Anales INAH* **1975**, *7*, 187–208.
10. López-Austin, A. Las cuatro columnas. *Arqueol. Mex.* **2019**, *83*, 20–29. Available online: https://arqueologiamexicana.mx (accessed on 19 May 2020).
11. Comisión Nacional de Áreas Naturales Protegidas. De Mitos y Leyendas, la Ceiba. Gobierno de México, 2021. Available online: https://www.gob.mx/conanp (accessed on 8 August 2021).

12. Hurtado, V. Datos Interesantes Sobre Algunos de los Símbolos Patrios de Guatemala. The Foundation for Latin American Anthropological Research (FLAAR Mesoamérica), 2021. Available online: https://flaar-mesoamerica.org (accessed on 16 May 2020).
13. Pezzini, F.F.; Dexter, K.G.; de Carvalho-Sobrinho, J.G.; Kidner, C.A.; Nicholls, J.A.; De Queiroz, L.P.; Pennington, R.T. Phylogeny and biogeography of *Ceiba* Mill. (Malvaceae, Bombacoideae). *Front. Biogeogr.* **2021**, *13*, 1–18. [CrossRef]
14. Gibbs, P.; Semir, J. A taxonomic revision of the genus *Ceiba* Mill. (Bombacaceae). *Anales Jard. Bot. Madrid* **2003**, *60*, 259–300. [CrossRef]
15. De Carvalho-Sobrinho, J.G.; de Queiroz, L.P. *Ceiba rubriflora* (Malvaceae: Bombacoideae), a new species from Bahia, Brazil. *Kew Bull.* **2008**, *63*, 649–653. [CrossRef]
16. Niembro-Rocas, A.; Vázquez-Torres, M.; Sánchez-Sánchez, O. *Árboles de Veracruz: 100 Especies Para la Reforestación Estratégica*; Universidad Veracruzana: Mexico City, Mexico, 2010.
17. Duno de Stefano, R.; Carnevali Fernández-Concha, G.; Ramírez Morillo, I.M.; Tapia Muñoz, J.L.; Can Itzá, L.L.; Hernández-Aguilar, S.; Embray, T. Flora de la Península de Yucatán. 2010. Revised on January 2022 Herbario CICY, Flora de la Península de Yucatán. Available online: https://www.cicy.mx/sitios/flora%20digital/ficha_virtual.php?especie=1746 (accessed on 10 January 2022).
18. Bullock, S.H.; Mooney, H.A.; Medina, E. *Seasonally Dry Tropical Forests*; Cambridge University Press: Cambridge, UK, 1995.
19. Dirzo, R.; Young, H.S.; Mooney, H.A.; Ceballos, G. *Seasonally Dry Tropical Forests: Ecology and Conservation*; Island Press: Washington, DC, USA, 2011; pp. xi–xiii. [CrossRef]
20. Linares-Palomino, R.; Oliveira-Filho, A.T.; Pennington, R.T. Neotropical Seasonally Dry Forests: Diversity, Endemism, and Biogeography of Woody Plants. In *Seasonally Dry Tropical Forests*; Dirzo, R., Young, H.S., Mooney, H.A., Ceballos, G., Eds.; Island Press: Washington, DC, USA, 2011; pp. 3–21. [CrossRef]
21. Lopezaraiza-Mikel, M.; Quesada, M.; Álvarez-Añorve, M.; Ávila-Cabadilla, L.; Martén-Rodríguez, S.; Calvo-Alvaado, J.; do Espírito-Santo, M.M.; Fernandes, G.W.; Sánchez-Azofeifa, A.; Aguilar-Aguilar, M.J.; et al. Phenological patterns of tropical dry forests along latitudinal and successional gradients in the neotropics. In *Tropical Dry Forests in the Americas*; Sanchez-Azofeifa, A., Powers, J.S., Fernandes, G.W., Quesada, M., Eds.; CRC Press: Boca Raton, FL, USA, 2013; pp. 119–146.
22. Maas, M.; Burgos, A. Water dynamics and the ecosystem level in seasonally dry tropical forests Seasonally Dry Tropical Forests: Ecology and Conservation. In *Seasonally Dry Tropical Forests*; Dirzo, R., Young, H.S., Mooney, H.A., Ceballos, G., Eds.; Island Press: Washington, DC, USA, 2011; pp. 141–156. [CrossRef]
23. Herrerías-Diego, Y.; Quesada, M.; Stoner, K.E.; Lobo, J.A. Effects of forest fragmentation on phenological patterns and reproductive success of the tropical dry forest tree *Ceiba aesculifolia*. *Conserv. Biol.* **2006**, *20*, 1111–1120. [CrossRef]
24. Valle-Díaz, O.; Blanco-García, A.; Bonfil, C.; Paz, H.; Lindig-Cisneros, R. Altitudinal range shift detected through seedling survival of *Ceiba aesculifolia* in an area under the influence of an urban heat island. *Forest Ecol. Manag.* **2009**, *258*, 1511–1515. [CrossRef]
25. Brondani, R.P.V.; Gaiotto, F.A.; Missiaggia, A.A.; Kirst, M.; Gribel, R.; Grattapaglia, D. Microsatellite markers for *Ceiba pentandra* (Bombacaceae), an endangered tree species of the Amazon Forest. *Mol. Ecol. Notes* **2003**, *3*, 177–179. [CrossRef]
26. IUCN 2022. The International Union for Conservation of Nature and Natural Resources Red List of Threatened Species. Version 2021-3. Available online: https://www.iucnredlist.org/ (accessed on 16 January 2022).
27. Cusack, D.F.; Karpman, J.; Ashdown, D.; Cao, Q.; Ciochina, M.; Halterman, S.; Lydon, S.; Neupane, A. Global change effects on humid tropical forests: Evidence for biogeochemical and biodiversity shifts at an ecosystem scale. *Rev. Geophys.* **2016**, *54*, 523–610. [CrossRef]
28. Velázquez-Rosas, N.; Ruiz-Guerra, B.; Sánchez-Coronado, M.E.; Gamboa de Buen, A.; Orozco-Segovia, A. Morphological variation in fruits and seeds of *Ceiba aesculifolia* and its relationship with germination and seedling biomass. *Bot. Sci.* **2017**, *95*, 81–91. [CrossRef]
29. Martínez-González, I.; Sánchez-Velázquez, L.R.; Ruiz-Guerra, B.; del Rosario Pineda-López, M.; Velázquez-Rosas, N. The role of seed size in the emergence and survival of seedlings in contrasting environments: The case of *Ceiba aesculifolia*. *New For.* **2021**, *52*, 493–507. [CrossRef]
30. Vargas-Rodriguez, Y.L.; Vázquez-García, J.A.; Williamson, G.B. Environmental correlates of tree and seedling–sapling distributions in a Mexican tropical dry forest. *Plant Ecol.* **2005**, *180*, 117–134. [CrossRef]
31. Pineda-García, F.; Paz, H.; Meinzer, F.C. Drought resistance in early and late secondary successional species from a tropical dry forest: The interplay between xylem resistance to embolism, sapwood water storage and leaf shedding. *Plant Cell Environ.* **2013**, *36*, 405–418. [CrossRef]
32. Mendoza, P.A.M.; Moya, E.G.; Rivera, J.R.A.; Xolocotzi, E.H. Regeneración natural de especies arbóreas en una selva mediana subperennifolia perturbada por extracción forestal. *Acta Bot. Mex.* **1995**, *32*, 11–23. [CrossRef]
33. Encino-Ruiz, L.; Lindig-Cisneros, R.; Gómez-Romero, M.; Blanco-García, A. Desempeño de tres especies arbóreas del bosque tropical caducifolio en un ensayo de restauración ecológica. *Bot. Sci.* **2013**, *91*, 107–114. [CrossRef]
34. Martínez-González, I.; Velázquez-Rosas, N.; del Rosario Pineda-López, M.; Ruiz-Guerra, B.; Sánchez-Velásquez, L.R. The role of seed size in the performance of *Ceiba aesculifolia* seedlings subjected to foliar damage. *Acta Oecol.* **2021**, *112*, 103772. [CrossRef]
35. Olvera-Mendoza, E.I.; Lara-Cabrera, S.I.; Sáenz-Romero, C.; Lindig-Cisneros, R. AFLP polymorphism in restored provenances of *Ceiba aesculifolia* within an urban heat island. *Phyton Inter. J. Exp. Bot.* **2016**, *85*, 169–175.
36. Gómez-Maqueo, X.; Soriano, D.; Velázquez-Rosas, N.; Alvarado-López, S.; Jiménez-Durán, K.; del Mar Garciadiego, M.; Gamboa-deBuen, A. The seed water content as a time-independent physiological trait during germination in wild tree species such as *Ceiba aesculifolia*. *Sci. Rep.* **2020**, *10*, 10429. [CrossRef] [PubMed]

37. Hending, D.; Randrianarison, H.; Holderied, M.; McCabe, G.; Cotton, S. The kapok tree (*Ceiba pentandra* (L.) Gaertn, Malvaceae) as a food source for native vertebrate species during times of resource scarcity and its potential for reforestation in Madagascar. *Austral Ecol.* **2021**, *46*, 1440–1444. [CrossRef]
38. Immanuel, R.R.; Ganapathy, M. Growth and physiological attributes of *Ceiba pentandra* (L.) Gaertn. seeds and seedlings under salt stress. *J. Agric. Biol. Sci.* **2007**, *2*, 12–16.
39. Immanuel, R.R.; Ganapathy, M. Agro-techniques for afforestation of degraded coastal agricultural lands with silk cotton (*Ceiba pentendra* (L.) Gaertn.). *J. Pharmacog. Phytochem.* **2019**, *8*, 1587–1590.
40. The University of Arizona, Campus Arboretum. 2021. Available online: https://apps.cals.arizona.edu/ (accessed on 26 November 2021).
41. Guillot, D.; Laguna, E.; Puche, C.; Meer, P. *Ceiba speciosa* 'Valencian Beauty'. *Bouteloua* **2016**, *23*, 3–13.
42. Dick, C.W.; Bermingham, E.; Lemes, M.R.; Gribel, R. Extreme long-distance dispersal of the lowland tropical rainforest tree *Ceiba pentandra* L.(Malvaceae) in Africa and the Neotropics. *Mol. Ecol.* **2007**, *16*, 3039–3049. [CrossRef] [PubMed]
43. Rahmah, A.U.; Abdullah, M.A. Evaluation of Malaysian *Ceiba pentandra* (L.) Gaertn. for oily water filtration using factorial design. *Desalination* **2011**, *266*, 51–55. [CrossRef]
44. Silitonga, A.S.; Ong, H.C.; Mahlia, T.M.I.; Masjuki, H.H.; Chong, W.T. Characterization and production of *Ceiba pentandra* biodiesel and its blends. *Fuel* **2013**, *108*, 855–858. [CrossRef]
45. Zhang, X.; Fu, W.; Duan, C.; Xiao, H.; Shi, M.; Zhao, N.; Xu, J. Superhydrophobicity determines the buoyancy performance of kapok fiber aggregates. *Appl. Surf. Sci.* **2013**, *266*, 225–229. [CrossRef]
46. Graham, B.P.; Haigler, C.H. Microtubules exert early, partial, and variable control of cotton fiber diameter. *Planta* **2021**, *253*, 47. [CrossRef]
47. Carranza-Nuñez, U.; Ramiro Vasquez-Garcia, S.; Flores-Ramirez, N.; Ahmed Abdel-Gawwad, H.; Luis Rico, J.; Arizbe Santiago, A.; Vargas, J.; Cruz-de-León, J. Physicochemical characterization of natural fibers obtained from seed pods of *Ceiba aesculifolia*. *BioResources* **2021**, *16*, 4200–4211. [CrossRef]
48. Anderson, D.B.; Kerr, T. Growth and structure of cotton fiber. *Ind. Eng. Chem.* **1938**, *30*, 48–54. [CrossRef]
49. Yan, J.; Xu, G.; Wang, F. A study on the quality of kapok blended yarns through different processing methods. *J. Tex. Ins.* **2013**, *104*, 675–681. [CrossRef]
50. Patwary, S. Clothing and Textile Sustainability: Current State of Environmental Challenges and the Ways Forward. *Text. Leather Rev.* **2020**, *3*, 158–173. [CrossRef]
51. Sartika, D.; Syamsu, K.; Warsiki, E.; Fahma, F.; Arnata, I.W. Nanocrystalline cellulose from kapok fiber (*Ceiba pentandra*) and its reinforcement effect on alginate hydrogel bead. *Starch-Stärke* **2021**, *79*, 2100033. [CrossRef]
52. Leal, M.R.; Flores-Sahagun, T.H.S.; Franco, T.S.; Muniz, G.I. *Ceiba speciosa* St. Hill fruit fiber as a potential source for nanocellulose production and reinforcement of polyvinyl acetate composites. *Polym. Compos.* **2021**, *42*, 397–411. [CrossRef]
53. Hori, K.; Flavier, M.E.; Kuga, S.; Lam, T.B.T.; Iiyama, K. Excellent oil absorbent kapok *Ceiba pentandra* (L.) Gaertn. fiber: Fiber structure, chemical characteristics, and application. *J. Wood Sci.* **2000**, *46*, 401–404. [CrossRef]
54. Thilagavathi, G.; Karan, C.P.; Thenmozhi, R. Development and investigations of kapok fiber based needle punched nonwoven as eco-friendly oil sorbent. *J. Nat. Fibers* **2018**, *17*, 18–27. [CrossRef]
55. Nkouam, G.B.; Adjoh, G.A.; Tchankou Leudeu, C.B.; Kouebou, C.; Tchiegang, C.; Kapseu, C. Physico-chemical properties of fruits, seed and oil of kapok (*Ceiba pentandra* Gaertn.) tree of different provenances from the northern part of Cameroon. *Int. J. Agric. Innov. Res.* **2017**, *6*, 275–278.
56. Berry, S.K. The characteristics of the kapok (*Ceiba pentadra* Gaertn.) seed oil. *Pertanika J. Trop. Agric. Sci.* **1979**, *21*, 1–4.
57. Rosselli, S.; Tundis, R.; Bruno, M.; Leporini, M.; Falco, T.; Gagliano Candela, R.; Badalamenti, N.; Loizzo, M.R. *Ceiba speciosa* (A. St.-Hil.) seeds oil: Fatty acids profiling by GC-MS and NMR and bioactivity. *Molecules* **2020**, *25*, 1037. [CrossRef]
58. Avendaño, A.; Casas, A.; Dávila, P.; Lira, R. Use forms, management and commercialization of "pochote" *Ceiba aesculifolia* (HB & K.) Britten & Baker f. subsp. parvifolia (Rose) PE Gibbs & Semir (Bombacaceae) in the Tehuacán Valley, Central Mexico. *J. Arid Environ.* **2006**, *67*, 15–35. [CrossRef]
59. Avendaño, A.; Casas, A.; Dávila, P.; Lira, R. In situ management and patterns of morphological variation of *Ceiba aesculifolia* subsp. parvifolia (Bombacaceae) in the Tehuacán-Cuicatlán Valley. *Econ. Bot.* **2009**, *63*, 138–151. [CrossRef]
60. Suastegui-Baylón, L.; Salazar, R.; Maldonado-Astudillo, Y.I.; Ramírez-Sucre, M.O.; Arámbula-Villa, G.; Flores-Casamayor, V.; Jiménez-Hernández, J. Physical, chemical and rheological characterization of tuber and starch from *Ceiba aesculifolia* subsp. *parvifolia*. *Molecule* **2021**, *26*, 2097. [CrossRef] [PubMed]
61. Marks, R.A.; Hotaling, S.; Frandsen, P.B.; VanBuren, R. Representation and participation across 20 years of plant genome sequencing. *Nat. Plants* **2021**, *7*, 1571–1578. [CrossRef] [PubMed]
62. Gao, Y.; Wang, H.; Liu, C.; Chu, H.; Dai, D.; Song, S.; Yu, L.; Han, L.; Fu, Y.; Tian, B.; et al. De novo genome assembly of the red silk cotton tree (*Bombax ceiba*). *GigaScience* **2018**, *7*, giy051. [CrossRef] [PubMed]
63. Gao, Y.; Wang, H.; Liu, C.; Chu, H.; Yan, Y.; Tang, L. Complete chloroplast genome sequence of the red silk cotton tree (*Bombax ceiba*). *Mitochondrial DNA Part B* **2018**, *3*, 315–316. [CrossRef]
64. Gao, Y.; Wang, H.; Liu, C.; Chu, H.; Yan, Y.; Tang, L. The complete mitochondrial genome of *Bombax ceiba*. *Mitochondrial DNA Part B* **2018**, *3*, 313–314. [CrossRef]

65. Wang, D.; Fan, W.; Guo, X.; Wu, K.; Zhou, S.; Chen, Z.; Li, D.; Wang, K.; Zhu, Y.; Zhou, Y. MaGenDB: A functional genomics hub for Malvaceae plants. *Nucleic Acids Res.* **2020**, *48*, D1076–D1084. [CrossRef]
66. Huang, S.; Zhu, Q.; Huang, G.; Han, B.; Zhou, Q.; Dai, J. The chloroplast genome of silk floss tree (*Ceiba speciosa*). *Mitochondrial DNA Part B* **2019**, *4*, 3632–3633. [CrossRef]
67. Nakabayashi, K.; Okamoto, M.; Koshiba, T.; Kamiya, Y.; Nambara, E. Genome-wide profiling of stored mRNA in *Arabidopsis thaliana* seed germination: Epigenetic and genetic regulation of transcription in seed. *Plant J.* **2005**, *41*, 697–709. [CrossRef] [PubMed]
68. Howell, K.A.; Narsai, R.; Carroll, A.; Ivanova, A.; Lohse, M.; Usadel, B.; Miller, A.H.; Whelan, J. Mapping metabolic and transcript temporal switches during germination in rice highlights specific transcription factors and the role of RNA instability in the germination process. *Plant Physiol.* **2009**, *149*, 961–980. [CrossRef] [PubMed]
69. Huang, G.; Huang, J.Q.; Chen, X.Y.; Zhu, Y.X. Recent advances and future perspectives in cotton research. *Annu. Rev. Plant Biol.* **2021**, *72*, 437–462. [CrossRef]
70. Wu, H.; Tian, Y.; Wan, Q.; Fang, L.; Guan, X.; Chen, J.; Hu, Y.; Ye, W.; Zhang, H.; Guo, W.; et al. Genetics and evolution of MIXTA genes regulating cotton lint fiber development. *New Phytol.* **2018**, *217*, 883–895. [CrossRef] [PubMed]
71. Gómez-Maqueo, X.M.; Gamboa-deBuen, A. (Universidad Nacional Autónoma de México, Ciudad de México, México). Personal communication, 2022.
72. Arellanes-Cancino, Y.; Romero-Sosa, M.Á.; Vega, E.; Maza-Villalobos, S.; Casas-Fernández, A. Ecological bases for sustainable management of Pochote (*Ceiba aesculifolia* subsp. parvifolia) through demographic analysis. *Econ. Bot.* **2018**, *72*, 20–37. [CrossRef]
73. Royal Botanic Gardens Kew. Plants of the World Online Database. Available online: https://powo.science.kew.org/ (accessed on 4 January 2022).
74. Tropicos.org. Missouri Botanical Garden. Version 3.3.1. Available online: https://tropicos.org (accessed on 5 January 2022).

plants

MDPI

Review

Edible Halophytes and Halo-Tolerant Species in Apulia Region (Southeastern Italy): Biogeography, Traditional Food Use and Potential Sustainable Crops

Rita Accogli [1], Valeria Tomaselli [2,*], Paolo Direnzo [3], Enrico Vito Perrino [2,4], Giuseppe Albanese [2], Marcella Urbano [3] and Gaetano Laghetti [3]

1 Department of Biological and Environmental Sciences and Technologies (DiSTeBA), Salento University, 73100 Lecce, Italy
2 Department of Biosciences, Biotechnologies and Environment, University of Bari "Aldo Moro", 70126 Bari, Italy
3 Institute of Biosciences and BioResources (IBBR), National Research Council (CNR), 70126 Bari, Italy
4 CIHEAM, Mediterranean Agronomic Institute of Bari, Via Ceglie 9, 70010 Valenzano, Italy
* Correspondence: valeria.tomaselli@uniba.it

Abstract: The Mediterranean basin is rich in wild edible species which have been used for food and medicinal purposes by humans throughout the centuries. Many of these species can be found near coastal areas and usually grow under saline conditions, while others can adapt in various harsh conditions including high salinity. Many of these species have a long history of gathering from the wild as a source of food. The aim of this contribution is an overview on the most important halophyte species (*Salicornia* sp. pl., *Arthrocaulon macrostachyum* (Moric.) Piirainen & G. Kadereit, *Soda inermis* Fourr., *Cakile maritima* Scop., *Crithmum maritimum* L., *Reichardia picroides* (L.) Roth., *Silene vulgaris* (Moench) Garcke subsp. *tenoreana* (Colla) Soldano & F. Conti, *Allium commutatum* Guss., *Beta vulgaris* L. subsp. *maritima* (L.) Arcang., *Capparis spinosa* L.) that traditionally have been gathered by rural communities in southern Italy, with special interest on their ecology and distribution, traditional uses, medicinal properties, marketing and early attempts of cultivation. It is worth noting that these species have an attractive new cash crop for marsh marginal lands.

Keywords: halophytes; edible wild species; coastal areas; *Salicornia; Arthrocaulon macrostachyum; Soda inermis; Cakile maritima; Crithmum maritimum; Reichardia picroides; Silene vulgaris* subsp. *tenoreana; Allium commutatum; Beta vulgaris* subsp. *maritima; Capparis spinosa*

Citation: Accogli, R.; Tomaselli, V.; Direnzo, P.; Perrino, E.V.; Albanese, G.; Urbano, M.; Laghetti, G. Edible Halophytes and Halo-Tolerant Species in Apulia Region (Southeastern Italy): Biogeography, Traditional Food Use and Potential Sustainable Crops. *Plants* **2023**, *12*, 549. https://doi.org/10.3390/plants12030549

Academic Editor: Oscar Vicente

Received: 30 December 2022
Revised: 15 January 2023
Accepted: 17 January 2023
Published: 25 January 2023

1. Introduction

The relationship between human civilizations and salinity has existed for thousands of years. The total area of saline and sodium lands is likely to be approximately 10% of arable land worldwide [1]. Many of the factors that lead to soil salinization are being exacerbated by climate change and it will get worse and worse over the next few years based on the indicators that the scientific world takes into consideration. In fact, considering the increase of global climate change and severe conditions prevailing all over the world, conventional crop cultivation is facing various limitations related to shortage of good quality water, rising temperature and the salinization and degradation of soil properties, especially in arid and semi-arid regions of the Mediterranean basin where the aforementioned problems intensify [2]. Therefore, it is of the utmost importance to propose alternative crop species able to adapt to difficult conditions in the framework of saline agriculture and constitute good candidates as potential food and medicinal crops [3], as already recently enacted in Apulia for some wild aromatic coastal species of *Lamiaceae* family [4–6] that can meet the future needs of populations who will face this challenge.

Halophytes are defined as plants capable of developing and completing their biological cycle in natural saline environments with concentrations greater than 200 mM NaCl [7].

They constitute a highly specialized flora adapted to hypersaline environments. The response to this type of stress is manifested through osmotic adjustments, morpho-anatomical and physiological adaptations. Most of these species are eu-halophytes, which are obligate halophytes because they grow and develop in media with high salt concentrations. Often, they are succulent plants, since they have large amounts of water in their tissues in addition to accumulating salts and ions, maintaining the osmotic balance [8]. In recent years, the nutritional interest of these genera has sparked an increase in research articles on elemental composition, minerals, bioactive compounds, polyphenols, fatty acids and flavonoids [9,10]. Halophytic species could be used for the saline agriculture proposed by [11]. The types of land proposed for use in halo-culture are generally internal and coastal marginal uncultivated lands, degraded saline lands, sandbanks and salinized lands in general which are not able to economically produce conventional agronomic crops.

There are two factors that make halophytes of special interest to be considered in the food industry: first, their economic potential, since their productivity in high-salinity and low-water intake environments is much higher than that of traditional edible species; and second, their nutritional value in terms of their protein, phenolic, lipid contents and the great quantity of minerals such as potassium, calcium, and magnesium and other bioactive compounds [12–14]. Several halophyte species are already being used as food, forage, oilseeds and medicines [15–18].

Examples of edible halophytes and halo-tolerant species include marine fennel (*Crithmum maritimum* L.) [19], Mediterranean saltwort (*Soda inermis* Fourr.) [20], glassworts (*Salicornia* sp. pl.; *Arthrocaulon macrostachyum* (Moric.) Piirainen & G. Kadereit) [21,22], sea rocket (*Cakile maritima* Scop.), sea beet (*Beta vulgaris* L. subsp. *maritima* (L.) Arcang.) [23], *Reichardia picroides* (L.) Roth. [24], caper (*Capparis spinosa* L.) [25], *Silene vulgaris* (Moench) Garcke subsp. *tenoreana* (Colla) Soldano & F. Conti and maritime wild leek (*Allium commutatum* Guss.) [26].

This review describes the edible halophytes and halo-tolerant species above reported in respects of their ecology and distribution, traditional uses, medicinal properties, marketing and early attempts of cultivation. Moreover, original new data are provided about the distribution of their use in the Apulia region (southeastern Italy) and about some early attempts of domestication/cultivation).

2. Materials and Methods

2.1. Study Area

The Apulia Region, located in the southeastern part of the Italian peninsula, has a surface area of more than 19,000 km^2, and is geologically characterized by Cretaceous limestones and calcarenites and by alluvial deposits (Pliocene–Pleistocene). It is the Italian region with the highest coastal extent (Figure 1), with about 1224 km of coastline (including the Tremiti Islands) stretched between the Adriatic and the Ionian Seas, from RodiGarganico to Ginosa Marina [27]. Throughout this paper we refer mostly to Gargano (northern Apulia) and Salento (southern Apulia) areas. This particular geographic conformation has induced uses and traditions in the Apulian people, who express a strong relationship of knowledge and uses of the coastal area not only for commercial exchanges but also for craft activities such as the extraction of building materials, the tanning of leather, wool washing and fabric dyeing [28] and, most importantly, the extraction of salt from sea water [29]. In relation to the latter, the most important extraction sites are (or have been) Salina di Margherita di Savoia (FG), Salina di Punta della Contessa (BR) and Salina dei Monaci di Manduria (TA); however, almost all the low Apulian coasts have established local resources for the supply of salt. The halophytes, growing in these coastal environments, have been an important asset for maritime populations, as source of mineral salts and numerous secondary metabolites precious for the human diet [30,31].

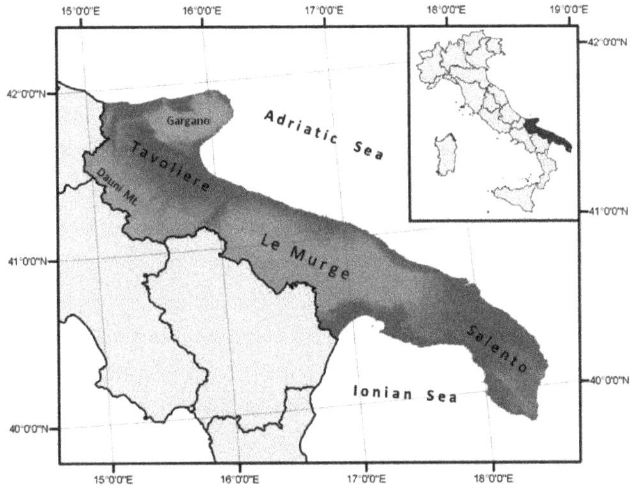

Figure 1. Study area.

The largest part of the region (over 80%) is used for agriculture and, as in other arid and semi-arid Mediterranean areas, has continuously been subjected to intensive management practices, potentially leading to land degradation [32,33]. As far as the plant biodiversity is concerned, Apulia is an area of high biogeographical interest and high floristic density [34,35], with a fairly large number of endemic taxa [36]. Similar diversity and richness are found in terms of plant communities, with a vegetational variability emphasized by a long-time exploitation of the area that has produced, in time, numerous, alternating and often contrasting landscapes. In particular, coastal areas, for their wide extent and geomorphological variability, are characterized by a high floristic richness [37] and variety of plant landscapes. Many phytosociological surveys have been carried out along the Apulian coast, on both rocky and sandy shores included salt marshes [38–52]. According to the most recent literature, the vegetation of the rocky coasts is classified in several associations of the orders *Crithmo-Staticetalia, Helichrysetalia italici* and *Senecionetalia cinerariae* (*Crithmo-Staticetea* class) [50] for perennial communities, and of *Saginetea maritimae* (*Saginetalia maritimae*) for annuals; salt marsh vegetation is framed, with numerous associations and alliances, within the classes *Thero-Salicornietea* (*Thero-Salicornietalia* and *Thero-Suaedetalia splendentis*), *Saginetea maritimae* (*Frankenietalia pulverulentae*), for the annual vegetation [51] and, for perennial vegetation, within the *Salicornietea fruticosae* and *Juncetalia maritimae* [45–48]. Finally, the vegetation of sandy coasts falls within the *Cakiletea maritimae* and the *Ammophiletea* classes [43,53]. These plant communities are included as habitats of community interest within 92/43 EEC "Habitat" Directive [54–59].

2.2. Bibliographic Review; New Data Collection in the Field

More than 50 bibliographic sources published in the last 40 years by Apulian publishers and referring to different uses and properties of local plants were retrieved and examined [60–68]. Information was derived from numerous surveys carried out in the last twenty years in various territories of the Apulia region, by interviews with the elderly people and with "connoisseurs" or "experts" of wild herbs. Similarly, in the last year, specific explorations in the most significant coastal areas of the region were carried out, to evaluate the presence of halophilic species of ethnobotanical interest. For each site, residents were contacted and site inspections were carried out in the coastal places they usually visit for herb collections, with the identification of the species known to them and recording, on semi-structured questionnaires, of information about: (a) "plant" (scientific and common name, habitat, collection period and site, used part, etc.); (b) "use" (e.g., human nutrition, animal food, folk medicine, etc.); and (c) "preparation, proverbs, recipes". The questionnaire structure is showed in Appendix B. Interviews were also addressed to restaurant managers in the most popular coastal locations and to fishermen in the most traditional fishing grounds. About eighty-five questionnaires were administered, and the collected data were then organized in a database. Moreover, extensive bibliographic research on food use and domestication/cultivation of the considered species also in other geographical areas, as well as other uses, has been carried out and reported in Results.

For each of the investigated species, we reported: brief description, distribution and ecology, food use, domestication/cultivation and other uses/properties.

As regards taxonomical nomenclature of the treated species, we followed the Italian checklist [69]. For the genus *Salicornia* (annual species), we followed "Flora d'Italia" [70]. For the syntaxonomical nomenclature of vegetation types mentioned in "distribution and ecology", we referred to the specialized literature reported in the previous section, "Study area".

2.3. Taxonomical Notes

Salicornioideae (Amaranthaceae/Chenopodiaceae *sensu* APG IV [71]; subfamily for succulent, articulated and apparently leafless species) stands out by comprising exclusively succulent hygro-halophytes with highly specialized morphological, anatomical and physiological adaptations to their environment in coastal and inland halophytic communities [72–74]. Within the *Salicornioideae*, the genus *Salicornia* (glassworts) displays a controversial taxonomical classification; this group has been thoroughly analyzed from both morphological and molecular traits. Common *Salicornioideae* characters are their succulent, articulated and apparently leafless stems, and the spike-like inflorescence of sessile, 3-flowered cymes, reduced flowers, usually consisting of a 2–4 lobed calyx tube with 1–2 stamens, and the sub-annular or curved embryo [75,76]. The genus *Salicornia* was for a long time considered as circumscribed to annual species, with the perennial species separated in other genera, but recent taxonomical revisions [74] provided a new framework, with the (perennial) *Sarcocornia* species included under *Salicornia* genus; thus, the *Salicornia* genus currently frames both annual and perennial species. In the same contribution, *Arthrocnemum macrostachyum* (Moric.) K. Koch is treated in a different genus (i.e., *Arthrocaulon*) as *Arthrocaulon macrostachyum* (Moric.) Piirainen & G. Kadereit.

In general, the taxonomic identity of the *Salicornia* species is a challenging issue. The complexity of the genus *Salicornia* in Europe has produced many reversals, alternating partitions within the genus. As regards the annual *Salicornia* species, two series have been traditionally identified, diploid and tetraploid, each of them with numerous species and micro-species. In the revision of [75], only three entities are recognized in the Mediterranean: *Salicornia procumbens* subsp. *procumbens* (tetraploid, distributed along Mediterranean and Atlantic coasts and including *Salicornia emerici*, *Salicornia veneta* and *Salicornia dolichostachya*), *Salicornia perennans* subsp. *perennans* (diploid, with Mediterranean and Eurasian distribution, including *Salicornia patula*) and *Salicornia europaea* subsp. *europaea* (diploid, western Mediterranean). As regards the Italian peninsula, only the two first taxa, *S. procumbens* subsp. *Procumbens* and *S. perennans* subsp. *perennans*, are recognized. Nevertheless, the existence of numerous microtaxa that are morphologically quite well-differentiated and characterized by particular ecological conditions would encourage them to maintain their use. For this reason, we refer to the classification proposed by [70], who recognizes the presence in the Italian peninsula (and the Apulia region) of the diploid *S. patula* and of the tetraploid *S. emerici*, *S. veneta* and *S. dolichostachya*.

Species of this group are characterized by the plasticity of their phenotype which, along with the reduction of their leaves and flowers and the difficulty of preserving the dry specimens in the herbaria, complicates the separation of the species with frequent confusion among *Sarcocornia* taxa [76]. The taxonomic complexity of this group implied that, in the course of our investigations, the identification of the correct taxonomic entity has often been problematic, especially when two or more related species coexist in the same site (e.g., *S. emerici* and *S. dolichostachya* or *Salicornia fruticosa* and *Salicornia perennis*).

3. Results

3.1. Salicornia sp. pl. (Annual)

3.1.1. Brief Description, Distribution and Ecology

Salicornia, also commonly known as pickleweed, glasswort, sea asparagus and samphire, derives its name from the Latin word meaning "salt" and includes strictly halophytes. These taxa accumulate inorganic salts and water in their stems [8]. The most common elements found are Na, Ca, K and Mg, among others, and are present in the stem and roots [9]. These species grow usually at the edges of wetlands, salt marshes and mudflats world-wide (except for Australia). Along with the perennial *Salicornia* sp.pl. and other perennial succulent Amaranthaceae/Chenopodiaceae, they thrive in littoral and coastal salt marsh or in inland salt pans, forming distinct vegetation types subjected to different flooding regimes with brackish-to-saline water (depending on tidal influences) and usually composed of almost monospecific plant communities. In general, annual *Salicornia* species occur in the innermost parts of the salt marshes, subject to longer periods of submersion. This group of species is characterized by succulent and articulated stems with opposite pairwise-fused leaves and bracts and inconspicuous flowers with three-flowered cymes in which the lateral flowers are in contact below the central flower. The green plant turns orange, pink to reddish in autumn, before dying in winter [75–77]. In the Apulia region, four species have been recognized: the diploid *S. patula* Duval-Jouve (Figure 2a), and the tetraploid *S. emerici* Duval-Jouve, *S. veneta* Pignatti & Lausi and *S. dolichostachya* Moss (Figure 3a) [70], often very similar morphologically to each other and distinguishable with certainty only if observed in full maturity (flowering and fruiting period). These species typically grow forming the pioneer coastal or continental vegetation of maritime and inland salt marshes, occupying those sites with the highest salt concentration, generally positioned in the first-line belt of the flooding zonation (Figures 2b and 3b) on raw soils from sandy to loamy or clay, depending on sedimentation conditions and in large part are poor in nutrients, temporarily flooded and drying out in summer (*Thero-Salicornietea* Tx. in Tx. & Oberd. 1958; *Salicornion patulae* Géhu & Géhu-Franck 1984; *Salicornion venetae* Tomaselli et al. 2020). The phenological optimum is late summer to autumn [51]. According to the 92/43/EEC "Habitat" Directive, this vegetation falls within the habitat of community interest 1310—"*Salicornia* and other annuals colonizing mud and sand".

Figure 2. A specimen of *Salicornia patula* (**a**); *S. patula* community ((**b**); the reddish glasswort carpet) at the edge of a coastal lagoon in Salento.

Figure 3. *Salicornia dolichostachya* (**a**); a plant community dominated by *S. dolichostachya* and *S. emerici* (**b**) at "Palude La Vela", Taranto, southern Apulia).

3.1.2. Food Use

The annual glassworts are almost all of food interest. Like other chenopods such as *Sarcocornia perennis*, *S. fruticosa* and *Arthrocaulon macrostachyum* recently introduced for human consumption [78], *S. patula* and other annuals are also an excellent candidate due to the mineral content. Actually, they are little known in the territories of southern Salento where, due to the rocky nature of the coasts, temporary salt ponds and salt marshes have sporadic presence and very limited extent. On the contrary, annual glassworts are widely used in the areas characterized by coastal salt marshes (e.g., Gargano, Capitanata, Brindisi and Taranto areas). The plants are low and herbaceous in consistency, and the tender tips of the young branches are easily broken by hand. The most abundant harvest is in spring, before the reproductive stage begins. The collected tips are cleaned, washed and boiled. After having drained them, one by one, the fleshy green part is removed (like a glove) and the central woody part is thrown away. They can be dressed with oil and lemon or vinegar, put in jars in oil or vinegar or frozen and used as needed. In the area of "Mar Piccolo", near Taranto, in summer it is used to mix the sea asparagus with leavened dough, which is then portioned and fried ("pettole"). Recently, in the Gargano area it is common to find pizzas or pasta topped with sea asparagus (Figures 4 and 5c; Table 1; Appendix A). In recent years, the sale of sea asparagus in oil or in vinegar has spread in various locations in northern Apulia (especially Gargano) and in the Bari area, even in large-scale distribution.

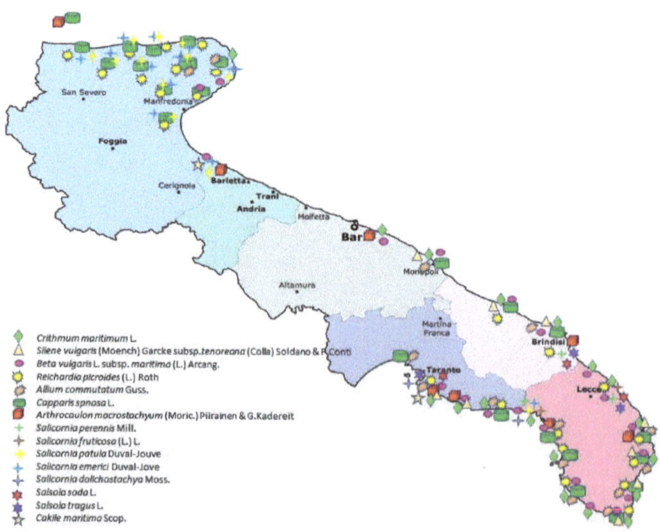

Figure 4. Distribution, across the Apulia region, of the food use of halophytes and halotolerant species treated in this contribution, according to our field observations and literature data.

Figure 5. *Salicornia* sp. sold at a fish shop (**a**) and at a superstore (**b**) in Bari; a dish of pasta with *Salicornia* sp. (**c**); Peschici, Gargano, Northern Apulia).

Table 1. Summary of the main features of the taxa considered in this contribution.

Taxon	Life Form	Geoelement	Protection	Use	Propagation
Salicornia patula	T scap	W-European	regional level	food/medical	generative
Salicornia emerici	T scap	Steno-Medit.	no	food/medical	generative
Salicornia dolichostachya	T scap	Steno-Medit.	no	food/medical	generative
Salicornia perennis	Ch succ	Euri-Medit.	regional level	food/medical	generative
Salicornia fruticosa	Ch succ	Euri-Medit.	no	food/medical	generative
Arthrocaulon macrostachyum	Ch succ/P succ	Medit.	no	food/medical	generative/ vegetative
Soda inermis	T scap	Paleotemp.	no	food/medical/ industrial	generative
Cakile maritima	T scap	Medit.-Atl. (Steno-)	no	food/medical	generative
Crithmum maritimum	Ch suffr	Euri-Medit./Steno-Medit.	regional level	food/medical	generative
Reichardia picroides	H scap	Steno-Medit.	no	food/medical	generative
Silene vulgaris subsp. *tenoreana*	H scap	Paleotemp./ Subcosmop.	no	food/medical/ phytoremediation	generative
Allium commutatum	G bulb	Steno-E-Medit.	LC (Least Concern)	food/medical	generative/ vegetative
Beta vulgaris subsp. *maritima*	H scap	Euri-Medit.	no	food/medical	generative
Capparis spinosa	NP	Eurasiat.	regional level	food/medical	generative/ vegetative

In the Iberian Peninsula and France, *S. emerici* has long been largely consumed as food [79]. In South Korea, tender *Salicornia* shoots are processed in drinks such as nuruk (a type of fermentation initiator), makgeolli (a Korean rice wine) or vinegar [80]; the aerial parts are used in salads, and their consumption is a source of salt in the diet [77].

3.1.3. Domestication/Cultivation

Being therophytes, the only method of propagation is by the generative way (from seeds). The seeds are located in a bract cavity, unprotected, so they fall as they reach maturity. It is necessary to be timely in collecting the parts that have already matured the seeds and to be careful not to drop them during the cutting and bagging operations. The quantity of seeds collected per plant is always minimal. Recently, interest towards *Salicornia* sp. has risen dramatically, especially in the Gargano area where some attempts of medium-scale cultivations have been run [22]).

In Israel, *Salicornia* and *Sarcocornia* cultivation for vegetable production is typically practiced under simple nets or in greenhouses on land areas of 0.5–1 ha. For the products of these crops to be successfully marketed as vegetables, the young shoots must be harvested manually, an intensive element critical to halophyte crop production. Different cultivation protocols, therefore, have been tested for *Salicornia* vegetable production. For example, the easiest and most straightforward way to produce *Salicornia* is to cultivate it in native sand dune soils watered with drip irrigation, as has been successfully accomplished by some farmers in the Dead Sea and Ramat HaNegev areas [81,82].

Salicornia sp. pl. is cultivated on a commercial scale in Israel and in Mexico, the United Arab Emirates, Saudi Arabia and India. In many parts of the Europe, the use of this halophyte is mainly based on the harvesting of wild plants, while only in the Netherlands, Portugal and France are the plants of *Salicornia* sp. pl. cultivated. In Italy, the first attempts of *S. patula* cultivation for family consumption started over 40 years ago in the gardens located around the Lesina lagoon (Apulia, Southern Italy) [22]. After only a few years, some local farmers started the cultivation of this wild grass as a cash crop, due to the increasing interest in this food, the attractive price and the possibility of economic profits from the cultivation of marginal lands [22].

In Apulia, harvesting in open fields takes place in July–August. In Israel, it is from August–September, and it is carried out manually in order to maintain the high quality of the final product; only fresh and tender parts can be sold (Figure 5a,b). *Salicornia* can be collected several times during the year in greenhouses. The harvesting can be repeated every two or three weeks (depending on the level of development). Plants are cut above 5 cm from the ground at a height of 10–15 cm. This repeated harvesting enables the same plant to be cut from three-to-four times, depending on the level of growth. The yield can reach 10–15 tons per hectare [9].

3.1.4. Other Uses/Properties

Like other halophilic species, even in glassworts the stress of the alkaline environment induces the synthesis of metabolites that can be useful to humans for their antioxidant and anti-inflammatory power [83]. For example, bioactive compounds such as phenols and fatty acids have been isolated for *S. patula*, and it has been found that the concentration of these compounds depends greatly on how close the plants are to the sea. Caffeic, coumaric, salicylic and trans-cinnamic acids and flavonoids, such as quercetin-3-O-rutinoside, kaempferol/luteolin, apigenin 7-glucoside and pelargonidin-3-O-rutinoside, have been isolated. Furthermore, high concentrations of palmitic, oleic and linoleic acids have been found [84].

3.2. Salicornia sp. pl. (Perennial) and Arthrocaulon macrostachyum (Moric.) Piirainen & G. Kadereit

3.2.1. Brief Description, Distribution and Ecology

This group includes species, commonly named as perennial glasswort or perennial marsh samphire, with erect-to-prostrate woody stems, sometimes creeping and rooting at the nodes; that are succulent, articulate and green but becoming reddish to maturity; that have opposite vestigial and long connate leaves clothing the internode, a reduction of vegetative and floral parts and 3–12 flowers hidden in cavities in the inflorescence axis. At present, the following taxa can be recognized in southeastern Italy: *S. perennis* Mill. subsp. *perennis*, characterized by a prostrate and decumbent habit with radicant branches, and a seed testa with scarce, patent, erect, long, fine and occasionally hooked hairs; *S. perennis* Mill. subsp. *alpini* (Lag.) Castrov. with prostrate-to-erect habit and dense, appressed, hooked, occasionally forked, long fine hairs that cover the entire seed testa (Figure 6c,d) and *S. fruticosa*, a 50–150 cm-tall woody shrub with an erect, branchy and pubescent seed exotesta covered mainly on the edges (Figure 6a,b) [76]. *Arthrocaulon macrostachyum* is a woody shrub usually erect, up to 150 cm tall, sometimes prostrate, usually richly branched, with fleshy segments that are glaucous or yellowish green (Figure 7a).

Figure 6. *Salicornia fruticosa* (**a**,**b**) at "Saline Margherita di Savoia" (FG, Northern Apulia); *S. alpini* (**c**,**d**) at "Saline di Punta della Contessa" (BR, Southern Apulia).

Figure 7. *Arthrocaulon macrostachyum* (**a**); *A. macrostachyum* communities at the edges of the "Capitanata" coastal lagoons ((**b**); Northern Apulia) and on the rocky coasts of Torre Guaceto ((**c**); Salento, Southern Apulia).

As for the annual *Salicornia* species, *S. perennis* subsp. *perennis*, *S. perennis* subsp. *alpini*, *S. fruticosa* and *A. macrostachyum* usually grow in coastal salt marsh-forming halophilous shrub plant communities developing on halomorphic soils and spatially arranged in distinct vegetation types (belts) depending on the different flooding period and the water salinity (*Salicornietea fruticosae* Br.-Bl. et Tx. ex A. Bolòs y Vayreda 1950; *Salicornion fruticosae* Br.-Bl. 1933; *Arthrocnemion glauci* Rivas-Mart. et Costa M. 1984; *Suaedion brevifoliae* Br.-Bl. et O. de

Bolòs 1958). According to the 92/43/EEC "Habitat" Directive, this vegetation falls within the habitat of community interest 1420—"Mediterranean and thermo-Atlantic halophilous scrubs (*Sarcocornetiea fruticosae*)". *A. macrostachyum* also grows extensively on rocky coasts forming, with *Crithmum maritimum* and other few species, the first belt of vegetation which is the most exposed to sea spray and waves (Figure 7b,c) [43,47,50,85].

3.2.2. Food Use

The bibliographic evidence on the use of perennial glassworts for food purposes dates back to the 18th century, but these halophytes have been collected and used in Apulia since the Middle Ages [29]. The young shoots are harvested by hand, in autumn or early spring, when they are still tender, breaking them at the point where they begin to lignify. As for the annual glassworts, the shoots are cleaned, washed and boiled; after having drained them, one by one, the central woody part is removed from the green fleshy part, which comes off easily. They can be dressed with oil and lemon or vinegar, put in jars in oil or vinegar or frozen; they are also used in dishes with soups and omelettes or in combination with fish and shellfish as a side dish.

S. perennis, *S. fruticosa* and *A. macrostachyum* are also widely consumed in other Mediterranean countries [78,79].

3.2.3. Domestication/Cultivation

In *A. macrostachyum* and perennial *Salicornia* species the seeds are placed in a bract cavity, unprotected, and the same harvesting and sowing techniques used for the annual *Salicornia* species must be used. However, since these are perennial species, with prostrate-ascending stems, it is also possible to adopt agamic propagation, using the branches at the base of the plant, which, in contact with the substrate, already tend to take root under the mother plant. *S. perennis* easily radicates at the nodes, and therefore it is possible to cut the stem before and after the node that generates the plant, eradicate the plant trying to extract the root with a clod of substrate and then transplant it into a nursery pot until it grows. It has been experimented that many halophilic species produce heteromorphic seeds not only in shape, color and size but also for physiological properties. Therefore, different seeds respond differently to environmental conditions, and this is an adaptive strategy that allows the species to adapt to fluctuating conditions. Furthermore, it has been ascertained that in many other halophytes there is no dormancy and that the different types of seeds respond with different germination percentages at the same salt concentration [86].

S. perennis is an economically valuable vegetable crop, which can be cultivated in soils where other vegetable plants cannot grow due to high salinity [87]. Some information about the cultivation of perennial *Salicornia* species and *A. macrostachyum* have been reported in the previous section.

3.2.4. Other Uses/Properties

The beneficial effects of *Salicornia* sp. pl. were already known throughout the Mediterranean Basin; for centuries, many populations have used it to treat gastrointestinal disorders, diabetes, hypertension and inflammation. Scientific studies in recent years have confirmed a high antioxidant and anti-radical activity for *Salicornia* sp.pl., thanks to the high content of flavonoids and polyphenolic compounds which have antidiabetic, neuroprotective, antibacterial, antihypertensive and antitumor activities [83,88].

A. macrostachyum could be a potential source of bioactive compounds that are useful for the treatment of several human diseases. It is traditionally used as an antibiotic and as an alexipharmic in Tunisia, and it plays a prominent role in traditional Indian medicine (Ayurveda). This species can tolerate and accumulate heavy metals and has a very high potential for phytoremediation [89]; moreover, the plant biomass has been used for fodder [90].

3.3. *Soda inermis Fourr. (=Salsola soda L.)*

3.3.1. Brief Description, Distribution and Ecology

The genus *Salsola* (from the Latin Salsus, meaning "salty"; Amaranthaceae/ Chenopodiaceae family) is widespread in salty areas of arid, semi-arid and temperate regions worldwide, with more than 140 species including both annual and perennial species [91]. *Soda inermis* (=*Salsola soda*; saltwort or barilla plant) is a succulent annual herb 20 to 120 cm high; its succulent leaves are linear, semi-cylindrical, 2–7 cm long and 2 mm wide, and shortly mucronate at the apex, green or reddish color (Figure 8); the flowers are small, sessile, axillary to the leaves, and conspicuously bibracteolate; it has five perianth segments, becoming hardened in fruit. It has wide Eurasian and African distribution and typically grows in correspondence with salt marshes and on soils with high organic content. This species colonizes maritime and inland salt marshes, usually grows in halo-nitrophilous and termophilous vegetation, on soils rich in organic content, and forms floristically very poor (almost mono-specific) communities dominated by S. inermis (*Salsoletum sodae* Pignatti 1953; *Thero-Suaedion splendentis* Br.-Bl. in Br.-Bl. & al. 1952). According to the 92/43/EEC "Habitat" Directive, this vegetation falls within the habitat of community interest 1310—"*Salicornia* and other annuals colonizing mud and sand".

Figure 8. *Soda inermis*.

3.3.2. Food Use

S. inermis is appreciated as a vegetable, especially in central Italy, mainly in the Latium region, where it is known as "agretti" because of the slightly bitter taste [92]. It is commonly sautéed with garlic and olive oil and served as a side dish. It can be used in salads, but another more common use is lightly steamed and dressed with lemon juice, olive oil, sea salt and fresh cracked black pepper.

In Salento, it is not collected and used as an edible species, whereas it is used in various localities of Gargano where the young seedlings are eaten both raw and boiled and dressed with vinegar and olive oil [67] (Figure 4; Table 1; Appendix A).

It is interesting to point out that the plant is used similarly, on a small scale, in Japan [93].

3.3.3. Domestication/Cultivation

S. inermis is cultivated and used in particular in Italy and in Spain (where it is known as "barilla"). This species can be cultivated in those salty lands where no other crops could give a good yield, or in those areas where irrigation is possible only with salty water. It is spread in south Europe, particularly in marginal areas near the coast [93,94]. *S. inermis* was cultivated also to reclaim brackish swamps [95].

Only for a few years has *S. inermis* been cultivated in Salento and other parts of Apulia. It should be sown in spring. The soil must be well worked and draining, exposed in a sunny position. The sowing must be done in March and the cultivation does not require fertilization. Germination is slow, it takes about a month for all the seeds to germinate and, between May and June, the young plants are ready to be harvested for food consumption.

3.3.4. Other Uses/Properties

In past times *S. inermis* had gained much attention due to its use as soda ash [93]. *S. inermis* is one of those plants that has lost economic importance as a consequence of new industrial processes. In fact, it was formerly highly demanded for the production of impure sodium carbonate that was used, e.g., for the production of soap and glass. This impure salt was called "barilla," hence the name of the plant. Barilla was obtained from the ashes of *S. inermis* and other halophilous plants able to produce sodium carbonate from the sodium chloride in the soil; it was also obtained from seaweeds. Barilla from *S. inermis* was preferred for the glass industry over that produced from other terrestrial plants because it was richer in potassium salts and therefore preferably employed in soap preparation. Nowadays it is very difficult to give an estimation of the former economic importance of *S. inermis* because soda is also obtained from the ashes of other plants. In fact, the Solvay process, which leads to the rather pure sodium carbonate used in all industrial processes, progressively replaced the production of barilla from plants since the beginning of the present century [95,96]. *S. inermis* is mentioned (as *Salsola soda*) by [97] among the Salento plants useful for obtaining sodium carbonate from the ashes (Appendix A).

S. inermis has also played a minor role in popular medicine: it was considered to be diuretic, aperitif and vermifuge [91]. This species can be used for bio-desalination of saline soils and as a companion plant with conventional crops [96,98].

3.4. *Cakile maritima Scop.*

3.4.1. Brief Description, Distribution and Ecology

Cakile maritima (sea rocket) is a succulent, annual species belonging to the Brassicaceae family, thriving along Mediterranean and European Atlantic sandy coasts, confined to maritime strandlines on sand or shingle, and associated fore dunes. The species is tolerant of salt spray and temporary seawater inundation. *C. maritima* is a succulent annual herb with prostrate or ascending stem, highly branched, fleshy leaves entire obovate or oblancelate-to-deeply pinnately lobed; its inflorescences are dense, many-flowered racemes, terminating the main stem and branches, with lilac to purple flowers (Figure 9). Together with *Salsola tragus*, it forms the typical vegetation of sandy and shingle beach drift lines (e.g., *Salsolo kali-Cakiletum maritimae* Costa & Mansanet 1981, corr. Rivas-Martínez et al. 1992; *Cakiletea maritimae* Tx. et Preising in Tx. ex Br.-Bl. et Tx. 1952 class). According to the 92/43/EEC "Habitat" Directive, this vegetation falls within the habitat of community interest 1210—"Annual vegetation of drift lines".

Figure 9. *Cakile maritima.*

3.4.2. Food Use

Harvested before flowering, the whole plant is boiled for about 15 min. Then drained and passed under hot water in order to eliminate the bitter part (this operation must be repeated 3/4 times). After it can be served with potatoes, stale bread softened in the same cooking water of the plant and then seasoned with olive oil (Margherita di Savoia sandy coast, Foggia).

C. maritima is used for food purposes in very few places in Salento (Figure 4; Table 1; Appendix A). The leaves, plump and with a spicy taste, can be added to raw salads and flavor cooked dishes [29].

3.4.3. Domestication/Cultivation

As a therophyte, *C. maritima* can only be multiplied generatively. No cultural protocols are known, nor if the seeds undergo a possible period of dormancy. However, it has been observed that, in conditions of salinity with NaCl equal to 75 mM, there is a significant delay in seed germination and that, in concentrations of NaCl equal to 100 mM, there is a greater vegetative development in the plant, with a greater number and size of leaves, greater water content and greater photosynthetic activity [99]. The response to the progressive salinization of coastal environments in *C. maritima* is immediate and positive, and for this reason it is proposed as a species to be included in agricultural systems as a companion crop or to increase its consumption for food purposes [100]. Several studies, especially from north Africa, have considered *C. maritima* as a promising species for domestication in the context of the biosaline agriculture approach relying on the domestication of halophytes [101,102].

3.4.4. Other Uses/Properties

C. maritima is a species of food interest for many Mediterranean populations and represents a promising species, owing its ecological plasticity and economic potential because of its ability to produce numerous secondary compounds and as an oilseed and energy crop. In recent years, its economic interest in its oilseeds and secondary metabolites, which are of therapeutic interest for humans, have been discovered. *C. maritima* has a high content of polyphenolic compounds that perform antioxidant activity and classify it among the species with antiallergenic, anti-inflammatory, anticancer, anticoagulant, antimicrobial, cardioprotective and vasodilatory properties [103–105]. In popular medicine *C. maritima* was considered an excellent anti-scurvy and anti-catarrhal, with diuretic and laxative properties, capable of counteracting jaundice and dropsy. In recent years, its microbial and antifungal activity has been evaluated, leading to industrial interest, as a substitute for the traditional antibiotics used for food preservation to which many bacterial and fungal species are now resistant [106].

Furthermore, it has been found that *C. maritima*, within 6 weeks, is able to reduce the phytotoxicity of phenanthrene, a Polycyclic Aromatic Hydrocarbon (PAH), by 75%; for this reason, it is thought to include *C. maritima* among the species to be used in phytoremediation interventions [107].

For all these reasons, the cultivation of this plant on salted marginal soil, in the context of the necessary development of bio-saline agriculture in the future, has been taken into consideration [102].

3.5. *Crithmum maritimum L.*

3.5.1. Brief Description, Distribution and Ecology

Marine fennel (*Crithmum maritimum*, Apiaceae), also known as crest marine, marine fennel, sea fennel, sampier and rock samphire, is a suffruticose chamaephyte, i.e., a perennial plant with woody basal stems and herbaceous upper branches, standing 20–60 cm tall and is glabrous, aromatic and pruinose. It has fleshy leaves, with lanceolate-linear leaflets and inflorescences in umbels with greenish-white flowers (Figure 10a). Its distribution area extents from the Mediterranean and Black Sea coasts, up to the Atlantic coast of Portugal and of south and south-west England, Wales and the Republic of Ireland [108]. This aromatic plant grows wild in rocky coastal environments, such as rock crevices, rocky shores and, sometimes, on shingle beaches. *C. maritimum* typically occurs in the halophytic and halotolerant, perennial plant communities which constitute the first vegetation belts on the rocky coasts in next proximity to the sea, directly exposed to the action of marine aerosol, wind and waves, and tolerating a high concentration of sodium chloride in the substrate (*Crithmo maritimi-Staticion* Molinier 1934; *Crithmo maritimi-Staticetea* Br.-Bl. in Br.-Bl. et al. 1952) [41,51]. According to the 92/43/EEC "Habitat" Directive, this vegetation falls within the habitat of community interest 1240—"Vegetated sea cliffs of the Mediterranean coasts with endemic *Limonium* spp.".

Figure 10. *Crithmum maritimum* (**a**); a typical side dish with sea fennel (**b**), consumed in Salento (Southern Apulia).

3.5.2. Food Use

Especially in the coastal villages of Salento, it is a tasty ingredient used to flavor fish-based dishes and for the preparation of many side dishes. The tender apical portions of the stems are harvested at the end of spring, peeled, blanched, drained and then seasoned with garlic and mint for long-term potting in oil or vinegar. In recent years, someone has experimented with freezing the freshly blanched plant, to be thawed and seasoned as a side dish (with oil, vinegar, garlic, mint and breadcrumbs (Figure 10b) or to be used as a flavoring [65,68]. In other parts of the Apulia region, fresh leaves are julienne cut, seasoned with oil and lemon (or vinegar) and consumed with salads; whole leaves can be used to make omelettes (Figure 4; Table 1; Appendix A).

Since ancient times, the fresh consumption of sea fennel has been recommended for its purifying, tonic, diuretic and purgative properties; the fruits of sea fennel were used in ancient times for reducing fermentation and intestinal spasms. Rich in vitamin C, sea fennel was also used as an anti-scurvy by sailors.

For promoting a full exploitation of this species, a new food product was obtained by drying sea fennel using different treatments (air-drying, microwave-drying, microwave-assisted air-drying and freeze-drying) [109]. Water activity, essential oil content, chlorophylls, surface color, coloring power and sensory evaluation were analyzed; the results indicated that microwaving and freeze-drying are optimal for preserving qualitative traits, including organoleptic properties, in dried sea fennel for food use. The culinary use of the sea fennel for several gastronomy products as a new spice-colorant has been also reported by [92].

3.5.3. Domestication/Cultivation

Marine fennel is cultivated in many areas across Europe for several economic and industrial purposes; in general, sea fennel is a strictly heliophile and therefore requires a planting pattern that ensures adequate inter-and intra-row spacing to maximize the leaf area exposed to the sun [106,108,110,111].

Despite this, species usually grow in close proximity to the seawater, salinities exceeding 50 mM NaCl were found to inhibit its germination [112,113]. According to [113], a useful approach to overcome the salt-induced seed dormancy observed in halophytes consists in the exogenous application of germination-promoting substances; in this way, nitrate, ammonium, and GA3 proved to significantly enhance seed germination of *C. maritimum* under salinities. Interestingly, red light application was also efficient for seed germination induction under salinity.

Propagation by softwood cuttings does not seem to be the appropriate method for mass propagation, as the mother plants of sea fennel provide a limited number of cuttings. In vitro culture techniques have been extensively used not only for rapid clonal propagation but also for the study of the salt tolerance mechanisms of many species; a first study on *C. maritimum* has been reported by [114].

In the Botanical Garden of the University of Salento, the multiplication of sea fennel has been carried out to be used for living collections and for interventions of environmental restoration on coastal habitats. The propagation by the vegetative method has not been successful, as the stem and the young branches have no rooting power, as said just above; on the other hand, the propagation by the generative method has provided an excellent result, certifying the germinative power around 70–75% and without germination-promoting substances. In this experiment, seeds were harvested between October and November, cleaned of impurities and stored in paper bags; then, they were sown in the second ten days of December on a substrate consisting of a mixture of peat and river sand in a 1:1 ratio, in honeycombed plateau placed in a cold tunnel. The emergence raised one month later, characterized by a slow scaling. In the plateau, the complete development of the root system took place after 5 months. Young plants tolerated well both transferring to larger pots and transplanting in the open field at the end of spring or in full autumn.

In the context of the use of sea fennel for the creation of green roofs, specific cultivation protocols have been provided. According to [115], sea fennel plants were satisfactorily grown on a soilless substrate, a mixture of grape marc compost, perlite and pumice, with a shallow depth of 15 cm. The better growth of sea fennel plants on a substrate consisting of pumice, perlite, compost, peat and zeolite, with a depth of 15 cm compared to a depth of 7.5 cm, is also reported by [116], while the same researchers found that the growth of plants was greater in the deficit irrigation treatment of 60% evapotranspiration (ETc) compared to the irrigation treatment of 30% ETc.

The use of *C. maritimum* in agricultural practices is an interesting possibility in areas where the growth of other species is limited. An evaluation of different agronomic protocols that have previously not been investigated for the cultivation of this species was proposed by [117].

3.5.4. Other Uses/Properties

In recent years, new scientific research has revived the interest in sea fennel, not only as a food plant due to its nutritional value, but also as a species of pharmacological interest [118,119]. *C. maritimum* is rich in phyto-compounds to which its numerous uses in medicine are owed. Roots, leaves and fruits are rich in several bioactive substances (essential oils, iodine, trace elements, beta carotene, proteins and mineral salts) with a wide range of uses as aromatic, medicinal, antimicrobial and insecticide. A large number of scientific studies show its effectiveness as an appetizer, tonic, digestive, carminative, diuretic, vermifuge and antimicrobial, as well as to treat kidney and heart disease [120–123]. Moreover, it has antiscorbutic properties, owing to the vitamin C content [124]. Recent studies have highlighted its properties as antioxidant, vasodilator, antibacterial, cytotoxic, cholinesterase inhibitory and anticancer [125–129]. Dehydrated and pulverized, it is an excellent flavoring and coloring agent for foods [92]. It was once used for the production of soda [110]. In temperate climates, the plant is used for ornamental decoration in rock gardens along the sea [129]. As a species that also tolerates strong insolation and long periods of drought, it is recommended in the construction of green roofs in the context of urban horticulture [115,116,130].

3.6. Reichardia picroides (L.) Roth

3.6.1. Brief Description, Distribution and Ecology

Reichardia picroides (Asteraceae) is a scapose hemicryptophyte, 20 to 40 cm long, with lush-green leaves spatulate to oblanceolate, entire to pennatopartite, gathered in a basal rosette; the branches of the flowering scape terminate with capitula of yellow ligulate flowers (Figure 11a,b). This species has a strictly Mediterranean distribution; it is a salt-tolerant species, with effective adaptation mechanism against saline conditions [131]. The var. *maritima* (Boiss) Fiori, no longer recognized as a valid taxon, with fleshy leaves and typically growing in coastal rocky habitats, has been considered as a diagnostic species of the *Crithmo-Staticetea* class [41,50]. According to the 92/43/EEC "Habitat" Directive, this vegetation falls within the habitat of community interest 1240—"Vegetated sea cliffs of the Mediterranean coasts with endemic *Limonium* spp.".

Figure 11. *Reichardia picroides*: flower heads (**a**) and basal rosette (**b**).

3.6.2. Food Use

R. picroides is among the best-known edible species in Italy, present both in coastal and inland habitats [132]. It rarely forms extensive and dense populations, and this makes its research for the collection more careful. In Puglia, it is used as only ingredient in soups. Generally, it is boiled and then dressed with olive oil, or sautéed in oil with onion and then enriched with cheese and other ingredients. In Salento, added to other wild herbs, it makes the "fojemmìsche" even more pleasant; in Martano (LE), it is referred to with the dialectal name "cannazzicula" to indicate its ability to delight the palate [65]. It has a sweet and delicate taste and can also be eaten raw in salads (Gargano) (Figure 4; Table 1; Appendix A). From November to July, the plant is always available; the basal rosette is collected, and the internal leaves ore those more edible. The still-closed flower heads can be used to flavor and decorate dishes or to add to salads. When the plant is old, it has a pungent taste, hence the epithet "picroides".

In the Umbria Region (Central Italy), its roots are roasted as a coffee substitute [133].

3.6.3. Domestication/Cultivation

This species is widely used as a wild vegetable and there is a hypothesis about some attempt at cultivation in the past [134]. Nevertheless, at present, no multiplication experiences by agronomic cultivation of *R. picroides* are reported.

Rich in phenolic compounds and tocopherols, the species can be cultivated under unfavorable conditions that can improve the bioactive properties through the increased phytochemicals content [24,131]). In a recent study [135], *R. picroides* was grown for four or six weeks under a greenhouse in a floating system. In order to improve the nutraceutical quality of the tissues, the plants were exposed to the following NaCl concentrations: 1.7 (control), 25, 50 and 100 mM. The results showed that a 4-week growing period in a floating system with 50 mM NaCl increased the content of bioactive molecules without affecting the fresh yield. After six weeks of cultivation, despite a decrease in biomass production as compared with the control, the leaves of salt-treated plants contained higher levels of bioactive molecules along with lower amounts of nitrate ion.

3.6.4. Other Uses/Properties

According to the popular pharmacopoeia of many countries, *R. picroides* has hypoglycaemic, diuretic, depurative, galactogenic and tonic properties [136,137]. In Sardinia, it was even used as a popular treatment against heart diseases such as angina pectoris [138]. Recent studies have confirmed that *R. picroides* extract has high antioxidant activities (at doses lower than 250 mg/kg) thanks to a high content of phenols and can have many therapeutic applications [30,139]. The ability of *R. picroides* extract to inhibit postprandial platelet aggregation in vitro has also been demonstrated [140].

R. picroides is also used for feeding rabbits in some areas of the Basilicata region (southern Italy) [141].

3.7. *Silene vulgaris (Moench) Garcke subsp. tenoreana (Colla) Soldano & F. Conti*

3.7.1. Brief Description, Distribution and Ecology

Silene vulgaris (Caryophyllaceae) is native to Eurasia. In Italy, it is present with six subspecies [69]. The subspecies differ from each other in their vegetative habit, leaf shape, leaf size, etc. The subspecies vulgaris is the most common and it grows in all Italian territory except in arid zones, and ranges from 0 to 1500 m asl (rarely up to 2400 m asl) [70]. It is a glabrous or poorly pubescent perennial herb, 30–70 cm high, generally with erect habit but sometimes prostrate; its leaves always sessile and opposed, the largest ones (the median) 12–18 mm × 40–60 mm are linear-lanceolate acute but not pointed; it has 3–9 flowers per plant, gathered in bunches, pendent on flexuous peduncles 5–15 cm long; it has an ovoid calyx (twice longer than wide), much wider than capsule, therefore apparently inflated around them; it has five white or lightly rosy petals with nails as long as the calyx; its anthers and style are purple; its capsule three times longer than the carpophore [70].

S. vulgaris subsp. *tenoreana* is similar to the related *S. vulgaris* subsp. *vulgaris*, but with a woody stem at the base and narrower, fleshy leaves. This taxon has an eastern Mediterranean distribution, preferentially growing in coastal areas, on rocky substrates (Figure 12), but also inland thanks to its great adaptability. It frequently appears, as "companion" species, in the communities of the *Crithmo-Staticetea* [41]. According to the 92/43/EEC "Habitat" Directive, this vegetation falls within the habitat of community interest 1240—"Vegetated sea cliffs of the Mediterranean coasts with endemic *Limonium* spp.". A recent study carried out to explain the current distribution of *S. vulgaris* on the European territory has shown that the species had a post-glacial distribution towards the North Pole ice cap from the refugia of southern Europe, contrary to the hypothesis that it has undergone a westward migration during agricultural expansion, such as most of the wild herbs that accompanied agricultural crops [142].

Figure 12. *Silene vulgaris* subsp. *tenoreana*.

3.7.2. Food Use

The tender leaves and the apical parts of the young stems are harvested, which are already abundant after the first autumn rains and in the spring. In Salento, numerous recipes mention it as a single ingredient (risotto with "strigoli", green pasta with "strigoli", sautéed "strigoli for filling focaccia", omelette with "strigoli", salads) [29], but it is also an ingredient in the so called "fojemmìsche". In the area of Monopoli (Bari), the leaves are cooked as part of an omelette with eggs and cheese (Figure 4; Table 1; Appendix A).

It is also culinarily used in other regions of Italy: the leaves are cooked and stewed in Amalfi coast (Campania Region) and the same use is reported in Tuscany [136], in the Marche region [143] and in Sicily [144]. In Veneto [145], its leaves are added to vegetable soups.

The subsp. *vulgaris* is also widely used in Italy as a vegetable and consumed in similar ways to subsp. *tenoreana*. In several localities of southern Italy, *S. vulgaris* subsp. *vulgaris* (named "culicid" in Apulia) are used for their more tender leaves which are fried with olive oil and eggs for preparing a sort of omelette. Shoots are in great demand by gourmets who use them for preparing vegetable soups, rice soups, minestrone or risotti. This is also true in Austria, Switzerland and Germany where folk names of the species often contain elements like "kohl" (cabbage) or "spinat" (spinach) [146,147].

3.7.3. Domestication/Cultivation

It is a rustic and very vigorous plant; it grows very well on any type of soil; exposure should be sunny. Generative propagation is possible. The seeds should be collected between June and July and selected, cleaned and stored in paper bags. Among the propagation techniques used at the Botanical Garden of Lecce, direct sowing on loose substrate (seed should be scattered sparsely by broadcasting or in rows 30–35 cm apart) enriched with organic fertilizer was adopted.

The case of *S. vulgaris* is an example of domestication processes from use to cultivation which is characteristic for many other vegetable plants in Italy [93,148] and which continues at the present time.

3.7.4. Other Uses/Properties

S. vulgaris has played a minor role in popular medicine: its rhizome contains saponin, alkaloids and tannins and it is considered to be a depurative of blood and anti-anaemic; its cultivation for producing those substances was attempted in southern Kazakhstan [93].

S. vulgaris is rich in mineral salts, and is therefore considered an excellent tonic [60,61]. This species is capable of colonizing contaminated and, because of its capacity to retain mercury and heavy metals, could be used in phytoremediation technologies [149,150].

3.8. Allium commutatum Guss

3.8.1. Brief Description, Distribution and Ecology

Allium commutatum (Amaryllidaceae), commonly named maritime wild leek, is a taxonomic entity typical of coastal rocky environments with central-eastern Mediterranean distribution. It has bulbous geophytes (bulbs are halo-tolerant) up to 100 cm tall and more, with showy spherical inflorescences (Figure 13a). It is usually found in open, rocky and/or stony places near the sea (Figure 13b), but also in fallow or abandoned fields and marginal areas. It grows from sea level to about 300 m asl and flowers from the end of June until the end of July [151]. In Italy, it is very common in coastal areas in the south (including Sardinia and Sicily), with the northern limit at the Tuscan archipelago (Tyrrhenian) and Marche (Adriatic). It is diagnostic species of the *Crithmo-Staticetea* class [41], that is, according to the 92/43/EEC "Habitat" Directive, habitat of community interest 1240—"Vegetated sea cliffs of the Mediterranean coasts with endemic *Limonium* spp.".

Figure 13. *Allium commutatum* inflorescence (**a**); *A. commutatum* community in the winter season (**b**).

3.8.2. Food Use

A. commutatum is a primary wild relative of and potential gene donor to leek (*A. porrum* L.) and a tertiary wild relative of and potential gene donor to a number of other crops in the *Allium* group, including onion (*A. cepa*), Welsh onion (*A. fistulosum*), garlic (*A. sativum*) and chives (*A. schoenoprasum*) [152,153].

As mentioned above, maritime wild leek is quite frequent on the Apulian coasts, especially on rocky coasts. As many other *Allium* species found in the inland area (*A. sativum* L., *A. vineale* L., *A. ampeloprasum* L., *A sphaerocephalon* L., *A. roseum* L., etc.), maritime wild leek is edible although not much used, probably because other wild and cultivated species are already abundantly available in the area. In Mediterranean and southern Italian cuisines, garlic is the main flavoring ingredient of numerous dishes: meat stews; a flavor enhancer for fish and grilled vegetables; greens and vegetables seasoned with mint and garlic before being preserved in oil or vinegar; marinades with breadcrumbs, vinegar, saffron, mint and garlic; oil flavored with garlic cloves [29,60,65] (Figure 4, Table 1).

The most commonly used parts of the plant are the bulbs, usually consumed raw or cooked. In the Gargano area, they is consumed raw in mixed salads or roasted and dressed with olive oil [66]; moreover, young leaves are used to flavor salads and soups and young seedlings are harvested and roasted. It is also used in sautéed foods, which are the basis of sauces, to season "friselle" (Monopoli, Bari) (Appendix A).

In traditional Sicilian cooking, a special dish, in the winter, is represented by the fried bulbs (the so called "purrietti") [154].

The consumption of the bulbs, eaten with bread, has been reported for various locations in eastern Mediterranean [155].

3.8.3. Domestication/Cultivation

A. commutatum can be propagated both by seeds and generatively. The ripe inflorescences, with the capsules containing mature seeds, are harvested between June and July; the sowing is to be done between October and November on a loose and draining substratum. Propagation by vegetative means involves the detachment of the largest bulbils from the main bulb and their transplantation into deep soil at the end of autumn.

Research on shoot and root regeneration from callus tissue of *A. commutatum* is reported by [156].

3.8.4. Other Uses/Properties

Like other *Allium* species, *A. commutatum* has numerous therapeutic properties. It can be used raw, in decoction or infusion for intestinal infections, for respiratory system diseases as an expectorant and anti-catarrhal, antibacterial, anthelmintic, hypotensive, antirheumatic and anti-inflammatory. Garlic poultices can be applied on inflamed or burned skin, on limbs and parts affected by bursitis, rheumatism and other strong inflammations of the tissues. For the populations settled between the Tigris and the Euphrates, a panacea for all ills was the decoction prepared with wine, garlic and calamus seeds. In addition to the nutraceutical and medicinal value, garlic was attributed a poisonous power against demons, evil spirits, the evil eye and witches [29,60,65].

Recent studies have evaluated the chemical profile of *A. commutatum* obtaining an alliin content of 31.5 mg/g in ethanolic extracts of bulbs (BE) and 38.8 mg/g in ethanolic extracts of aerial parts (APE); in the latter, quercetin (38.5 mg/g) and luteolin (31.8 mg/g) were also present. The bulbs and leaves are a precious source of organic compounds that fight obesity and perform antioxidant activities [157].

In other species of the *Allium* genus, studies of organic compounds confirm antioxidant, antiseptic and anticancer activities thanks to the high content of reduced glutathione, flavonoids, soluble proteins, vitamin C, carotenoids, chlorophylls a and b and malonyldialdehyde [158,159].

3.9. *Beta vulgaris L. subsp. maritima (L.) Arcang*

3.9.1. Brief Description, Distribution and Ecology

Beta vulgaris subsp. *maritima* (sea beet or wild beet) is a plant species belonging to the Amaranthaceae/Chenopodiaceae family (subfamily *Betoideae*). Scapose chamaephyte, with a dense basal rosette of leaves of spatulate shape (but variable in shape and size), is shiny green with inconspicuous greenish flowers in groups of two-to-five flowers in long spikes (Figure 14a). It has a wide distribution ranging from the Canary Islands in the west, northward along Europe's Atlantic coast and Baltic Seas, extending eastward through the Mediterranean basin up to the coasts of the Black Sea. It usually grows on stony and pebbly soils along the coasts but also, more rarely, inland on clays; it is commonly cultivated from sea level to about 600 m. This species has a large environmental adaptability to conditions such as high salinity and poor soil, which is related to its extreme genotypic and phenotypic variation [160].

Figure 14. *Beta vulgaris* subsp. *maritima* (**a**); a typical side dish with sea beet (**b**).

3.9.2. Food Use

Sea beet is one of the most appreciated wild herbs for the preparation of typical dishes. It is abundantly available not only in coastal environments but also in those immediately adjacent to. The tender leaf lamina is scarcely fibrous, and, after a very short blanching, it is pounded and added to flour mixture to make the so called "green pasta". In Salento, the sea beet is a fundamental ingredient for the "fojemmische", a typical dish which has as ingredients a mixture of 10–20 wild herbs, depending on the season. Used as a single ingredient in almost all the sited visited in the Apulia region (Figure 4; Table 1; Appendix A), the methods of preparation are simple: boiled and seasoned with oil and lemon (Figure 14b) or sautéed with onion and then enriched with pancetta, speck, ricotta, cheese, to be used as a filling for focaccia [61,161]. It is also commonly used in broad bean and legume soups, or in the preparation of omelettes. In the Gargano area, it is used to enrich a local dish called "pancotto", a dish in which the plant is boiled and with the addition of tomatoes and spices and the use of stale bread, was a typical dish in the tradition of the peasant people.

In Sicily, traditionally, in order to avoid constipation, it is recommended to consume the leaves of sea beet (locally called "gira"), which are also considered refreshing [162].

3.9.3. Domestication/Cultivation

Sea beet is considered the progenitor of all the crop varieties of beet [64,163–165]. Its domestication is traced back to the Egyptians and Babylonians. The Greeks and Romans made extensive use of it and selected a large number of varieties handed down to the present and all with an annual or biennial cycle.

Sea beet is a perennial species; propagation by the vegetative method is not known whilst its propagation by the generative method is possible. The seeds are difficult to release because they are enclosed inside hard glomeruli which are soaked for 24 h and then sown on a soft and draining substrate.

Sea beet survives in extreme conditions, from brackish marshes to cliffs continually beaten by the waves; it has the ability to tolerate high salt concentrations and aridity, unlike all the other cultivated varieties which instead show signs of suffering and morpho-functional changes such as growth retardation, wilting of the leaves, reduction of photosynthetic speed and stomatal conductance [166].

3.9.4. Other Uses/Properties

The food value of sea beet is to be attributed to the high content of fibers, mineral salts, vitamins (A, B1, B2, PP, C) and proteins that provide disinfectant, diuretic, refreshing, purifying and laxative and anti-anemic properties [60,65]. In the first century BC, the Greek physician and botanist Dioscorides proposed using sea beet against earache, against lice and dandruff and to apply boiled root on pustules and burns [163]. An essential oil was distilled from the aerial part of sea chard which revealed antioxidant, anticholinesterase, anti-tyrosinase and cytotoxic properties on the A549 cell line [167]. A recent study showed that the leaves' extracts are usefulness to prevent diabetes complications and have promising chemo-preventive properties [23].

3.10. *Capparis spinosa* L.

3.10.1. Brief Description, Distribution and Ecology

Capparis spinosa (Capparaceae), commonly known as flinders rose or caper bush, is a procumbent shrub with semi-prostrate (sometimes pendulous) unramified (or scarcely branching) branches, 40 to 80 cm long, green sometimes to reddish or yellowish. Leaf stipules may be formed into spines, granting it the name "spinosa". Its leaves are usually rounded to ovate; its flowers are solitary, with four white-pinkish obovate or rounded-ovate petals, and numerous stamens (Figure 15a,b). The fruits are like small cucumbers when immature, and when ripe they enlarge and turn purplish-red-brown with numerous seeds plunged into a white mucilaginous placenta. The species is native to dry regions of western and central Asia; at present, it grows naturally from the Atlantic coast of the Canary Islands and Morocco to the Black Sea and to the east side of the Caspian Sea; it is also spread in north Africa, Europe, west Asia, and Australia [168,169]. It requires average annual temperatures above 14 °C and average annual rainfall of no less than 200 mm; moreover, it resists strong winds [170]. The drought and salt tolerance of *C. spinosa* allows it to persist in a wide range of habitats, even on nutrient-poor, rocky and gravelly soils [170]. The species has a wide distribution in anthropogenic wall and rock crevices chasmophyte nitrophilous vegetation (*Cymbalario-Parietarietea diffusae* Oberdorfer 1969 = *Parietarietea judaicae* Oberdorfer 1977) of both coastal and inland areas.

3.10.2. Food Use

The consumption of the caper buds and fruits was well known already in Greek and Roman times, both as a condiment and for therapeutic use, and for this reason caper cultivation was widespread in the conquered territories. Commonly, the closed buds (capers) are harvested for use and consumption, but not too far into their development because when small they are crunchier and tastier. Capers can be kept for long periods, even 2–3 years if they are harvested, put immediately in a glass jar and covered with vinegar, or made into a pickle (that is covered with coarse salt and turned over for 4–5 days, then put into jars and covered with the same salt; Figure 15c). In some localities of Salento, the tender tips of the stems with the last two-to-three very tender leaves are also collected and transformed, with the same dialectal name that is given to the "chiapparara" plant; however, the removal of the apex implies the arrest of the elongation of the stem, and for this reason the producers avoid doing it. Capers are used as a condiment in raw salads, in caponata, with fish, in fillings and in sauces. Throughout the Apulia (Figure 4; Table 1; Appendix A), they are essential ingredients of the characteristic "rustic pizza", stuffed with onions, black olives, tomatoes and capers [65]. The bibliographic sources do not refer to the use of freshly picked capers, but only after maceration in vinegar or pickle, probably due to the strong bitter taste of the raw buds.

Figure 15. *Capparis spinosa* (**a**) and a detail of the flower (**b**); caper buds and fruits in salt (**c**).

The plant starts producing flower buds in the month of June and continues until the month of August by lengthening the stems, on which ripening fruit, fruit setting and new flowers successively appear. The aerial part loses its leaves and dries between October–November; the growing season re-starts in February–March.

3.10.3. Domestication/Cultivation

C. spinosa has always been considered of minor importance among the plants of agricultural interest. However, on the island of Pantelleria and in the Aeolian archipelago (and, in particular, on the island of Salina), the caper has reached an increasing economic importance, which has led to the developing of specialized cultivation [171,172]. It is also widely diffused in cultivated form, especially in Southern Europe (Italy, Spain and Greece), in North Africa (Tunisia, Morocco and Egypt) and in the Middle East (Syria and Turkey) [173]. In Linosa (Sicily), island cultivation practices provide rooted cuttings placed in holes 30 cm deep with blond peat to increase soil water-holding capacity. During the first year of growth, five rescue irrigations were carried out in summer to encourage establishment of the young plantings. Pruning was carried out at the end of each year during the autumn–winter period (November–December) by cutting branches to approximately 6–10 cm from the base (long pruning). Subsequently, three-to-four lava stones were placed around the plantings to protect them from the wind and to limit water loss from evaporation; crop care included manual weeding five times and hoeing three times [174]. Additional information on the cultivation of capers in the Pantelleria and Pelagie archipelagos is reported by [175].

C. spinosa is a dominant plant element on the limestone cliffs of the Apulian coast. It adapts very easily to harsh environments, from the coastal cliffs to the urban walls and monuments. The root system of *C. spinosa* creeps into the rock fissures and develops to make up 62.5% of the total plant biomass already 4–5 months after germination; the cortical layers of the roots, both fictive and fibrous, are able to store large quantities of water and to overcome water crises [176].

Its seeds have low germinative power and, for this reason, various pre-germinative treatments are used in generative propagation: mechanical scarification of the integuments, cold stratification, immersion of the seeds in a 0.2% H_2SO_4 solution and seed treatment with gibberellins (G4+7 and GA3). The vegetative propagation consists of the realization of woody or semi-woody cuttings, 15–30 cm long, with a diameter of 1–2.5 cm, to be taken between February and March, then stratified in sand and placed in a cold box at an ambient temperature of 3–4 °C [169,177].

Propagation experiments carried out in the Botanical Garden of Lecce have shown a good germination percentage (75–80%) for seeds of *C. spinosa* collected in Marina Serra di Tricase (LE) in the month of October and sown in the second ten days of December in an alveolar plateau (kept in a cold tunnel), on a substrate made up of a mix of peat and agri-perlite in a 1:1 ratio. No treatments were carried out to accelerate or improve germination, which began 40 days after sowing, continuing until the month of May.

3.10.4. Other Uses/Properties

C. spinosa has a long history as an archaeophyte. This species has been used in medicine since ancient times as testified by the Bible and the writings of Hippocrates, Aristotle and Pliny the Elder. Its numerous pharmacological effects are due to the richness of secondary metabolites. Traditional medicines have handed down different uses: in Iran, traditional medicine recommended the bark of the roots and fruits against malaria, hemorrhoids and as a diuretic; in Pakistan, the plant and roots were the parts used for the preparation of decoctions or infusions with purifying and analgesic power to relieve rheumatic pains, toothaches and coughs; in China, it was used against dropsy, anemia and stomach pain; in Morocco, it is recommended for conjunctivitis, for the expulsion of kidney stones, for diabetes and for gastrointestinal disorders [25,169,178]. Recent scientific studies attest to the presence of very important phytochemicals such as phenols, flavonoids, tocopherols, terpenes which attribute important therapeutic effects to the caper: antihypertensive, antiobesity, anti-inflammatory, hypoglycemic, antibiotic, anti-allergic and antihistamine, hepatoprotective and antitumor. Almost all parts of the caper plant contain high quantities of glucosinolates (84–89%). Glucocapperin (methyl glucosinolate) is present above all in the buds and shoots and glucocleomine is present in the seeds and leaves. In the lipid fraction of the seeds, a high content of linoleic and oleic acid, sterols and tocopherols, aliphatic and triterpenic alcohol was detected [179–183].

No acute or chronic toxicity has been detected for the use of *C. spinosa* and/or its extracts [179,184–186].

The people of Israel produced wine from flower buds, and one of the uses of this wine was for preparing incense for the Temple [187].

C. spinosa is used as a fodder and ornamental plant, too [186].

This plant, thanks to its drought-resistance and strong root system, has high potential against desertification and soil erosion [169].

4. Conclusions

Coastal areas have an important ecological value, and are among the most threatened environments, both in the Mediterranean region and worldwide. In fact, especially in recent decades, they are undergoing rapid anthropogenic development. Increasing human pressure (e.g., urbanization, exploitation of natural resources, plant invasion) is causing degradation of coastal areas, along with the reduction, fragmentation and isolation of their habitats [57,188]. Land claims, agricultural intensification and hydrological modifications are the main drivers of changes [189–192]. Marine and coastal ecosystems provide a wide range of services to human society including supporting, regulating, cultural and provisioning services; among the provisioning services, the provisioning of cultivated crops ("nutrition") and of genetic materials ("materials"), especially of genetic resources for new crops, are included [193–196].

The changing climate by global warming and the increasing aridity, along with sea-level rising, contribute to the salinization of the hydromorphic soils, especially in coastal areas, and determining, among others, serious environmental hazards in agriculture with which Mediterranean policymakers and scientists are beginning to interact through integrated initiatives such as REstoration ACTions for the MEDiterranean (REACT4MED) (https://react4med.eu, accessed on 30 November 2022). In this framework, and in view of the subtropical conditions to which many territories in the Mediterranean region (and, among these, many areas in southern Italy) are approaching due to climate change, halophytes become a valid alternative to conventional vegetable crops, and a strong point for agricultural reconversion and regeneration programs, as they offer the possibility of selecting those more tolerant to dry farming and saline stress, and those which under stress conditions produce greater amounts of secondary metabolites useful for human health [86,167,197]. Focusing on the Apulia region, in Salento, the problem of salinization of groundwater is becoming increasingly serious, especially in coastal areas due to the ingression of sea water. Therefore, halophytes may be used as both associated crops and for human consumption. Moreover, the growing halophytes can also be used in phytoremediation processes of soils polluted by heavy metals or polycyclic aromatic hydrocarbons.

In the framework of the Integrated Coastal Zone Management (ICZM) [198], the creation of buffer zones surrounding protected and sensitive coastal sites, such as those in spatial contact to agricultural areas subject to intensive exploitation, may contribute to both mitigate the effects of agricultural practices on coastal environments and recovery of marginal areas [199]; the latter may be abandoned lands due to soil salinization, which could be subject to dry farming with halophyte cash crops. This practice, apart from preventing possible conflicts with stakeholders (e.g., farmers) standing in or around the protected areas, could also mitigate the effects of gathering wild halophyte species; in fact, as highlighted in the Results, many of the considered species in the wild fall within vegetation types that are protected by the EU Directive as habitat of community interest and, in those places where gathering occurs extensively, natural populations, along with habitat structure, may be put at risk.

Among the most recently interesting cases, *Salicornia* sp. pl. is a good candidate for dryland farming in the presence of salinized soils, and also for reclamation of barren lands and salt flats; to be an economically viable cash crop, this emerging crop should ensure high-yield production [77,200]. Current knowledge suggests that sea fennel also has good potential as an emerging cash crop, even in the context of a saline agriculture regime; nevertheless, this halophyte plant is underutilized for commercial cultivation, possibly by the lack of consumer demand [109,129].

For a full exploitation of these species, further studies on domestication and cultivation practices, as well as on possible product transformation, are needed. Moreover, an overall promotion campaign of these products and their nutritional and healthy virtues should be planned, also in terms of new food products.

Author Contributions: Conceptualization, V.T., R.A. and G.L.; methodology, R.A., V.T. and G.L.; bibliographic research, R.A., V.T., G.L., P.D., M.U. and E.V.P.; field investigation, R.A., V.T., G.L., E.V.P. and G.A.; writing—original draft preparation, V.T., R.A., G.L., P.D., E.V.P. and M.U.; writing—review and editing, V.T., R.A. and G.L.; visualization, V.T. and R.A.; supervision, V.T. All authors have read and agreed to the published version of the manuscript.

Funding: This research received no external funding.

Data Availability Statement: Not applicable.

Acknowledgments: The authors wish to thank Domenico Nardone and Nunzia Maria Di Tonno, Gina Negro and Fabiana Renzo, Francesco Chetta, and Patrizia Pagliara, for the precious information provided. We also thank the two anonymous reviewers for their useful comments and constructive suggestions.

Conflicts of Interest: The authors declare no conflict of interest.

Appendix A

Table A1. a–c: Distribution, across the Apulia region, of the use (food use; sale; domestication; cultivation) of the halophytes and halotolerant species treated in this contribution, according to our field observations and the literature data.

(a)

	loc name 1	loc name 2	loc name 3	loc name 4	loc name 5	loc name 6	loc name 7	loc name 8	loc name 9	loc name 10	loc name 11	loc name 12	loc name 13
	Isole Tremiti	Lesina	Apricena	Sannicandro Garganico	Cagnano	Ischitella	Vico del Gargano	Vieste	Peschici	S. Giovanni Rotondo	Manfredonia	Mattinata	Margherita di Savoia
Salicornia patula		up; cu; hs	up; cu	up; cu; hs	up; cu; hs	up; cu;	up; cu; hs	up	up; cu; hs	up; cu;	up		up; cu
Salicornia emerici		up; cu; hs	up; cu; hs	up; cu; hs	up; cu; hs	up; cu;	up; cu; hs	up	up; cu; hs	up; cu;	up		
Salicornia dolichostachya													
Salicornia perennis													up; cu
Salicornia fruticosa													
Arthrocaulon macrostachyum	up; cu; hs												up; cu
Soda inermis													
Cakile maritima													up
Crithmum maritimum								up; cu					
Reichardia picroides		up; cu	up; cu	up; cu	up; cu	up; cu	up; cu	up; cu	up; cu	up; cu	up; cu	up; cu	
Silene vulgaris subsp. *tenoreana*													
Allium commutatum							up; cu	up; cu	up; cu				
Beta vulgaris subsp. *maritima*								up; cu	up; cu			up; cu	up; cu
Capparis spinosa	up; cu; hs	up; cu; hs	up; cu; hs	up; cu; hs	up; cu; hs	up; cu; hs	up; cu; hs	up; cu; hs	up; cu; hs	up; cu; hs; cul	up; cu; hs	up; cu; hs; cul	

(b)

	loc name 14	loc name 15	loc name 16	loc name 17	loc name 18	loc name 19	loc name 20	loc name 21	loc name 22	loc name 23	loc name 24
	Mola di Bari	Monopoli	Carovigno	Brindisi	Casalabate	San Cataldo	Torre dell'Orso	Otranto	Santa Cesarea	Tricase	S. Maria di Leuca
Salicornia patula											
Salicornia emerici											
Salicornia dolichostachya				up			up				
Salicornia perennis				up; cu		up; cu					

Table A1. *Cont.*

Salicornia fruticosa								up; cu					
Arthrocaulon macrostachyum	up; cu; hs					up; cu; hs			up; cu				
Soda inermis						up		up					
Cakile maritima													
Crithmum maritimum	up; cu		up; cu	up; cu		up; cu	up; cu	up; cu	up; cu	up; cu	up; cu	up; cu	up; cu
Reichardia picroides				up; cu		up; cu		up; cu	up; cu	up; cu	up; cu	up; cu	up; cu
Silene vulgaris subsp. *tenoreana*			up	up; cu		up; cu			up; cu				
Allium commutatum			up; cu	up; cu		up; cu			up; cu	up; cu	up; cu	up; cu	up; cu
Beta vulgaris subsp. *maritima*	up; cu; hs		up; cu; hs	up; cu		up; cu	up; cu	up; cu	up; cu	up; cu	up; cu	up; cu	up; cu
Capparis spinosa			up; cu; hs	up; cu		up; cu			up; cu	up; cu	up; cu	up; cu	up; cu
(c)													
	loc name 25	loc name 26	loc name 27		loc name 28		loc name 29	loc name 30	loc name 31	loc name 32	loc name 33	loc name 34	
	Pescoluse	Torre Suda	Gallipoli		S. Maria al Bagno		S. Isidoro	Porto Cesareo	Salina di Monaci	San Pietro in Bevagna	S.giorgio jonico	Taranto (la Vela)	
Salicornia patula							up						
Salicornia emerici							up					up; cu	
Salicornia dolichostachya							up			up		up; cu	
Salicornia perennis							up		up; cu			up; cu	
Salicornia fruticosa							up						
Arthrocaulon macrostachyum					up					up		up; cu	
Soda inermis										up			
Cakile maritima										up			
Crithmum maritimum	up; cu	up; cu	up; cu		up; cu		up; cu	up; cu	up; cu	up; cu			
Reichardia picroides		up; cu	up; cu		up; cu			up; cu		up; cu			
Silene vulgaris subsp. *tenoreana*									up	up			
Allium commutatum		up; cu	up; cu		up; cu			up; cu		up; cu	up; cu		
Beta vulgaris subsp. *maritima*	up; cu	up; cu	up; cu		up; cu		up; cu	up; cu	up; cu	up; cu			
Capparis spinosa		up; cu; hs	up; cu		up; cu			up; cu			up; cu		

up = used in the past; cu = currently used; hs = harvested for sale; dom = domestication attempts; cul = cultivation attempts.

Appendix B

Type of ethnobotanic sheet used in the interviews with people residing in the investigated areas.

ETHNOBOTANIC FORM

PLANT
Family
Scientific name
Common name
Habitat
Information on vegetation type, soil, degree of anthropic disturbance

---

Used part
Collection period
Collection site
Municipality
Geographic coordinates

USE
Human feeding Animal feeding Folk medicine
Cosmetics Craft
Other uses

Used at present?
Knowledge about its past use in the local community/family

PREPARATION, RECIPES AND OTHER INFORMATION

ATTACHMENTS
Photo Video Audio

References

1. Shahid, S.A.; Zama, M.; Heng, L. Soil Salinity: Historical Perspectives and a World Overview of the Problem. In *Guideline for Salinity Assessment, Mitigation and Adaptation Using Nuclear and Related Techniques*; Zaman, M., Shahid, S.A., Heng, L., Eds.; Springer International Publishing: Cham, Switzerland, 2018; pp. 43–53. [CrossRef]
2. Slama, I.; Abdelly, C.; Bouchereau, A.; Flowers, T.; Savouré, A. Diversity, distribution and roles of osmoprotective compounds accumulated in halophytes under abiotic stress. *Ann. Bot.* **2015**, *115*, 433–447. [CrossRef] [PubMed]
3. Panta, S.; Flowers, T.; Lane, P.; Doyle, R.; Haros, G.; Shabala, S. Halophyte agriculture: Success stories. *Environ. Exp. Bot.* **2014**, *107*, 71–83. [CrossRef]
4. Perrino, E.V.; Valerio, F.; Jallali, S.; Trani, A.; Mezzapesa, G.N. Ecological and Biological Properties of *Satureja cuneifolia* Ten. and Thymus spinulosus Ten.: Two Wild Officinal Species of Conservation Concern in Apulia (Italy). A Preliminary Survey. *Plants* **2021**, *10*, 1952. [CrossRef] [PubMed]
5. Valerio, F.; Mezzapesa, G.N.; Ghannouchi, A.; Mondelli, D.; Logrieco, A.F.; Perrino, E.V. Characterization and antimicrobial properties of essential oils from four wild taxa of Lamiaceae family growing in Apulia. *Agronomy* **2021**, *11*, 1431. [CrossRef]
6. Casella, F.; Vurro, M.; Valerio, F.; Perrino, E.V.; Mezzapesa, G.; Boari, A. Phytotoxic effects of essential oils from six *Lamiaceae* species. *Agronomy* **2023**, *13*, 257. [CrossRef]
7. Flowers, T.J.; Colmer, T.D. Salinity tolerance in halophytes. *New Phyt.* **2008**, *179*, 954–963. [CrossRef]
8. Grigore, M.N.; Toma, C. Definition and classification of halophytes. In *Anatomical Adaptations of Halophytes*; Springer: Cham, Switzerland, 2017; pp. 3–28.
9. Ventura, Y.; Wuddineh, W.A.; Myrzabayeva, M.; Alikulov, Z.; Khozin-Goldberg, I.; Shpigel, M.; Samocha, T.M.; Sagi, M. Effect of seawater concentration on the productivity and nutritional value of annual Salicornia and perennial *Sarcocornia* halophytes as leafy vegetable crops. *Sci. Hortic.* **2011**, *128*, 189–196. [CrossRef]
10. Loconsole, D.; Cristiano, G.; De Lucia, B. Glassworts: From wild salt marsh species to sustainable edible crops. *Agriculture* **2019**, *9*, 14. [CrossRef]
11. Glenn, E.P.; Brown, J.J.; Blumwald, E. Salt Tolerance and Crop Potential of Halophytes. *Crit. Rev. Plant Sci.* **1999**, *18*, 227–255. [CrossRef]

12. Ksouri, R.; Megdiche, W.; Falleh, H.; Trabelsi, N.; Boulaaba, M.; Smaoui, A.; Abdelly, C. Influence of biological, environmental and technical factors on phenolic content and antioxidant activities of Tunisian halophytes. *Comptes Rendus Biol.* **2008**, *331*, 865–873. [CrossRef]
13. Benhammou, N.; Bekkara, F.A.; KadifkovaPanovska, T. Antioxidant activity of methanolic extracts and some bioactive compounds of *Atriplex halimus*. *Comptes Rendus Chim.* **2009**, *12*, 1259–1266. [CrossRef]
14. Mishra, A.; Patel, M.K.; Jha, B. Non-targeted metabolomics and scavenging activity of reactive oxygen species reveal the potential of *Salicornia brachiata* as a functional food. *J. Funct. Foods* **2015**, *13*, 21–31. [CrossRef]
15. Arya, S.S.; Devi, S.; Ram, K.; Kumar, S.; Kumar, N.; Mann, A.; Chand, G. Halophytes: The plants of therapeutic medicine. In *Ecophysiology, Abiotic Stress Responses and Utilization of Halophytes*; Hasanuzzaman, M., Nahar, K., Öztürk, M., Eds.; Springer: Singapore, 2019; pp. 271–287.
16. del Rosario Marinoni, L.; Zabala, J.M.; Taleisnik, E.L.; Schrauf, G.E.; Richard, G.A.; Tomas, P.A.; Giavedoni, J.A.; Pensiero, J.F. Wild halophytic species as forage sources: Key aspects for plant breeding. *Grass Forage Sci.* **2019**, *74*, 321–344. [CrossRef]
17. Castañeda-Loaiza, V.; Oliveira, M.; Santos, T.; Schüler, L.; Lima, A.R.; Gama, F. Wild vs cultivated halophytes: Nutritional and functional differences. *Food Chem.* **2020**, *333*, 127536. [CrossRef] [PubMed]
18. Joshi, A.; Kanthaliya, B.; Arora, J. Halophytes: The nonconventional crops as source of biofuel production. In *Handbook of halophytes: From Molecules to Ecosystems towards Biosaline Agriculture*; Grigore, M.N., Ed.; Springer: Cham, Switzerland, 2020; pp. 1–28.
19. Montesano, F.; Gattullo, C.; Parente, A.; Terzano, R.; Renna, M. Cultivation of potted sea fennel, an emerging mediterranean halophyte, using a renewable seaweed-based material as a peat substitute. *Agriculture* **2018**, *8*, 96. [CrossRef]
20. Karakas, S.; Cullu, M.A.; Dikilitas, M. Comparison of two halophyte species (*Salsola soda* and *Portulaca oleracea*) for salt removal potential under different soil salinity conditions. *Turk. J. Agric. For.* **2017**, *41*, 183–190. [CrossRef]
21. Ventura, Y.; Eshel, A.; Pasternak, D.; Sagi, M. The development of halophyte-based agriculture: Past and present. *Ann. Bot.* **2014**, *115*, 529–540. [CrossRef]
22. Urbano, M.; Tomaselli, V.; Bisignano, V.; Veronico, G.; Hammer, K.; Laghetti, G. Salicornia patula Duval-Jouve: From gathering of wild plants to some attempts of cultivation in Apulia region (southern Italy). *Genet. Resour. Crop Evol.* **2017**, *64*, 1465–1472. [CrossRef]
23. Bouchmaa, N.; Mrid, R.B.; Kabach, I.; Zouaoui, Z.; Karrouchi, K.; Chtibi, H.; Zyad, A.; Cacciola, F.; Nhiri, M. *Beta vulgaris* subsp. *maritima*: A Valuable Food with High Added Health Benefits. *Appl. Sci.* **2022**, *12*, 1866. [CrossRef]
24. Maggini, R.; Benvenuti, S.; Leoni, F.; Pardossi, A. Terracrepolo (*Reichardia picroides* (L.) Roth.): Wild food or new horticultural crop? *Sci. Hortic.* **2018**, *240*, 224–231. [CrossRef]
25. Rivera, D.; Inocencio, C.; Obón, C.; Alcarez, F. Review of Food and Medicinal Uses of *Capparis* L. subgenus *Capparis* (*Capparidaceae*). *Econ. Bot.* **2003**, *57*, 515–534. [CrossRef]
26. Rivera, D.; Obón, C.; Heinrich, M.; Inocencio, C.; Verde, A.; Fajardo, J. Gathered Mediterranean food plants—Ethnobotanical investigations and historical development. *Forum Nutr.* **2006**, *59*, 18–74.
27. Lavarra, P.; Angelini, P.; Augello, R.; Bianco, P.M.; Capogrossi, R.; Gennaio, R.; La Ghezza, V.; Marrese, M. *Il Sistema Carta Della Natura Della Regione Puglia*; Serie Rapporti: Ispra, Italy, 2014; p. 204.
28. Accogli, R. *Piante da Fibra e Piante Tintorie Spontanee Nell'economia del Salento (Puglia)*; Organo Ufficiale Associazione Geografi Italiani (AGEI)—Geotema Patron Editore: Bologna, Italy, 2009; Volume 35–36, pp. 99–102.
29. Ditonno, N.; Lamusta, S.; Nardone, D. *Sapori e Aromi da Piante e Frutti Spontanei della Puglia Peninsulare*; Amici della "A. De Leo": Brindisi, Italy, 1997; p. 794.
30. Savo, V.; Salomone, F.; Mattoni, E.; Tofani, D.; Caneva, G. Traditional Salads and Soups with Wild Plants as a Source of Antioxidants: A Comparative Chemical Analysis of Five Species Growing in Central Italy. *Evid.-Based Complement. Altern. Med.* **2019**, *5*, 6782472. [CrossRef]
31. Tava, A.; Biazzi, E.; Mella, M.; Doria, A.; Accogli, R.; Argentieri, M.P.; Avato, P. Triterpenic saponins from *Medicago marina* L. *Phytochemistry* **2019**, *174*, 112333. [CrossRef]
32. Shelef, O.; Stavi, I.; Zdruli, P.; Rachmilevitch, S. Land use change, a case study from southern Italy: General implications for agricultural subsidy policies. *Land Degrad. Dev.* **2015**, *27*, 868–870. [CrossRef]
33. Perrino, E.V.; Ladisa, G.; Calabrese, G. Flora and plant genetic resources of ancient olive groves of Apulia (southern Italy). *Genet. Resour. Crop Evol.* **2014**, *61*, 23–53. [CrossRef]
34. Wagensommer, R.P.; Fröhlich, T.; Fröhlich, M. First record of the southeast European species *Cerinthe retorta* Sibth. & Sm. (Boraginaceae) in Italy and considerations on its distribution and conservation status. *Acta Bot. Gall. Bot. Lett.* **2014**, *161*, 111–115. [CrossRef]
35. Di Pietro, R.; Wagensommer, R.P. A new *Sesleria juncifolia* association from south-eastern Italy and its position in the amphi-Adriatic biogeographical context. *Acta Bot. Croat.* **2014**, *73*, 171–207. [CrossRef]
36. Wagensommer, R.P.; Medagli, P.; Albano, A.; Peruzzi, L.; Bartolucci, F.; Villani, M.; Conti, F.; Passalacqua, N.G.; Alessandrini, A.; Barberis, G.; et al. Loci classici of the Italian endemic vascular plants described for Apulia. *Inf. Bot. Ital.* **2014**, *46*, 323–369.
37. Perrino, E.V.; Signorile, G.; Marvulli, M. A first checklist of the vascular flora of the "Polignano a Mare" coast (Apulia, southern Italy). *Nat. Croat.* **2013**, *22*, 295–318.

38. Corbetta, F. Lineamenti della vegetazione macrofitica dei Laghi di Lesina e di Varano [Outlines of the macrophytic vegetation of the Lesina and Varano lakes]. *Giorn. Bot. Ital.* **1970**, *104*, 165–191. [CrossRef]
39. Géhu, J.-M.; Costa, M.; Scoppola, A.; Biondi, E.; Marchiori, S.; Peris, J.B.; Frank, J.; Caniglia, G.; Veri, L. Essai synsystématique et synchorologique sur les végétations littorals italiennes dans un but conservatoire. I. Dunes et vasessalees [Synsystematic and synchorological analysis on Italian coastal vegetation for a conservatory purpose. I. Dunes and salt marshes]. *Doc. Phytosoc.* **1984**, *8*, 394–474.
40. Corbetta, F.; Gratani, L.; Moriconi, M.; Pirone, G. Lineamenti vegetazionali e caratterizzazione ecologica delle spiagge dell'arco jonico da Taranto alla foce del Sinni [Vegetation features and ecological characterization of the Ionian beaches from Taranto to the Sinni mouth]. *Coll. Phytosoc.* **1989**, *19*, 55–81.
41. Bartolo, G.; Brullo, S.; Signorello, P. La classe *Crithmo-Limonietea* nella penisola Italiana. *Colloq. phytosoc.* **1992**, *19*, 55–81.
42. Brullo, S.; De Marco, G. *Antyllidion barbae-jovis* new alliance of *Crithmo-Limonietea*. *Arch. Bot. Ital.* **1989**, *65*, 109–120.
43. Biondi, E.; Casavecchia, S.; Guerra, V. Analysis of vegetation diversity in relation to the geomorphological characteristics in the Salento coasts (Apulia-Italy). *Fitos* **2006**, *43*, 25–38.
44. Tomaselli, V.; Perrino, E.V.; Cimmarusti, G. Paludi Sfinale e Gusmay, due aree umide di rilevante interesse naturalistico nel Parco Nazionale del Gargano [Sfinale and Gusmay, two wetlands of significant naturalistic interest in the Gargano National Park]. *Inf. Bot. Ital.* **2008**, *40*, 183–192.
45. Biondi, E.; Casavecchia, S. The halophilous retro-dunegrasslands of the Italian Adriatic coastline. *Braun-Blanquetia* **2010**, *46*, 11–127.
46. Tomaselli, V.; Di Pietro, R.; Sciandrello, S. Plant communities structure and composition in three coastal wetlands in southern Apulia (Italy). *Biologia* **2011**, *66*, 1027–1043. [CrossRef]
47. Sciandrello, S.; Tomaselli, V. Coastal salt marshes plant communities of the *Salicornietea fruticosae* class in Apulia (Italy). *Biologia* **2014**, *69*, 53–69. [CrossRef]
48. Tomaselli, V.; Sciandrello, S. Contribution to the knowledge of the coastal vegetation of the Zone Umide dellaCapitanata (Apulia, Italy). *Plant Biosyst.* **2017**, *151*, 673–694. [CrossRef]
49. Veronico, G.; Sciandrello, S.; Medagli, P.; Tomaselli, V. Vegetation survey and plant landscape mapping of the SCI IT9140002 "LitoraleBrindisino" (Puglia, Southern Italy). *Plant Sociol.* **2017**, *54*, 89–106.
50. Tomaselli, V.; Terzi, M. Rocky coastal vegetation of the class *Crithmo-Staticetea* in the south-east of Italy. *Acta Bot. Croat.* **2019**, *78*, 46–56. [CrossRef]
51. Tomaselli, V.; Veronico, G.; Sciandrello, S.; Forte, L. Therophytic halophilous vegetation classification in South-Eastern Italy. *Phytocoenologia* **2020**, *50*, 187–209. [CrossRef]
52. Margiotta, B.; Colaprico, G.; Urbano, M.; Veronico, G.; Tommasi, F.; Tomaselli, V. Halophile wheatgrass *Thinopyrum elongatum* (Host) D.R. Dewey (Poaceae) in three Apulian coastal wetlands: Vegetation survey and genetic diversity. *Plant Biosyst.* **2020**, *156*, 1–15. [CrossRef]
53. Brullo, S.; Giusso Del Galdo, G.; Siracusa, G.; Spampinato, G. Considerazioni fitogeografiche sulla vegetazione psammofila dei litorali italiani. *Biogeographia* **2001**, *22*, 93–137.
54. Perrino, E.V.; Wagensommer, R.P. Aggiornamenti floristici per il Gargano (Puglia) con riferimento agli habitat della Direttiva 92/43/EEC. *Inform. Bot. Ital.* **2012**, *44*, 163–170.
55. Perrino, E.V.; Tomaselli, V.; Costa, R.; Pavone, P. Conservation status of habitats (Directive 92/43 EEC) of coastal and low hill belts in a Mediterranean biodiversity hot spot (Gargano-Italy). *Plant Biosyst.* **2013**, *147*, 1006–1028. [CrossRef]
56. Perrino, E.V.; Wagensommer, R.P.; Silletti, G.N.; Signorile, G.; Angiulli, F. Nuovi dati distributivi e relazione con la Direttiva 92/43/CEE di taxa critici pugliesi dalla Provincia di Bari. *Inform. Bot. Ital.* **2013**, *45*, 53–62.
57. Tomaselli, V.; Tenerelli, P.; Sciandrello, S. Mapping and quantifying habitat fragmentation in small coastal areas: A case study of three protected wetlands in Apulia (Italy). *Environ. Monit. Assess.* **2012**, *184*, 693–713. [CrossRef]
58. Adamo, M.; Tarantino, C.; Tomaselli, V.; Veronico, G.; Nagendra, H.; Blonda, P. Habitat mapping of coastal wetlands using expert knowledge and Earth observation data. *J. Appl. Ecol.* **2016**, *53*, 1521–1532. [CrossRef]
59. Gavish, Y.; O'Connell, J.; Marsh, C.J.; Tarantino, C.; Blonda, P.; Tomaselli, V.; Kunin, W.E. Comparing the performance of flat and hierarchical habitat/land-cover classification models in a NATURA 2000 site. *ISPRS J. Photogramm. Remote Sens.* **2018**, *136*, 1–12. [CrossRef]
60. Palombi, D. *Erbario Salentino*; Edizioni del Grifo: Lecce, Italy, 2005; p. 527.
61. Accogli, R.; Mele, C.; Minonne, F.; Medagli, P.; Marchiori, S. Utilizzo delle piante spontanee mangerecce nel Salento. 2° Convegno Nazionale "Piante Mediterranee". *Atti Convegno* **2007**, *I*, 520–528.
62. Leporatti, M.L.; Guarrera, P.M. Ethnobotanical remarks in Capitanata and Salento areas (Puglia Southern Italy). *Etnobiología* **2005**, *5*, 51–64.
63. Accogli, R.; Marchiori, S. Ricerche etnobotaniche nell'agro di Fasano (Brindisi Puglia). In Proceedings of the IV Convegno Nazionale "Piante Mediterranee. Le Potenzialità del Territorio e dell'Ambiente", Marina di Nova Siri (MT), Italy, 7–10 October 2009; p. 165.
64. Accogli, R.; Medagli, P.; Marchiori, S. Conservazione della biodiversità vegetale: Indagini e reperimento di ancestori spontanei. *Ital. J. Agric.* **2009**, *4*, 225–230.

65. Accogli, R.; Medagli, P. *Erbe Spontanee Salentine. Guida al Riconoscimento e all'uso Delle Piante Alimentari Tradizionali*; Edizioni Grifo: Lecce, Italy, 2014; p. 198, ISBN 9788898175642.
66. Biscotti, N.; Pieroni, A. The hidden Mediterranean diet: Wild vegetables traditionally gathered and consumed in the Gargano area, Apulia, SE Italy. *Acta Soc. Bot. Pol.* **2015**, *84*, 327–338. [CrossRef]
67. Biscotti, N.; Bonsanto, D.; Del Viscio, G. The traditional food use of wild vegetables in Apulia (Italy) in the light of Italian ethnobotanical literature. *Ital. Bot.* **2018**, *5*, 1–24. [CrossRef]
68. Nardone, D.; Ditonno, N.; Lamusta, S. *Fave e Favelle. Piante della Puglia Peninsulare Nelle voci Dialettali in Uso e di Tradizione. Aforismi. Modi di Dire. Farmaci. Cosmetici*; Centro Studi Salentini: Lecce, Italy, 2012; p. 587.
69. Bartolucci, F.; Peruzzi, L.; Galasso, G.; Albano, A.; Alessandrini, A.; Ardenghi, N.; Astuti, G.; Bacchetta, G.; Ballelli, S.; Banfi, E.; et al. An updated checklist of the vascular flora native to Italy. *Plant Bios.* **2018**, *152*, 179–303. [CrossRef]
70. Pignatti, S.; Guarino, R.; La Rosa, M. *Flora D'italia 2017–2019*, 2nd ed.; Edagricole–Edizioni Agricole di NewBusiness Media: Bologna, Italy, 2019; Volume 1–4.
71. The Angiosperm Phylogeny Group; Chase, M.W.; Christenhusz, M.J.M.; Fay, M.F.; Byng, J.W.; Judd, W.S.; Soltis, D.E.; Mabberley, D.J.; Sennikov, A.N.; Soltis, P.S.; et al. An update of the Angiosperm Phylogeny Group classification for the orders and families of flowering plants: APG IV. *Bot. J. Linn. Soc.* **2016**, *181*, 1–20. [CrossRef]
72. Kadereit, G.; Borsch, T.; Weising, K.; Freitag, H. Phylogeny of *Amaranthaceae* and *Chenopodiaceae* and the evolution of C4 photosynthesis. *Intern. J. Plant Sci.* **2003**, *164*, 959–986. [CrossRef]
73. Rufo, L.; Fuente, V.; Sánchez-Mata, D. Sarcocornia plant communities of the Iberian Peninsula and the Balearic Islands. *Phytocoenologia* **2016**, *46*, 383–396. [CrossRef]
74. Piirainen, M.; Liebisch, O.; Kadereit, G. Phylogeny, biogeography, systematics and taxonomy of Salicornioideae (Amaranthaceae/ Chenopodiaceae)—A cosmopolitan, highlyspecializedhygrohalophyte lineage dating back to the Oligocene. *Taxon* **2017**, *66*, 109–132. [CrossRef]
75. Kadereit, G.; Piirainen, M.; Lambinon, J.; Vanderpoorten, A. Cryptic taxa should have names: Reflections in the glasswort genus *Salicornia* (Amaranthaceae). *Taxon* **2012**, *61*, 1227–1239. [CrossRef]
76. De La Fuente, V.; Rufo, L.; Rodríguez, N.; Sánchez-Mata, D.; Franco, A.; Amils, R. A study of *Sarcocornia* A.J. Scott (*Chenopodiaceae*) from Western Mediterranean Europe. *Plant Biosyst.* **2016**, *150*, 343–356. [CrossRef]
77. Patel, S. Salicornia: Evaluating the halophytic extremophile as a food and a pharmaceutical candidate. *Biotech* **2016**, *6*, 104. [CrossRef]
78. Barreira, L.; Resek, E.; Rodrigues, M.J.; Rocha, M.I.; Pereira, H.; Bandarra, N.; Moreira da Silva, M.; Varela, J.; Custódio, L. Halophytes: Gourmet food with nutritional health benefits? *J. Food Comp. Anal.* **2017**, *59*, 35–42. [CrossRef]
79. Chevalier, A. Les Salicornes et leuremploi dans l'alimentation: Etude historique, botanique, economique. *Jour. D'agric. Trad. Bot. Appl.* **1922**, *2*, 697–785.
80. Kim, S.M. Quality Characteristics of Low-Salt Kimchi with Salt Replaced by *Salicornia herbacea* L. Powder. *J. Korean Soc. Food Cult.* **2013**, *28*, 674–683. [CrossRef]
81. Ventura, Y.; Sagi, M. Halophyte crop cultivation: The case for Salicornia and Sarcocornia. *Environ. Exp. Bot.* **2013**, *92*, 144–153. [CrossRef]
82. Gunning, D. *Cultivating Salicornia europaea (Marsh Samphire)*; Daithi O' Murchu Marine Research Station & University College Cork: Dublin, Ireland, 2016; pp. 1–50.
83. Limongelli, F.; Crupi, P.; Clodoveo, M.L.; Corbo, F.; Muraglia, M. Overview of the polyphenols in *Salicornia*: From recovery to health-promoting effect. *Molecules* **2022**, *27*, 7954. [CrossRef]
84. Sánchez-Gavilán, I.; Ramírez, E.; de la Fuente, V. Bioactive Compounds in *Salicornia patula* Duval-Jouve: A Mediterranean Edible Euhalophyte. *Foods* **2021**, *10*, 410. [CrossRef] [PubMed]
85. Tomaselli, V.; Adamo, M.; Veronico, G.; Sciandrello, S.; Tarantino, C.; Dimopoulos, P.; Medagli, P.; Nagendra, H.; Blonda, P. Definition and application of expert knowledge on vegetation pattern, phenology and seasonality for habitat mapping, as exemplified in a Mediterranean coastal site. *Plant Biosyst.* **2017**, *151*, 887–899. [CrossRef]
86. Nisar, F.; Gul, B.; Khan, M.A.; Hamed, A. Heteromorhic seed of coastal halophytes *Arthrocnemum macrostachyum* and *indicum* display differential patterns of hydrogen peroxide accumulation, lipid peroxidation and antioxidat activities under increasing salinity. *Plant PhysiolBiochem* **2019**, *144*, 58–63. [CrossRef]
87. Antunes, M.D.; Gago, C.; Guerreiro, A.; Sousa, A.R.; Julião, M.; Miguel, M.G.; Faleiro, M.L.; Panagopoulos, T. Nutritional Characterization and Storage Ability of *Salicornia ramosissima* and *Sarcocornia perennis* for Fresh Vegetable Salads. *Horticulturae* **2021**, *7*, 6. [CrossRef]
88. Castagna, A.; Mariottini, G.; Gabriele, M.; Longo, V.; Souid, A.; Dauvergne, X.; Foggi, G.; Conte, G.; Santin, M.; Ranieri, A. Nutritional composition and bioactivity of *Salicornia europaea* L. plants grown in monocolture or intercropped with tomato plants in salt-affected soils. *Horticulturae* **2022**, *8*, 828. [CrossRef]
89. Navarro-Torre, S.; Mateos-Naranjo, E.; Caviedes, M.A.; Pajuelo, E.; Rodríguez-Llorente, I.D. Isolation of plant-growth-promoting and metal-resistant cultivable bacteria from *Arthrocnemum macrostachyum* in the Odiel marshes with potential use in phytoremediation. *Mar. Pollut. Bull.* **2016**, *110*, 133–142. [CrossRef]
90. El-Naker, N.A.; Yousef, A.F.; Yousef, L.F. A review of *Arthrocnemum* (*Arthrocaulon*) *macrostachyum* chemical content and bioactivity. *Phytochem. Rev.* **2020**, *19*, 1427–1448. [CrossRef]

91. Hanif, Z.; Hafiz Haider, A.; Ghulam, R.; Asif, T.; Bhagirath Singh, C. Genus *Salsola*: Its Benefits, Uses, Environmental Perspectives and Future Aspects—A Review. *J. Rangel. Sci.* **2018**, *8*, 315–328.
92. Renna, M.; Gonnella, M. Ethnobotany, Nutritional Traits, and Healthy Properties of Some Halophytes Used as Greens in the Mediterranean Basin. In *Handbook of Halophytes*; Grigore, M.N., Ed.; Springer: Cham, Switzerland, 2020. [CrossRef]
93. Hammer, K. Chenopodiaceae. In *J. Schultze-Motel, Rudolf MansfeldsVerzeichnislandwirtschaftlicher und gartnenscherKulturpflanzen (ohneZierpflanzen)*; Akademie-Verl.: Berlin, Germany, 1986; pp. 145–170.
94. Hammer, K.; Pignone, D.; Cifarelli, S.; Perrino, P. Barilla (*Salsola soda, Chenopodiaceae*). *Econ. Bot.* **1990**, *44*, 410–412. Available online: http://www.jstor.org/stable/4255259 (accessed on 25 November 2022). [CrossRef]
95. Tesi, R. Salsola. In *Enciclopedia Agraria Italiana*; REDA: Roma, Italy, 1980; p. 1008.
96. Centofanti, T.; Bañuelos, G. Evaluation of the halophyte *Salsola soda* as an alternative crop for saline soils high in selenium and boron. *J Environ. Manag.* **2015**, *157*, 96–102. [CrossRef]
97. Bisceglia, V. *Sulle Piante Utili*; Marelli: Milano, Italy, 1807.
98. Colla, G.; Rouphael, Y.; Fallovo, C.; Cardarelli, M.; Graifenberg, A. Use of *Salsola soda* as a companion plant to improve greenhouse pepper (*Capsicum annuum*) under saline conditions. *N. Zeal. J. Crop hortic. Sci.* **2006**, *34*, 283–290. [CrossRef]
99. Debez, A.; Braun, H.P.; Pich, A.; Taamalli, W.; Koyro, H.W.; Abdelly, C.; Huchzermeyer, B. Proteomic and physiological response of the halophyte *Cakile maritima* to moderate salinity at the germinative and vegetative stages. *J. Protemics* **2012**, *75*, 5667–5694. [CrossRef]
100. Augedelo, A.; Carvajal, M.; Martinez-Ballesta, M.d.C. Halophytes of the Mediterranean Basin-Underutilized Species with the Potential to Be Nutritions Crops in the Scenario of the Climate Change. *Foods* **2021**, *10*, 119. [CrossRef]
101. Gandour, M.; Tamaalli, W.; Trabelsi, N.; Hessini, K.; Sebei, K.; Debez, A.; Abdelly, C. How to optimize the seed and seed-oil production in the cash crop halophyte *Cakile maritima*? *J. Med. Plants Res.* **2011**, *5*, 5982–5987. [CrossRef]
102. Arbelet-Bonnin, D.; Ben-Hamed-Louati, I.; Laurenti, P.; Abdelly, C.; Ben-Hame, K.; Bouteau, F. *Cakile maritima*, a promising model for halophyte studies and a putative cash crop for saline agriculture. *Adv. Agric.* **2019**, *155*, 45–78. [CrossRef]
103. Farhat, N.; Hichri, S.T.M.; Hildebrandt, A.; Debez, A.; Braun, H.P. Composition and Stability of the Oxidative Phosphorylation System in the Halophile Plant *Cakile maritima*. *Front. Plant Sci.* **2019**, *10*, 1010. [CrossRef]
104. Fuochi, V.; Barbagallo, I.I.; Distefano, A.; Puglisi, F.; Palmieri, R.; Di Rosa, M.; Giallongo, C.; Longhitano, L.; Fontana, P.; Sferrazzo, G.; et al. Biological properties of *Cakile maritima* Scop. (*Brassicaceae*) extracts. *Eur. Rev. Med. L Pharm. Sci.* **2019**, *23*, 2280–2292. [CrossRef]
105. Ksouri, R.; Megdiche, W.; Debez, A.; Falleh, H.; Grignon, C.; Abdelly, C. Salinity effects on polyphenol content and antioxidant activities in leaves of the halophyte *Cakile maritima*. *Plant Physiol. Biochem.* **2007**, *45*, 244–249. [CrossRef]
106. Meot-Duros, L.; Le Floch, G.; Magné, C. Radical scavenging, antioxidant and antimicrobial activities of halophytic species. *J. Ethnopharmacol.* **2008**, *116*, 258–262. [CrossRef]
107. Shiri, M.; Rabbi, M.; El Amrani, A.; Abdelly, A. The Halophyte Cakile maritima Reduces Phenanthrene Phytotoxicity. *J. Phytorimediation* **2015**, *17*, 925–928. [CrossRef]
108. Atia, S.; Barhoumi, Z.; Mokded, R.; Abdelly, C.; Smaoui, A. Environmental eco-physiology and economical potential of the halophyte *Crithmum maritimum* L. (*Apiaceae*). *J. Med. Plants Res.* **2011**, *5*, 3564–3571. [CrossRef]
109. Renna, M.; Gonnella, M.; Caretto, S.; Mita, G.; Serio, F. Sea fennel (*Crithmum maritimum* L.): From underutilized crop to new dried product for food use. *Genet. Resour. Crop Evol.* **2017**, *64*, 205–216. [CrossRef]
110. Franke, W. Vitamin C in sea fennel (*Crithmum maritimum*), an edible wild plant. *Econ. Bot.* **1982**, *36*, 163–165. [CrossRef]
111. Renna, M.; Gonnella, M. The use of the sea fennel as a new spice-colorant in culinary preparations. *Int. J. Gastron. Food Sci.* **2012**, *1*, 111–115. [CrossRef]
112. Atia, A.; Ben, H.K.; Debez, A.; Abdelly, C. Salt and seawater effects on the germination of *Crithmum maritimum*. In *Biosaline Agriculture and Salinity Tolerance in Plants*; Öztürk, M., Waisel, Y., Khan, M.A., Görk, G., Eds.; Birkhäuser Verlag: Basel, Switzerland, 2006; pp. 29–33.
113. Atia, A.; Debez, A.; Barhoumi, Z.; Smaoui, A.; Abdelly, C. Interactive effects of salinity, nitrate, light, and seed weight on the germination of the halophyte *Crithmum maritimum*. *Acta Biol. Hung* **2009**, *60*, 433–439. [CrossRef]
114. Grigoriadou, K.; Maloupa, E. Micropropagation and salt tolerance of in vitro grown *Crithmum maritimum* L. *Plant Cell Tiss. Organ. Cult.* **2008**, *94*, 209–217. [CrossRef]
115. Martini, A.N.; Papafotiou, M.; Evangelopoulos, K. Effect of substrate type and depth on the establishment of the edible and medicinal native species *Crithmum maritimum* on an extensive urban Mediterranean green roof. *Acta Hortic.* **2017**, *1189*, 451–454. [CrossRef]
116. Nektarios, P.A.; Nydrioti, E.; Kapsali, T.; Ntoulas, N. *Crithmum maritimum* growth in extensive green roof systems with different substrate type, depth and irrigation regime. *Acta Hortic.* **2016**, *1108*, 303–308. [CrossRef]
117. Zenobi, S.; Fiorentini, M.; Zitti, S.; Aquilanti, L.; Foligni, R.; Mannozzi, C.; Mozzon, M.; Orsini, R. *Crithmum maritimum* L.: First results on phenological development and biomass production in Mediterranean areas. *Agronomy* **2021**, *11*, 773. [CrossRef]
118. Pereira, C.G.; Barreira, L.; da Rosa Neng, N.; Nogueira, J.M.F.; Marques, C.; Santos, T.; Varela, J.; Custodio, L. Searching for new sources of innovative products for the food industry within halophyte aromatic plants: In vitro antioxidant activity and phenolic and mineral contents of infusions and decoctions of *Crithmum maritimum* L. *Food Chem. Toxicol.* **2017**, *107*, 581–589. [CrossRef]

119. Pereira, A.G.; Fraga-Corral, M.; García-Oliveira, P.; Jimenez-Lopez, C.; Lourenço-Lopes, C.; Carpena, M.; Otero, P.; Gullón, P.; Prieto, M.A.; Simal-Gandara, J. Culinary and nutritional value of edible wild plants from northern Spain rich in phenolic compounds with potential health benefits. *Food Funct.* **2020**, *11*, 8493. [CrossRef]
120. Atia, A.; Debez, A.; Barhoumi, Z.; Abdelly, C.; Smaoui, A. Histochemical localization of essential oils and bioactive substances in the seed coat of the halophyte *Crithmum maritimum* L. (*Apiaceae*). *J. Plant Biol.* **2009**, *52*, 448–452. [CrossRef]
121. Cornara, L.; D'Arrigo, C.; Pioli, F.; Borghesi, B.; Bottino, C.; Patrone, E.; Mariotti, M.G. Micromorphological investigation on the leaves of the rock samphire (*Crithmum maritimum* L.): Occurrence of hesperidin and diosmin crystals. *Plant Biosyst.* **2009**, *143*, 283–292. [CrossRef]
122. Meot-Duros, L.; Magné, C. Antioxidant activity and phenol content of *Crithmum maritimum* L. leaves. *Plant Physiol. Biochem.* **2009**, *47*, 37–41. [CrossRef]
123. Meot-Duros, L.; Cérantola, S.; Talarmin, H. New antibacterial and cytotoxic activities of falcarindiol isolated in *Crithmum maritimum* L. leaf extract. *Food Chem. Toxicol.* **2010**, *48*, 553–557. [CrossRef]
124. Generalić Mekinić, I.; Blažević, I.; Mudnić, I.; Burčul, F.; Grga, M.; Skroza, D.; Jerčić, I.; Ljubenkov, I.; Boban, M.; Miloš, M.; et al. Sea fennel (*Crithmum maritimum* L.): Phytochemical profile, antioxidative, cholinesterase inhibitory and vasodilatory activity. *J. Food Sci. Technol.* **2016**, *53*, 3104–3112. [CrossRef]
125. Gil, L.; Pinya, S.; Tejada, S.; Capó, X.; Sureda, A. Antioxidant Defenses in Wild Growing Halophyte *Crithmum maritimum* from Inland and Coastline Populations. *Chem. Biodiv.* **2019**, *16*, e1800448. [CrossRef]
126. Jallali, I.; Zaouali, Y.; Missaoui, I.; Smeoui, A.; Abdelly, C.; Ksouri, R. Variability of antioxidant and antibacterial effects of essential oils and acetonic extracts of two edible halophytes: *Crithmum maritimum* L. and *Inula crithmoïdes* L. *Food Chem.* **2014**, *145*, 1031–1038. [CrossRef]
127. Gnocchi, D.; Cesari, G.; Calabrese, G.J.; Capone, R.; Sabbà, C.; Mazzocca, A. Inhibition of Hepatocellular Carcinoma Growth by Ethyl Acetate Extracts of Apulian *Brassica oleracea* L. and *Crithmum maritimum* L. *Plant Foods Hum. Nutr.* **2020**, *75*, 33–40. [CrossRef]
128. D'Agostino, G.; Giambra, B.; Palla, F.; Bruno, M.; Badalamenti, N. The Application of the Essential Oils of *Thymus vulgaris* L. and *Crithmum maritimum* L. as Biocidal on Two TholuBommalu Indian Leather Puppets. *Plants* **2021**, *10*, 1508. [CrossRef]
129. Renna, M. Reviewing the Prospects of Sea Fennel (*Crithmum maritimum* L.) as Emerging Vegetable Crop. *Plants* **2018**, *7*, 92. [CrossRef] [PubMed]
130. Azeñas, V.; Janner, I.; Medrano, H.; Gulías, J. Performance evaluation of five Mediterranean species to optimize ecosystem services of green roofs under water-limited conditions. *J. Environ. Manag.* **2018**, *212*, 236–247. [CrossRef] [PubMed]
131. Alexopoulos, A.A.; Assimakopoulou, A.; Panagopoulos, P.; Bakea, M.; Vidalis, N.; Karapanos, I.C.; Petropoulos, S.A. Impact of Salinity on the Growth and Chemical Composition of Two Underutilized Wild Edible Greens: *Taraxacum officinale* and *Reichardia picroides*. *Horticulturae* **2021**, *7*, 160. [CrossRef]
132. Guarrera, M.P.; Savo, V. Wild food plants used in traditional vegetable mixtures in Italy. *J. Ethnopharmacol.* **2016**, *185*, 202–234. [CrossRef] [PubMed]
133. Ranfa, A.; Bodesmo, M. An Ethnobotanical investigation of traditional knowledge and uses of edible wild plants in the Umbria Region, Central Italy. *J. Appl. Bot. Food Qual.* **2017**, *90*, 246–258. [CrossRef]
134. Tittel, C. Compositae. In *Rudolf Mansfeld Verzeichnis landwirtschaftelicher und gartnerishcer Kulturpflanzed (ohne Zierpflanzen)*; Schultze-Motel, J., Ed.; Akademie-Verlag: Berlin, Germany, 1986; pp. 1261–1335.
135. Maggini, R.; Benvenuti, S.; Leoni, F.; Incrocci, L.; Pardossi, A. Effects of NaCl on hydroponic cultivation of *Reichardia picroides* (L.) roth. *Agronomy* **2021**, *11*, 2352. [CrossRef]
136. Pieroni, A. Medicinal plants and food medicines in the folk traditions of the upper Lucca 471 Province, Italy. *J. Ethnopharmacol.* **2000**, *70*, 235–273. [CrossRef]
137. Loi, M.C.; Poli, F.; Sacchetti, G.; Selenu, M.B.; Ballero, M. Ethnopharmacology of Ogliastra (Villagrande Strisaili, Sardinia, Italy). *Fitoterapia* **2004**, *75*, 277–295. [CrossRef]
138. Atzei, A.; Orioni, D.S.; Sotgiu, R. Contributo alla conoscenza degli usi etnobotanici nella 362 Gallura (Sardegna). *Boll. Della Soc. Sarda Sci. Nat.* **1991**, *28*, 137–177.
139. Aouachria, S.; Boumerfeg, S.; Benslama, A.; Benbacha, F.; Guemmez, T.; Khennouf, S.; Arrar, L.; Baghiani, A. Acute, sub-acutetoxicity and antioxidant activities (in vitro and in vivo) of *Reichardia picroides* crude estract. *J. Ethnopharmacol.* **2017**, *208*, 105–116. [CrossRef]
140. Fragopulou, E.; Detopoulou, P.; Nomikos, T.; Pliakis, E.; Panagiotakos, D.B.; Antonopoulou, S. Mediterranean wild plants reduce prostprandial platelet aggregation in patients with metabolic syndrome. *Metabolism* **2012**, *61*, 325–334. [CrossRef]
141. Guarrera, P.M.; Salerno, G.; Caneva, G. Food, Flavouring and feed plant traditions in the Tyrrhenian sector of Basilicata, Italy. *J Ethnobiol. Ethnomed.* **2006**, *2*, 37. [CrossRef]
142. Sebasky, M.; Keller, S.R.; Taylor, D.R. Investigating past range dynamics for a weed of cultivation Silene vulgaris. *Ecol. Evol.* **2016**, *6*, 4800–4811. [CrossRef]
143. Taffetani, F. Rugni, Speragne e Crispigne. Piante Spontanee negli usi del Territorio maceratese. In *Fondazione Cassa di Risparmio della Provincia di Macerata. Carime Arte srl*; Pigini Group: Loreto, Italy, 2005.
144. Arcidiacono, S.; Napoli, M.; Oddo, G.; Pavone, P. Piante selvatiche d'uso popolare nei territori di Alcara li Fusi e Militello Rosmarino (Messina, N-E Sicilia). *Quad. Bot. Amb.* **2007**, *18*, 105–146.

145. Zampiva, F. Le erbe nella cucina popolare veneta. *Erbor. Domani* **1983**, *4*, 60–69.
146. Laghetti, G.; Hammer, K.; Perrino, P. Collecting in north-west Italy. *FAO/IBPGR Newsl.* **1993**, *91/92*, 23.
147. Laghetti, G.; Perrino, P.; Hammer, K. Utilization of *Silene vulgaris* (Moench) Garcke in Italy. *Econ. Bot.* **1994**, *48*, 337–339. [CrossRef]
148. Hammer, K.; Kniipffer, H.; Laghetti, G.; Perrino, P. *Seeds from the Past: A Catalogue of the Crop Germplasm of South Italy and Sicily*; Germplasm Institute: Bari, Italy, 1992; p. 174.
149. Pacwa-Płociniczak, M.; Płociniczak, T.; Yu, D.; Kurola, J.M.; Sinkkonen, A.; Piotrowska-Seget, Z.; Romantschuk, M. Effect of Silene vulgaris and Heavy Metal Pollution on Soil Microbial Diversity in Long-Term Contaminated Soil. *Water Air Soil Pollut.* **2018**, *229*, 13. [CrossRef]
150. Antonidiadis, V.; Shaheen, S.M.; Stark, H.J.; Wennrich, R.; Levizou, E.; Merbach, I.; Rinklebe, J. Phytoremediation potential of twelve wild plants species for toxic elements in a contaminated soil. *Environ Int.* **2021**, *146*, 106233. [CrossRef]
151. Tison, J.M.; Jauzein, P.; Michaud, H. *Flore de France Méditerranéenne Continentale*; Naturalia Publications: Turriers, Italy, 2014.
152. GRIN. Germplasm Resources Information Network—(GRIN) [Online Database]. National Germplasm Resources Laboratory, Beltsville, Maryland. 2010. Available online: http://www.ars-rin.gov/cgibin/npgs/html/index.pl (accessed on 30 November 2022).
153. Véla, E.; Kell, S.P. *Allium commutatum. IUCN Red List Threat. Species* **2018**, e.T172281A19174707.
154. Geraci, A.; Amato, F.; Di Noto, G.; Bazan, G.; Schicchi, R. The wild taxa utilized as vegetables in Sicily (Italy): A traditional component of the Mediterranean diet. *J. Ethnob. Ethnomed.* **2018**, *14*, 14. [CrossRef]
155. Mathew, B. *A review of* Allium *section* Allium; Royal Botanic Gardens, Kew: Richmond, UK, 1996; ISBN 9780947643935.
156. Vujević, M.; Pevalek-Kozlina, B.; Pavlica, M.; Šolić, M.E. Shoot and root regeneration from callus tissue of Allium commutatum-Guss. *Acta Bot. Croat.* **1999**, *58*, 57–64.
157. Loizzo, M.R.; Tundis, R.; Sut, S.; Dall'Acqua, S.; Ilardi, V.; Leporini, M.; Falco, T.; Sicari, V.; Bruno, M. High-Performance Liquid Chromatography/Electrospray Ionization Tandem Mass Spectrometry (HPLC-ESI-MSn) Analysis and Bioactivity Useful for Prevention of "Diabesity" of *Allium commutatum* Guss. *Plant Foods Hum. Nutr.* **2020**, *75*, 124–130. [CrossRef] [PubMed]
158. Demirtas, I.; Erenler, R.; Elmastas, M.; Goktasoglu, A. Studie on the antioxidant potential of flavones of *Allium vineale* isolated from its water-soluble fraction. *Food Chem.* **2013**, *136*, 34–40. [CrossRef] [PubMed]
159. Stajner, D.; Igic, R.; Popovic, B.M.; Malencic, D.j. Comparative study of antioxidant properties of wild growing and cultivated *Allium* species. *Phytother. Res.* **2007**, *22*, 113–117. [CrossRef]
160. Leys, M.; Petit, E.J.; El-Bahloul, Y.; Liso, C.; Fournet, S.; Arnaud, J.F. Spatial genetic structure in *Beta vulgaris* subsp. *maritima* and *Beta macrocarpa* reveals the effect of contrasting mating system, influence of marine currents, and footprints of postglacial recolonization routes. *Ecol. Evol.* **2014**, *4*, 1828–1852. [CrossRef]
161. Bianco, V.; Mariani, R.; Santamaria, P. *Piante Spontanee Nella Cucina Tradizionale Molese*; Levante Ed: Bari, Italy, 2009; pp. 84–89.
162. Lentini, F.; Venza, F. Wild food plants of popular use in Sicily. *J. Ethnob. Ethnomed.* **2007**, *3*, 15. [CrossRef]
163. Goldman, I.L.; Janick, J. Evolution of root morphology in table beet: Historical and iconographic. *Front. Plant. Sci.* **2021**, *12*, 689926. [CrossRef]
164. Perrino, E.V.; Wagensommer, R.P. Crop Wild Relatives (CWR) Priority in Italy: Distribution, Ecology, In Situ and Ex Situ Conservation and Expected Actions. *Sustainability* **2021**, *13*, 1682. [CrossRef]
165. Perrino, E.V.; Perrino, P. Crop wild relatives: Know how past and present to improve future research, conservation and utilization strategies, especially in Italy: A review. *Genet. Resour. Crop Evol.* **2020**, *67*, 1067–1105. [CrossRef]
166. Yolcu, S.; Alavilli, H.; Ganesh, P.; Panigrahy, M.; Song, K. Salt and drought stress responses in cultivated beets (*Beta vulgaris* (L.) andwild beet (Beta maritima L.). *Plants* **2021**, *10*, 1843. [CrossRef]
167. Zardi-Bergaoui, A.; Nejma, A.B.; Harzallah-Skhiri, F.; Flamini, G.; Ascrizzi, R.; Jannet, H.B. Chemical composition and biological studies of the essential oil from aerial parts of *Beta vulgaris* subsp. *maritima* (L.) Arcang. Growing in Tunisia. *Chem Biodivers.* **2017**, *14*, e1700234. [CrossRef]
168. Inocencio, C.; Alcaraz, F.; Calderon, F.; Obon, C.; Rivera, D. The use of floral characters in *Capparis* sect. *Capparis* to determine the botanical and geographical origin of capers. *Eur. Food Res. Technol.* **2022**, *214*, 335–339. [CrossRef]
169. Chedraoui, S.; Abi-Rizk, A.; El-Beyrouthy, M.; Chalak, L.; Ouaini, N.; Rajjou, L. *Capparis spinosa* L. in a systematic review: A xerophilous species of multi values and promising potentialities. *Front. Plant Sci.* **2017**, *8*, 1845. [CrossRef]
170. Ashraf, U.; Chaudhry, M.N.; Ahmad, S.R.; Ashraf, I.; Arslan, M.; Noor, H.; Jabbar, M. Impacts of climate change on Capparis spinosa L. based on ecological niche modeling. *PeerJ* **2018**, *6*, e5792. [CrossRef]
171. Barbera, G.; Di Lorenzo, R. The caper culture in Italy. *Acta Hortic.* **1984**, *144*, 167–172. [CrossRef]
172. Barbera, G.; Di Lorenzo, R.; Barone, E. Observations on *Capparis* populations cultivated in Sicily and on their vegetative and productive behavior. *Agric. Mediterr.* **1991**, *121*, 32–39.
173. Sottile, F.; Caltagirone, C.; Peano, C.; Del Signore, M.B.; Barone, E. Can the Caper (Capparis spinosa L.) Still Be Considered a Difficult-to-Propagate Crop? *Horticulturae* **2021**, *7*, 316. [CrossRef]
174. La Bella, S.; Rossini, F.; Licata, M.; Virga, G.; Ruggeri, R.; Iacuzzi, N.; Leto, C.; Tuttolomondo, T. Four-Year Study on the Bio-Agronomic Response of Biotypes of *Capparis spinosa* L. on the Island of Linosa (Italy). *Agriculture* **2021**, *11*, 327. [CrossRef]
175. Laghetti, G.; Hammer, K.; Perrino, P. Plant genetic resources in Pantelleria and Pelagie archipelago, Italy: Collecting and conservation of local crop germplasm. *FAO/IBPGR Plant Gen. Res. Newsl.* **1996**, *108*, 17–25.

176. Gan, L.; Zhang, C.; Yin, Y.; Lin, Z.; Huang, Y.; Xiang, J.; Fu, C.H.; Li, M.T. Anatomical adaptation of the xerophilus medicinal plant, *Capparis spinosa*, to drougth conditions. *Hortic. Environ. Biotechnol.* **2013**, *54*, 156–161. [CrossRef]
177. Santo, A.; Dessì, L.; Ucchesu, M.; Bou Dagher Kharrat, M.; Charbel Sark, R.; Accogli, R.; Buhagiar, J.; Kyratsis, A.; Fournakaru, C.; Bacchetta, G. Seed germination ecology and salt stress response in eight Mediterranean populations of Sarcopoterium spinosum (L.) Spach. *Plant Species Biol.* **2019**, *34*, 110–121. [CrossRef]
178. Annaz, H.; Sane, Y.; Bitchagno, G.T.M.; Ben Bakrim, W.; Drissi, B.; Mahdi, I.; El Bouhssini, M.; Sobeh, M. Caper (Capparis spinosa L.): An Update Review on Its Phytochemistry, Nutritional Value, Traditional Uses and Therapeutic Potential. *Front. Pharmacol.* **2022**, *13*, 878749. [CrossRef] [PubMed]
179. Nabavi, S.F.; Maggi, F.; Daglia, M.; Habetmariam, S.; Rastrelli, L.; Nabavi, S.M.P. Pharmacological effects of Capparis spinosa L. *Phytother. Res.* **2016**, *30*, 1733–1744. [CrossRef]
180. Petropoulos, S.A.; Karkanis, A.; Martins, N.; Ferreira, C.F.R. Edible halophytes of the Mediterranean basin: Potential candidates for novel food products. *Trends Food Sci. Technol.* **2018**, *74*, 69–84. [CrossRef]
181. Vahid, H.; Rakhandeh, H.; Ghorbani, A. Antidiabetic proprierties of *Capparis spinosa* L. and its components. *Biomed Pharmacother.* **2017**, *92*, 293–302. [CrossRef]
182. Zhang, H.; Ma, Z.F. Phytochemical and Pharmacological Properties of *Capparis spinosa* as a Medicinal Plant. *Nutrients* **2018**, *10*, 116. [CrossRef]
183. Zhu, X.; Yang, Y.; Gao, W.; Jiang, B.; Shi, L. Capparis spinosa Alleviates DSS-Induced Ulcerative Colitis via Regulation of the Gut Microbiota and Oxidative Stress. Evid. *Based Complement. Altern. Med.* **2021**, *15*, 1227876. [CrossRef]
184. Rahnavard, R.; Razavi, N. A review on the medical effects of *Capparis spinosa* L. *Adv. Herb. Med.* **2016**, *2*, 44–53.
185. Saleem, H.; Khurshid, U.; Sarfraz, M.; Ahmad, I.; Alamri, A.; Anwar, S.; Alamri, A.S.; Locatelli, M.; Tartaglia, A.; Mahomoodally, M.F.; et al. Investigation into the biological properties, secondary metabolites composition, and toxicity of aerial and root parts of *Capparis spinosa* L.: An important medicinal food plant. *Food Chem. Toxicol.* **2021**, *155*, 112404. [CrossRef]
186. Kdimy, A.; El Yadini, M.; Guaadaoui, A.; Bourais, I.; El Hajjaji, S.; Le, H.V. Phytochemistry, Biological Activities, Therapeutic Potential, and Socio-Economic Value of the Caper Bush (*Capparis Spinosa* L.). *Chem. Biod.* **2022**, *19*, e202200300. [CrossRef]
187. Faran, M. Capparis spinosa-The Plant on the Wall. In *Medicinal and Aromatic Plants of the Middle-East. Medicinal and Aromatic Plants of the World*; Yaniv, Z., Dudai, N., Eds.; Springer: Dordrecht, The Netherlands, 2014; Volume 2.
188. Stinca, A.; Chianese, G.; D'Auria, G.; Del Guacchio, E.; Fascetti, S.; Perrino, E.V.; Rosati, L.; Salerno, G.; Santangelo, A. New alien vascular species for the flora of southern Italy. *Webbia* **2017**, *72*, 295–301. [CrossRef]
189. Gibbs, J.P. Wetlands loss and biodiversity conservation. *Cons. Biol.* **2000**, *14*, 314–317. [CrossRef]
190. van der Maarel, E. Some remarks on the functions of European coastal ecosystems. *Phytocoenologia* **2003**, *33*, 187–202. [CrossRef]
191. Underwood, E.C.; Viers, J.H.; Klausmeyer, K.R.; Cox, R.L.; Shaw, M.R. Threats and biodiversity in the Mediterranean biome. *Divers. Distrib.* **2009**, *15*, 188–197. [CrossRef]
192. Tomaselli, V.; Veronico, G.; Adamo, M. Monitoring and recording changes in natural landscapes: A case study from two coastal wetlands in SE Italy. *Land* **2021**, *10*, 50. [CrossRef]
193. Maes, J.; Fabrega, N.; Zulian, G.; Barbosa, A.; Vizcaino, P.; Ivits, E.; Polce, C.; Vandecasteele, I.; Marí Rivero, I.; Guerra, C.; et al. *Mapping and Assessment of Ecosystems and their Services. JRC Science and Policy Reports*; European Commission: Brussels, Belgium, 2015.
194. Liquete, C.; Piroddi, C.; Drakou, E.G.; Gurney, L.; Katsanevakis, S.; Charef, A.; Egoh, B. Current Status and Future Prospects for the Assessment of Marine and Coastal Ecosystem Services: A Systematic Review. *PLoS ONE* **2013**, *8*, e67737. [CrossRef]
195. Perrino, E.V.; Wagensommer, R.P. Crop Wild Relatives (CWRs) Threatened and Endemic to Italy: Urgent Actions for Protection and Use. *Biology* **2022**, *11*, 193. [CrossRef]
196. Pisani, D.; Pazienza, P.; Perrino, E.V.; Caporale, D.; De Lucia, C. The Economic Valuation of Ecosystem Services of Biodiversity Components in Protected Areas: A Review for a Framework of Analysis for the Gargano National Park. *Sustainability* **2021**, *13*, 11726. [CrossRef]
197. Dimita, R.; Min Allah, S.; Luvisi, A.; Greco, D.; De Bellis, L.; Accogli, R.; Mininni, C.; Negro, C. Volatile Compounds and Total Phenolic Content of *Perilla frutescens* at Microgreens and Mature Stages. *Horticulturae* **2022**, *8*, 71. [CrossRef]
198. Dronkers, J. Integrated Coastal Zone Management (ICZM). Available online: http://www.coastalwiki.org/wiki/Integrated_Coastal_Zone_Management_(ICZM) (accessed on 23 December 2022).
199. Tomaselli, V.; Mantino, F.; Tarantino, C.; Albanese, G.; Adamo, M. Changing landscapes: Habitat monitoring and landtransformation in a long-time used Mediterranean coastal wetland. *Wetl. Ecol Manag.* **2022**. [CrossRef]
200. O'leary, J.W.; Glenn, E.P. Global distribution and potential for halophytes. In *Halophytes as a Resource for Livestock and for Rehabilitation of Degraded Lands*; Squires, V.R., Ayoub, A.T., Eds.; Tasks for Vegetation Science; Springer: Dordrecht, The Netherlands, 1994; Volume 32. [CrossRef]

MDPI
St. Alban-Anlage 66
4052 Basel
Switzerland
www.mdpi.com

Plants Editorial Office
E-mail: plants@mdpi.com
www.mdpi.com/journal/plants